Digital Tools for Academic Branding and Self-Promotion

Marga Cabrera
Polytechnic University of Valencia, Spain

Nuria Lloret
Polytechnic University of Valencia, Spain

A volume in the Advances in Educational Marketing, Administration, and Leadership (AEMAL) Book Series

IGI GLOBAL
DISSEMINATOR OF KNOWLEDGE

www.igi-global.com

Published in the United States of America by
IGI Global
Information Science Reference (an imprint of IGI Global)
701 E. Chocolate Avenue
Hershey PA, USA 17033
Tel: 717-533-8845
Fax: 717-533-8661
E-mail: cust@igi-global.com
Web site: http://www.igi-global.com

Library of Congress Cataloging-in-Publication Data

Names: Cabrera, Marga, editor. | Lloret Romero, Nuria, 1966- editor.
Title: Digital tools for academic branding and self-promotion / Marga Cabrera and Nuria Lloret, editors.
Description: Hershey PA : Information Science Reference, 2017. | Series: Advances in educational marketing, administration, and leadership | Includes bibliographical references and index.
Identifiers: LCCN 2016033383| ISBN 9781522509172 (hardcover) | ISBN 9781522509189 (ebook)
Subjects: LCSH: College teachers--Professional relationships. | Digital communications. | College teachers--Blogs. | Branding (Marketing) | Reputation.
Classification: LCC LB1778 .D53 2017 | DDC 378.1/2--dc23 LC record available at https://lccn.loc.gov/2016033383

This book is published in the IGI Global book series Advances in Educational Marketing, Administration, and Leadership (AEMAL) (ISSN: 2326-9022; eISSN: 2326-9030)

British Cataloguing in Publication Data
A Cataloguing in Publication record for this book is available from the British Library.

All work contributed to this book is new, previously-unpublished material. The views expressed in this book are those of the authors, but not necessarily of the publisher.

For electronic access to this publication, please contact: eresources@igi-global.com.

Advances in Educational Marketing, Administration, and Leadership (AEMAL) Book Series

Siran Mukerji
IGNOU, India
Purnendu Tripathi
IGNOU, India

ISSN:2326-9022
EISSN:2326-9030

MISSION

With more educational institutions entering into public, higher, and professional education, the educational environment has grown increasingly competitive. With this increase in competitiveness has come the need for a greater focus on leadership within the institutions, on administrative handling of educational matters, and on the marketing of the services offered.

The **Advances in Educational Marketing, Administration, & Leadership (AEMAL) Book Series** strives to provide publications that address all these areas and present trending, current research to assist professionals, administrators, and others involved in the education sector in making their decisions.

COVERAGE

- Enrollment Management
- Governance in P-12 and Higher Education
- Academic Administration
- Marketing Theories within Education
- Students as Consumers
- Direct marketing of educational programs
- Educational Marketing Campaigns
- Educational Leadership
- Faculty Administration and Management
- Technologies and Educational Marketing

IGI Global is currently accepting manuscripts for publication within this series. To submit a proposal for a volume in this series, please contact our Acquisition Editors at Acquisitions@igi-global.com or visit: http://www.igi-global.com/publish/.

The Advances in Educational Marketing, Administration, and Leadership (AEMAL) Book Series (ISSN 2326-9022) is published by IGI Global, 701 E. Chocolate Avenue, Hershey, PA 17033-1240, USA, www.igi-global.com. This series is composed of titles available for purchase individually; each title is edited to be contextually exclusive from any other title within the series. For pricing and ordering information please visit http://www.igi-global.com/book-series/advances-educational-marketing-administration-leadership/73677. Postmaster: Send all address changes to above address.

Titles in this Series

For a list of additional titles in this series, please visit: www.igi-global.com

Assessing the Current State of Education in the Caribbean
Charmaine Bissessar (Independent Researcher, Trinidad and Tobago)
Information Science Reference • copyright 2017 • 367pp • H/C (ISBN: 9781522517009) • US $175.00 (our price)

World University Rankings and the Future of Higher Education
Kevin Downing (City University of Hong Kong, China) and Fraide A. Ganotice, Jr. (The University of Hong Kong, China)
Information Science Reference • copyright 2017 • 534pp • H/C (ISBN: 9781522508199) • US $190.00 (our price)

Handbook of Research on Administration, Policy, and Leadership in Higher Education
Siran Mukerji (Indira Gandhi National Open University, India) and Purnendu Tripathi (Indira Gandhi National Open University, India)
Information Science Reference • copyright 2017 • 678pp • H/C (ISBN: 9781522506720) • US $295.00 (our price)

Student Experiences and Educational Outcomes in Community Engagement for the 21st Century
Cathryn Crosby (Teachers College Columbia University, USA) and Frederick Brockmeier (Northern Kentucky University, USA)
Information Science Reference • copyright 2017 • 306pp • H/C (ISBN: 9781522508748) • US $175.00 (our price)

Literacy Program Evaluation and Development Initiatives for P-12 Teaching
Salika A. Lawrence (Medgar Evers College, City University of New York, USA)
Information Science Reference • copyright 2017 • 309pp • H/C (ISBN: 9781522506690) • US $180.00 (our price)

Community Engagement Program Implementation and Teacher Preparation for 21st Century Education
Cathryn Crosby (Teachers College Columbia University, USA) and Frederick Brockmeier (Northern Kentucky University, USA)
Information Science Reference • copyright 2017 • 330pp • H/C (ISBN: 9781522508717) • US $185.00 (our price)

Systemic Knowledge-Based Assessment of Higher Education Programs
Edgar Oliver Cardoso Espinosa (Instituto Politécnico Nacional, Mexico)
Information Science Reference • copyright 2016 • 264pp • H/C (ISBN: 9781522504573) • US $170.00 (our price)

Applied Chaos and Complexity Theory in Education
Şefika Şule Erçetin (Hacettepe University, Turkey)
Information Science Reference • copyright 2016 • 299pp • H/C (ISBN: 9781522504603) • US $170.00 (our price)

IGI GLOBAL
DISSEMINATOR OF KNOWLEDGE
www.igi-global.com

701 E. Chocolate Ave., Hershey, PA 17033
Order online at www.igi-global.com or call 717-533-8845 x100
To place a standing order for titles released in this series, contact: cust@igi-global.com
Mon-Fri 8:00 am - 5:00 pm (est) or fax 24 hours a day 717-533-8661

Editorial Advisory Board

Table of Contents

Detailed Table of Contents

Many researchers are uncomfortable with the idea of promoting themselves or their work, but it is as important to report on it as it is to do it, no one will know about your latest work unless you tell them about it. It is very important to be able to describe what you are good at and what you are best at, not just to let others talk about you. As a career researcher, it is absolutely essential to promote yourself. In addition to your Google profile and your Google Scholar profile you are strongly advised to have your own web page or blog, where you can post your work history, latest developments, report on goals, projects, news and a lot more. Why wait until someone else talks about your news if you can do it yourself? In this chapter we offer an overview of new digital tools that can help to improve your online academic reputation.

The Internet have led to the emergence of a new communication ecosystem that is not restricted to the online context and in which professional and social media intersect and cohabitate to compete for the attention of audiences. This ecosystem occurs in cyberspace; a placeless space where new forms of power, influence, control, and management of collective perceptions have developed. These emerging macro social platforms are giving rise to a new paradox: people need social media to express their opinions and creativity, while the major platforms tend to view users as products. A new model is presented to describe and explain this new communication ecosystem that has put an end to the mediating exclusivity of professional media and maximizes collective interpersonal communication on one and the same social continuum. It is a new interpersonal and collective communication ecosystem whose deep logic is necessary to understand and in which academics must compete.

Nowadays, many professionals of different areas of expertise have found themselves at the needs to incorporate in the digital world, their personal digital branding as a result of entrepreneurship projects, or as a crisis of business sectors. A different phenomenon occurs in the academic field; the researchers are forced to incorporate themselves in the digital field. The research professors have a responsibility to society, popularizing science, this represents to value their own work and contribute to its visibility, and the starting point is to have "digital academic Branding. This chapter explains step-by-step techniques that develop a digital academic branding, starting from an interior analysis of the academic up to the formulation of the visual and conceptual brand; it is important is to make in a sequential manner every one of the activities described here.

Academics have always been in the business of producing and promoting ideas. The configuration of their identities or brands can play a central role in the successful realization of this ambition. This chapter uses the participative stance in netnography to discuss academic personal branding, but also interweaves advice, core principles, and a series of short inductive case studies about personal academic brands and fields. Featuring cases like those of Plato, Timothy Leary, John Lilly, and others, the chapter outlines a theory of personal academic branding, relating core identities to academic branding practices and purposes. In particular, this chapter suggests the relationship between thinking innovative thoughts, representing them to various audiences, and having one's idea shared and built upon by a network of interested others. Core to the successful realization of this endeavor are the following nine elements: unconventionality, creativity, internal and external audiences, segmentation, core constituents, co-branding, simplicity, uniqueness, and catchiness.

The present chapter provides a clear vision for the social networks environment from the self-promotion point of view. Chapter focuses on organizing tools, audience, and type of publications. Tools are organized to contextualize their use and to give a proper understanding of the relevant contents that can be published. Audience is presented according to the relations and interests with the teacher and researcher. Simultaneously, chapter gives a vision of the privacy scope or the publications, and provides an evaluation mechanism to distinguish the most convenient area of publication depending of the message content. Following submission of these analyses, chapter focuses on the teacher and research activity and how to promote these activities through social networks. The chapter ends with a set of suggestions to make a strategic use of new media with the goal of promoting efficiently personal brand as a teacher and researcher.

The Academic environment is changing, especially with the nominated web 2.0 and science 2.0. The increase in new academic social network used for researchers to publicize their knowledge, research and contributions in science, can not take lightly. For this and everything that comes with being a researcher, the academic community manager comes to manage the social networks and academic social network, care workflow, generate networking and disseminate results, among others. Increasingly universities seeking to spread the results generated by their students, teachers, masters, doctors or research groups. This research is intended to show the important role that will have the academic community manager inside the group of researchers.

The book chapter seeks to highlight the importance of brand journalism and personal branding within the university context. First, the focus is on determining how universities are adapting to the changing needs of their audience by utilising new narrative techniques. Second, we must determine whether brand journalism in turn allows for the construction of personal branding. If so, we would like to know whether teachers coming from the Schools of Communication / Information of universities in Spain have created their personal brands, in addition to what criteria and tools had to be taken into consideration for its creation. In conclusion, a lack of progress in regards to the development of a personal brand as well as personal branding in academia is observed.

There are numerous assumptions on research evaluation in terms of quality and relevance of academic contributions. Researchers are becoming increasingly acquainted with bibliometric indicators, including; citation analysis, impact factor, h-index, webometrics and academic social networking sites. In this light, this chapter presents a review of these concepts as it considers relevant theoretical underpinnings that are related to the content marketing of scholars. Therefore, this contribution critically evaluates previous papers that revolve on the subject of academic reputation as it deliberates on the individual researchers' personal branding. It also explains how metrics are currently being used to rank the academic standing of journals as well as higher educational institutions. In a nutshell, this chapter implies that the scholarly impact depends on a number of factors including accessibility of publications, peer review of academic work as well as social networking among scholars.

In publishing their education background together with the professional experience, users make LinkedIn a privileged web source for understanding "University-Industry" connections. Precisely, the main goal of this study is to test LinkedIn as a valid source for analyses oriented to the quantification of the university-industry interactions. To this end, the authors propose two different procedures (method A: direct through the URL mentions between LinkedIn profiles; and Method B: indirect through the information from LinkedIn University Pages), comparing them against the direct procedure based on URL mentions between official websites (Method C). To do this, the authors have selected the whole Spanish academic system. The results show that method A is unusable yet due to the low web connectivity between LinkedIn profiles, while method B provides reliable though too volatile data that complements method C, which reveal in turn relations of different nature.

According to Klamm (2011), some of the best practices for universities seeking to build effective social media engagement are: 1) to develop a strategy and set goals; 2) to pick and choose the platforms; 3) to empower and support individual departments; 4) to put guidelines in place; 5) to develop a consistent voice across platforms; and 6) to communicate across campus. In the following pages, we will delve more deeply into Twitter for universities. After checking 10 reasons why universities should take Twitter seriously, we will offer some good practices for universities willing to use Twitter in a way that optimizes the tool; that is, not only for disseminating information, but also for talking with the community and taking advantage of its multimedia potential. We will illustrate these good practices with screenshots to better show what we mean and also to reinforce their utility with contributions from some of the early theorists on what the proper use of social media should be.

This document aims to show how teachers can make their mark in the territory through the appropriate use of their personal brand and geomarketing techniques, with the example of the author, and how this methodology can be transferred to students in their connection with the professional world. It is a process of structuring and decentralization of personal branding in the territory, linking physical space with online space and where geography acquires a new integral dimension through SoLoMo methodology (Social, Local, and Mobile).

Preface

Traditional methods of marketing for academics work include attending conferences and publishing in journals and books. These are still very important, but personal branding increasingly takes place on the Internet. The reputation management is a relatively new field for academic research and it is a topic for study in fields as diverse as marketing, management, economics, accounting, finance, communications, sociology, political science and cultural studies. It improves the ability to process and manage information and knowledge-related processes in order to create new knowledge to increase personal strategic capability and organizational performance. In the fields of information studies, information systems, knowledge management, etc., there exists a need for an edited collection of articles in this area.

The rapid progress on computer networks and pervasive computing has offered the base conditions for the establishment of a new Academic Branding scene and they have emerged during the last years as a result of the challenges faced by the business, social, and scientific worlds and enabled by the fast progress in the information and communication technologies. Advanced and highly integrated social networks, virtual organizations, virtual (professional) communities, virtual laboratories / e-science, are illustrations of a major trend in which entities seek complementarities and join efforts that allow them to better participate in challenging and competitive opportunities. However, no one will know about your work unless you tell them about it. As a career researcher, it is absolutely essential to have in mind the academic branding, both professionally and personally.

While in academia remains vital publishing in the impact publications increasingly open access journals and social media are making more visible the work of researchers beyond the publication in a journal with impact. Although a factor does not exclude the other relevant publications if it is true that renowned authors have seen increased its visibility through good management of their digital image even this visibility has served to promote their publications with impact revealing all his scientific production.

We have shown in this book the unstoppable force of digital brand in the academic environment and how it is here to stay as a new way to manage our image as a researchers and teachers. This book tries to give lines about important aspects in the digital brand to improve their personal brand from different points of view. His greatest contribution is the specialization in the field of academia since the digital brand management is not at all similar in the business environment that in the academic environment.

Science should be for and by the citizens, so that researchers should disclose their research for the common good, especially because in many cases the financing of their projects is public so the moral accountability front citizens should be paramount when disseminate their work. The digital personal brand of the researchers not only serves to publicize their work but to approach colleagues who may be interested in working in the same field. The platforms 2.0 allow us to meet colleagues who are working in similar issues as us and follow their work, they can create now what in other decades was called "invisible colleges". It is essential fot the sciences to know each other researchers working on similar themes to collaborate in their research which will provide more contrasted and expanded results. For all researchers the digital presence is presented as a tool in their research not only for the dissemination of its work and in creating networks among colleagues who research in the same field.

OBJECTIVE OF THE BOOK

This book will aim to provide relevant theoretical frameworks and the latest empirical research findings in the area. It is written for professionals who want to improve their understanding of the strategic role of trust at different levels of the information and knowledge society.

A growing number of personal branding marketing can be observed in many domain areas. However, the developments and even the understanding of these cases have suffered from ad-hoc approaches, being urgent to establish a proper theoretical foundation for the area of the academic world. Furthermore, the fast developments in the area and the nature of the paradigm configure the emergence of a new discipline, which needs to be built on a sounder theoretical basis.

TARGET AUDIENCE

The target of the book is researchers, academics, practitioners working in the field of information and communication technologies and knowledge management in various disciplines (e.g., education, library science, sociology, information and communication sciences, computer science and information technology), educational technologists and individuals who are interested in how the academic branding and emerging technologies will impact on actual society and the social implications that surround the reformulation and fluidity of online academic reputation.

ORGANIZATION OF THE BOOK

The book is organized into 12 chapters. Here you can see a brief description of each of them:

Chapter 1: New Tools for Online Academic Dissemination – Let the World Know Your Best

The chapter identifies new ways for academic dissemination. Many researchers are uncomfortable with the idea of promoting themselves or their work, but it is as important to report on it as it is to do it, no one will know about your latest work unless you tell them about it. The chapter helps the authors to identify in what are they good at and what are they best at, not just to let others talk about them. As a career researcher, it is absolutely essential to promote yourself. The chapter analyzes the importance of the Google profile as well as the Google Scholar profile and the importance of having an own web page or blog. In this chapter the authors offer an overview of some new digital tools that can help to improve the online academic reputation.

Chapter 2: We the New Media: The Disruption of Social Media in Interpersonal and Collective Communication

The disruption of social media in interpersonal and collective communication. The Internet have led to the emergence of a new communication ecosystem that is not restricted to the online context and in which professional and social media intersect and cohabitate to compete for the attention of audiences. This chapter presents the new ecosystem in cyberspace; a placeless space where new forms of power, influence, control, and management of collective perceptions have developed. The author shows the new emerging macro social platforms that are giving rise to a new paradox: people need social media to express their opinions and creativity, while the major platforms tend to view users as products. Finally the chapter suggests a new model to describe and explain this new communication ecosystem that has put an end to the mediating exclusivity of professional media and maximizes collective interpersonal communication on one and the same social continuum.

Chapter 3: Building Academic Branding – The Digital Branding as Academic Footprint

This chapter present the background for the academics. The researchers are forced to incorporate themselves in the digital field. The research and professors have a responsibility to society, popularizing science, this represents to value their own work and contribute to its visibility, and one starting point is to have "digital academic branding". This chapter explains step-by-step techniques that develop a digital academic branding, starting from an interior analysis of the academic up to the formulation of the visual and conceptual brand; it is important to do it in a sequential manner every one of the activities described here.

Chapter 4: Flow My Bits, the Professor Screened – Netnography, Academic Micro-celebrity, and Personal Branding

This chapter uses the participative stance in netnography to discuss academic personal branding, but also interweaves advice, core principles, and a series of short inductive case studies about personal academic brands and fields. Featuring cases like those of Plato, Timothy Leary, John Lilly, and others, the chapter outlines a theory of personal academic branding, relating core identities to academic branding practices and purposes. In particular, this chapter suggests the relationship between thinking innovative thoughts, representing them to various audiences, and having one's idea shared and built upon by a network of interested others. Core to the successful realization of this endeavor are the following nine elements: unconventionality, creativity, internal and external audiences, segmentation, core constituents, co-branding, simplicity, uniqueness, and catchiness.

Chapter 5: Social Media as a Tool, and Tools Offered by Social Media, to Teachers and Researchers – Personal or Professional Use?

The chapter provides a clear vision for the social networks environment from the self-promotion point of view. Chapter focuses on organizing tools, audience, and type of publications. Tools are organized to contextualize their use and to give a proper understanding of the relevant contents that can be published. Audience is presented according to the relations and interests with the teacher and researcher. Simultaneously, this chapter gives a vision of the privacy scope or the publications, and provides an evaluation mechanism to distinguish the most convenient area of publication depending of the message content. Following submission of these analyses, chapter focuses on the teacher and research activity and how to promote these activities through social networks. The chapter ends with a set of suggestions to make a strategic use of new media with the goal of promoting efficiently personal brand as a teacher and researcher.

Chapter 6: Academic Community Manager – Manager of the Academic Community

The chapter shows the important role that will have the academic community manager inside the group of researchers. The academic environment is changing, especially with the nominated web 2.0 and science 2.0. The increase in new academic social network used for researchers to publicize their knowledge, research and contributions in science, cannot take lightly. For this and everything that comes with being a researcher, the academic community manager comes to manage the social networks and academic social network, care workflow, generate networking and disseminate results, among others. Increasingly universities seeking to spread the results generated by their students, teachers, masters, doctors or research groups.

Chapter 7: Brand Journalism and Personal Branding in Academia

The chapter seeks to highlight the importance of brand journalism and personal branding within the university context. First the chapter focus is on determining how universities are adapting to the changing needs of their audience by utilizing new narrative techniques and analyze whether brand journalism in turn allows for the construction of personal branding. In the chapter the author suggests how to know whether teachers coming from the Schools of Communication / Information of universities in Spain have created their personal brands, in addition to what criteria and tools had to be taken into consideration for its creation. In conclusion, a lack of progress in regards to the development of a personal brand as well as personal branding in academia is observed in this chapter.

Chapter 8: Utilizing Content Marketing Metrics and Social Networks for Academic Visibility – Content Marketing for Academic Impact

There are numerous assumptions on research evaluation in terms of quality and relevance of academic contributions. Researchers are becoming increasingly acquainted with bibliometric indicators, including; citation analysis, impact factor, h-index, webometrics and academic social networking sites. In this light, this chapter presents a review of these concepts as it considers relevant theoretical underpinnings that are related to the content marketing of scholars. Therefore, this contribution critically evaluates previous papers that revolve on the subject of academic reputation as it deliberates on the individual researchers' personal branding. It also explains how metrics are currently being used to rank the academic standing of journals as well as higher educational institutions. In a nutshell, this chapter implies that the scholarly impact depends on a number of factors including accessibility of publications, peer review of academic work as well as social networking among scholars.

Chapter 9: From Universities to Private Companies – A Measurable Route of LinkedIn Users

The main goal of this chapter is to present the study to test LinkedIn as a valid source for analyses oriented to the quantification of the university-industry interactions. To this end, the authors propose two different procedures (method A: direct through the URL mentions between LinkedIn profiles; and Method B: indirect through the information from LinkedIn University Pages), comparing them against the direct procedure based on URL mentions between official websites (Method C). To do this, the authors have selected the whole Spanish academic system. The results show that method A is unusable yet due to the low web connectivity between LinkedIn profiles, while method B provides reliable though too volatile data that complements method C, which reveal in turn relations of different nature.

Chapter 10: Good Practices for Universities on Twitter

The chapter presents a study made after checking 10 reasons why universities should take Twitter seriously, the authors offer in this chapter some good practices for universities willing to use Twitter in a way that optimizes the tool; that is, not only for disseminating information, but also for talking with the community and taking advantage of its multimedia potential. The author illustrates these good practices with screenshots to better show what he mean and also to reinforce their utility with contributions from some of the early theorists on what the proper use of social media should be.

Chapter 11: The Use of Geomarketing to Promote Personal Branding – Make Your Mark as a teacher in the Territory of Internet

This chapter aims to show how teachers can make their mark in the territory through the appropriate use of their personal brand and geomarketing techniques, with the example of the author, and how this methodology can be transferred to students in their connection with the professional world. It is a process of structuring and decentralization of personal branding in the territory, linking physical space with online space and where geography acquires a new integral dimension through SoLoMo methodology (Social, Local and Mobile).

Chapter 12: Audiobranding and Its Importance to Your Personal Brand – New Ways to Communicate Our Messages through Sound

In this chapter the author tries to discover the importance of audio and voice as a key element in building a personal brand. Podcast, audioblogs and web tools that reference the sound are analyzed based on their degree of influence on personal branding. By using the voice we can differentiate ourselves from our competitors, so the audio branding, or sonic branding as it is also known, shows a verbal identity of his own personality. The sound has always been known for being a great tool to convey memorable messages to consumers, in fact we are able to consume the sound since we are in the womb. Having an audio brand is one way to ensure that no other company uses a similar audio and make our brand consistent. The radio language, consisting of voice, music, silence and the special effects are responsible for generating emotions in the receiving public. The audio conveys information, entertains consumers and in the long term, helps build a positive image that reinforces the brand values. The design of an audio brand goes beyond creating a catchy jingle or a piece of melody.

Acknowledgment

We would like to thank the authors for their contributions to this monograph, all of which have been made on a voluntary basis. And of course we would like to thank all those who made it possible for us to finish this book, particularly our families and our editor, Courtney Tychinski, who has had a bunch of patience and has done a great work along the project.

Chapter 1
New Tools for Online Academic Dissemination:
Let the World Know Your Best

Margarita Cabrera Mendez
Polytechnic University of Valencia, Spain

Nuria Lloret-Romero
Polytechnic University of Valencia, Spain

Rebeca Díez-Somavilla
Polytechnic University of Valencia, Spain

ABSTRACT

Many researchers are uncomfortable with the idea of promoting themselves or their work, but it is as important to report on it as it is to do it, no one will know about your latest work unless you tell them about it. It is very important to be able to describe what you are good at and what you are best at, not just to let others talk about you. As a career researcher, it is absolutely essential to promote yourself. In addition to your Google profile and your Google Scholar profile you are strongly advised to have your own web page or blog, where you can post your work history, latest developments, report on goals, projects, news and a lot more. Why wait until someone else talks about your news if you can do it yourself? In this chapter we offer an overview of new digital tools that can help to improve your online academic reputation.

INTRODUCTION

This chapter puts forward some recommendations for knowing who is who in academic networks and in the online world relating to specialist fields of knowledge. It will deal with the basic tools for discovering the current position of academics and what people are saying about their work, in order to report more widely on advances in research, to facilitate access and academic exchange with other researchers worldwide, creating your own brand as a specialized academic.

DOI: 10.4018/978-1-5225-0917-2.ch001

What happens when someone looks for your name on Google? What about when we look for our lines of research? Are we or our research group in there? Are they the desired results? Let's support the development of a vision of your unique contribution (your "brand"). Create a plan to update your brand, and offer guidance on creating a platform consistent with your brand.

Many researchers are uncomfortable with the idea of promoting themselves or their work, but it is as important to report on it as it is to do it, no one will know about your latest work unless you tell them about it. It is very important to be able to describe what we are good at and what we are best at, not just to let others talk about us.

As a career researcher, it is absolutely essential to promote yourself. In addition to your Google profile and your Google Scholar profile you are strongly advised to have your own website page or blog, where you can post your work history, latest developments, report on goals, projects, news and a lot more. Why wait till someone talks about your news if you can do it yourself? If you can, it is very useful to buy your own domain with your name or a name identifying your highly specialized work. With that domain you can publish your blog or webpage, or if you prefer you can use a basic personal page as www.about.me to report about you. It allows you to have a personal landing page, you can add widgets to all your social media outlets and web presences. It is easy to format, no coding needed, simply drop in a background, pick a colour scheme, and write your story. It is the perfect home base to put your professional skills on. This kind of tools helps a lot with your online personal brand and they are very easy to use. Furthermore, you can look for other professionals like you worldwide. Remember to link all your profiles on different networks, your page will be like your digital home where anybody can find all the information about you, most of it: the search engines will do!! Don't forget to keep your content updated with your latest professional news. Remember: you need a home page, it doesn't matter if it is a blog, a webpage or an about.me page, you need a place to talk about yourself and to let others know who you are.

BACKGROUND

"Personal branding is how we define ourselves in the work space while at the same time, incorporating the personal elements that make us who we are" (Rasmussen, 2012). We also agree with the definition of personal branding in Wikipedia: "personal branding is the practice of people marketing themselves and their careers as brands. While previous self-help management techniques were about self-improvement, the personal-branding concept suggests instead that success comes from self-packaging. Personal branding also involves creating an asset by defining an individual's body, clothing, physical appearance, digital and online presence and areas of knowledge in a way leading to a uniquely distinguishable, and ideally memorable, impression. The term is thought to have been first used and discussed in a 1997 article by Tom Peters (1997). Personal Branding is essentially; the ongoing process of establishing a prescribed image or impression in the mind of others about an individual, group or organization. Personal branding can often involves the application of one's name to various products or services" (Peters, 1997).

Everyone has personal branding; it is intrinsically part of all of our personality and ourselves. The way we talk, the way we act, how we do things and how we think, our style of dress, our lifestyle, how we sit, move, and relate to other persons, how we shake hands and even who our friends are or about our partner… each of our roles and movements are important, they define us and our personal branding. A teacher's personal brand precedes them before they enter the classroom; it is all about what people are saying about other people even before they know them. Not having a reputation is as bad as having a bad one. Academics must leave a trace, they and their work must have a good reputation. Including their relationship with the students in networks as Facebook, to improve the engagement on the class (Junco 2012).

Dilip. S. Mutom wrote in his ebook about social media for researchers. "Personal branding is not just for celebrities or those in the corporate sector. As a researcher, adopting the techniques and practices developed in business schools to create your own personal brand can be a powerful tool for marketing yourself. Many researchers are uncomfortable with the idea of promoting themselves or their work. The general opinion seems to be "let your work do the talking". However, no one will know about your work unless you tell them about it. As an early career researcher, it is absolutely essential to promote yourself" (Mutom, 2011).

OBJECTIVES

Within this chapter the aims are to:

- Identify what matters for online visibility.
- Know the importance of what people say about others or their work online.
- Know who is who in their international academic field across the social media community.
- Develop a clear, concise dissemination of research and teaching contributions.
- Create an academic "brand" consistent with academic contributions.
- Think through the communication changes needed to fully disseminate academic work.
- Know the tools for updating a brand through a teaching and research portfolio.
- Design and develop Internet visibility through publications, ePortfolio, website, or academic blog.
- Select the most adequate format for disseminating your work.

THE IMPORTANCE OF ACADEMIC BRANDING

As Dilip. S. Mutom said: "Recognition by your peers is likely to be taken into account in the Research Excellence Framework as well as university appointments and promotions. Universities are increasingly looking at the contribution of academics in terms of their public profile and the impact of their research beyond academia. Traditional methods of marketing your work include attending conferences and publishing in journals and books. These are still very important, but personal branding increasingly takes place on the Internet" (Mutom, 2011).

The Center for Corporate Reputation Management is the first centre at a U.S. university to focus exclusively on issues related to reputation management. The Centre serves as a focal point for the exchange of theory and best practices among business leaders, professors and researchers. They say "reputation management is a relatively new field for academic research and it is a topic for study in fields as diverse as marketing, management, economics, accounting, finance, communications, sociology, political science and cultural studies. Today, damage to reputation is one of the greatest concerns of CEOs. Trust is critical in a market economy. Without trust, people do not invest; the best talent does not join a company; good employees leave; and the competitive position of the company erodes" Heath & Ni, 2008). It is as important for companies to be trusted as for Universities or Research centres, they need people with a great reputation, and nowadays there is no better way to improve it than by using a communication plan to promote our team's personal brand, if we are good we have to let the world know we are (of course we have to be good to say something interesting). Everything we do to improve our personal brand will help to protect and enhance the reputation and brand of our organization, company, university or researcher centre.

Personal reputation is an opinion about a person, typically as a result of social evaluation on a set of criteria. It is important in business, education, online communities and many other fields. Reputation may be considered, as a component of identity as defined by others, it is the result of what you do, what you say, and what others say about you. There is another definition that says "it is what people are saying around you when you are out", our personality footprint.

Our personal reputation in the academic field is for example what students, colleagues or former students, say about us, not just as teachers, but also as people. If we are good at listening or lecturing, if we contribute to students' enjoyment, growth, and understanding, that will get around by word of mouth to future generations. It is the same in our research field, if we do a good job and we are also able to communicate it through our publications and presenting it in workshops and events, our professional reputation will be enhanced.

Wikipedia has a clear and very easy explanation of academic visibility: it relates to dissemination, accessibility, and recognition of scholarship produced by university teaching staff, especially in terms of promotion and tenure criteria and academic rank. Traditionally, "visibility" was a function of printed output like books, book chapters, and peer-reviewed journal articles primarily available in university libraries. This is rapidly changing with the meteoric rise in electronic and online indexing of these materials. Scholarly recognition and reputation grows as visibility increases from works being frequently referred to as authoritative.

Digital Branding

It is very important for you to be able to describe what you are good at and what you are best at, not just to let others talk about you. Many researchers are uncomfortable with the idea of promoting themselves or their work, but it is as important to report it as it is to do it, no one will know about your latest work unless you tell them about it. As an early career researcher, it is absolutely essential to promote yourself. In addition to your Google profile and your Google Scholar profile you are strongly advised to have your own web page or blog, where you can post your work history, latest developments, report on goals, projects, news and a lot more. Why wait until someone talks about your news if you can do it yourself?

As we said earlier the importance of the Internet in our personal brand is increasing, we agree with the Wikipedia explanation which says that information and communications technologies are rapidly changing how academic achievements and reputation can be assessed. The Internet is becoming an all-purpose source for scholarship. "Web mentions" and URL citations are analogous to journal citations for scholarly work appearing or referenced on web sites or blogs. Like article citations, web citations can represent the noteworthiness of a scholar's contributions. And, like frequent academic journal citations, a wide range of web mentions add to an academic's reputation and prestige.

The advantage of assessing a wider range of academic output conveys breadth and scope; it takes a more holistic view of an academic's body of work, or figuratively, an overall "academic footprint". An assessment of someone's academic footprint and/or visibility is similar to that of the actual tenure review process because it can include nearly all of the activities undertaken by a faculty member, including but not limited to: dissertation, book reviews, conference proceedings, conference presentations, research reports, funded grants, and teaching activities. Webometrics can be used to assess an academic's visibility based on "web mentions" or references to academic work by web pages. Metrics include the number of URLs, Google Pagerank, and the number of sources linking to associated pages (backlinks). The importance of all of our actions on the web is crucial to our effort to improve our personal branding and academic visibility.

Today, our names are constantly being looked up online: whether we are looking for employ mentor looking for a specialist, whenever we meet new people or when we are just engaging in online networking. The famous fashion designer Coco Chanel said "no one has a second chance at a first impression" and it's true, the impression left with someone when we meet them for the very first time is very important and is usually unforgettable. Today this first impression is moving online too, when someone searches for us or our work, when the results are about our research, publications or lessons, the searcher will form a strong opinion about us and our work. That is the impression we make on the Internet and on the mind of others, it is our digital personal brand. Do we know what is said about us online? Do you "google" yourself often?

Effective Online Tools

Social media tools have the power to put you in contact with millions of people around the world. In order to build your online identity you need to know all of the possibilities, as well as how all of your activities are working together. It is important to be consistent so that each part of your social media network is contributing positively to the brand you are trying to build. Of course there are a lot of tools, and the more you use them, the more it increases the visibility of your work. But let's focus on the most important ones.

Google Alerts

It is important to know what is said about you or your brand, where it is said and how influential those people are who are talking about you. The easiest way to do it is with a Google search using different variations of your name: your forename + surname(s), married surname + name, nickname + surname… This will give you an estimate of the position of your brand on the Internet, and will show the 'possible' competitors with the same name as you. If we have a common name or a famous name (because it is the same name as an influential or famous person) it will be very difficult to appear among the top places in the results. But of course we cannot do this every day, so the best way to have information about our reputation online is to use the Google Alerts tool (www.google.com/alerts). With this tool you can receive an email with an alert every time someone writes your name on the Internet; it is useful to have an overview of what is happening around you.

Google Alerts are, as you can see from their help page (https://support.google.com/alerts), emails sent to you when Google finds new results, such as web pages, newspaper articles, or blogs, that match your search term. You can use Google Alerts to monitor anything on the web. For example, people use Google Alerts to:

1. Find out what is being said about their university.
2. Monitor developing new research.

3. Keep up to date on a competitor or sector.
4. Get the latest news on a research group.
5. Find out what's being said about themselves.
6. Monitor their research.

To find out what's being said about you or your work, you just have to enter your full name in quotes to receive an email when new stories, blogs or web pages mention your name. Example: ["Marga Cabrera"]. If you have a common name, you can use negative terms to remove irrelevant results. For example, if you have the same name as a professional football player, try adding -football to your query. Example: ["Marga Cabrera" -football]. If you get a lot of results from a site that you aren't interested in, you can exclude results from that site. For example, if you don't want results from twitter.com, add [-site:twitter.com] to your query. Example: ["Marga Cabrera" -site:twitter.com].

Google Account

A Google account lets you access a variety of Google products such as Gmail, Google+, YouTube, Google docs and many more. A Google account includes a Gmail address and a Google profile, which will help you have a richer, more personalized experience across Google products. You can create a Google account by going to the Account Creation page, or by clicking the Create an account button on the top-right corner of any Google sign-in page. You can use your Google account with all Google products, so once you create an account – regardless of whether you do it through the Account Creation page or through a specific product's sign-in page – you can use that username and password with any Google product.

But remember, we do so in particular to improve the information about our work and us when someone searches for us. It is important as well to have a Google account (www.google.com/account) if you want to tell Google about yourself. It helps if you give your own information and not just what the searcher can find, so if someone searches for you, Google will have its own information provided by you. For example it is useful to give your name and surname plus your areas of interest such as your research specialism, your university or your research group.

You can choose how much or how little to share with people, the name is the only part of a profile that must be public. If you choose to share other information, such as interests, hometown or email address, you can decide who gets to see it. On your profile and privacy settings page, you can change the information on your profile, as well as who can see what details. You can prevent your profile from showing up in Google web search results by changing your settings, though it may show up in other contexts.

Personal Domain and Landing Page

If you can do it, it is very useful to buy your own domain with your name or a name identifying your highly specialized work. With that domain you can publish your blog or a webpage, or if you prefer you can use a basic personal page such as www.about.me to provide information about yourself.

About.me allows you to have a personal landing page, you can add widgets to all your social media outlets and web presences. It is easy to format; no coding needed, simply drop in a background, pick a colour scheme, and write your story. It is the perfect home base to put your professional skills on. And Google recognizes it as important data associated with your name, so it will be shown in the top results.

This kind of tools helps a lot with online personal branding and they are very easy to use. Furthermore, you can look for other existing professional people in your area worldwide in about.me. Remember to link all your profiles across different networks, your page is like your digital home where anybody can find all your information, most of it: the search engines will do!! Don't forget to keep your content updated with your latest professional news. Remember: you need a home page, it doesn't matter if it is a blog, a webpage or an about.me page, you need a place to talk about yourself and to provide information on who you are. And you can use this page and your domain as your email signature; with just one address you are delivering all your background.

Google Scholar

This is an essential tool for many academic purposes. As they say on their website (http://scholar.google.com/intl/en/scholar/about.html) Google Scholar provides a simple way to broadly search for scholarly literature. From one place, you can search across many disciplines and sources: articles, theses, books, abstracts and court opinions, from academic publishers, professional societies, online repositories, universities and other web sites. Google Scholar helps you find relevant work across the world of scholarly research.

The features of Google Scholar are:

1. Search all scholarly literature from one convenient place
2. Explore related works, citations, authors, and publications
3. Locate the complete document through your library or on the web
4. Keep up with recent developments in any area of research
5. Check who's citing your publications, create a public author profile

Google Scholar aims to rank documents the way researchers do, weighing the full text of each document, where it was published, who it was written by, as well as how often and how recently it has been cited in other scholarly literature.

Google Scholar Citations provide a simple way for authors to keep track of citations to their articles. You can check who is citing your publications, graph citations over time, and compute several citation metrics. You can also make your profile public, so that it may appear in Google Scholar results when people search for your name or your publications. Best of all, it's quick to set up and simple to maintain – even if you have written hundreds of articles, and even if your name is shared by several different scholars. You can add groups of related articles, not just one article at a time; and your citation metrics are computed and updated automatically as Google Scholar finds new citations to your work on the web. You can choose to have your list of articles updated automatically or review the updates yourself, or to manually update your articles at any time.

Google Scholar Metrics provide an easy way for authors to quickly gauge the visibility and influence of recent articles in scholarly publications. Scholar Metrics summarize recent citations to many publications, to help authors as they consider where to publish their new research.

To get started, you can browse the top 100 publications in several languages, ordered by their five-year h-index and h-median metrics. To see which articles in a publication were cited the most and who cited them, click on its h-index number to view the articles as well as the citations underlying the metrics.

You can also explore publications in research areas of interest to you. To browse publications in a broad area of research, select one of the areas in the left column. For example: Engineering & Computer Science or Health & Medical Sciences. To explore specific research areas, select one of the broad areas, click on the "Subcategories" link and then select one of the options. For example: Databases & Information Systems or Development Economics.

Browsing by research area is, as yet, available only for English publications. You can, of course, search for specific publications in all languages by words in their titles.

To start you can search for any article at http://scholar.google.com, but it is much more useful if you have your own account and update your articles, books and publications. You can follow other authors or items as well. We strongly recommend you to use this tool to establish your online brand and improve your reputation, as well as to meet other professionals within your area and network with them.

Linkedin

This is a business-oriented social networking site where the users can join and follow groups and companies. It is very useful to be in touch with colleagues around the world and to discuss issues of common interest.

Academic professionals can greatly benefit from LinkedIn as can business people, and groups on the site offer a great opportunity for collaboration, discussion, and sharing between academics. There are so many academic groups available on LinkedIn, we can't possibly list them all, but OnlineUniversities.com has collected these 50 as a great place to start. Check them out to find networking groups, news, professional organizations, and more: http://www.onlineuniversities.com/50-great-linkedin-groups-for-academics

Also how you show your curriculum is a very good method of first presentation, millions of people use this tool to look for workers or colleagues. It allows you to build an online profile that features your experience and skills as well as to network with other users in a professional environment, so it provides a great way to connect with contacts from your research world. Take advantage of the many opportunities that this network can provide you with by stating who you are and who others are.

Academia

Academia.edu focuses on academic connections; it is essentially a sort of LinkedIn focusing on the academic world, so it provides improved features for things like listing publications. Also, going back to our personal brand online, LinkedIn and Academia.edu profiles tend to feature high in Google searches, so a well-constructed profile can be a great way to develop your online brand.

At "23 things for research" on the online learning programme for researchers, students and staff at the University of Oxford it says that Academia.edu has less functionality, but it is geared towards academic activities (McCarthy & Lindsay, 2014). You can write update posts on your activities, upload papers and other documents, which might include 'grey' literature such as conference papers as well as links to your journal articles. It can also tell you how many people have viewed your profile, what keywords they used to find you, and who is following your work. You can also follow the profiles of other scholars, which is useful for keeping up to date with people's publications.

Twitter

Twitter is a micro-blogging platform, allowing you to share updates with your followers in 140 characters or less. It is a great way to share content about your interests and research. You can search just about anything you are interested in and join conversations in real time, find the latest news, and connect to users with similar interests.

Twitter is what you make of it, and its flexibility is one of its greatest strengths. It's important to have a hashtag or a keyword, that others can search for when you want to communicate with networks about a subject, you can find hashtags used by other users, or you can invent your own. The right way to act is to share and participate with value-added content, not just promoting your own work but to disseminate the work of others. It is very important to tweet with added value content related to your research area of interest; this will help with the positioning of your brand.

The Twitter search engine is very powerful for finding content or people discussing a concrete area of interest. It helps you to get to know who is who in your area all over the world, and the latest news around an issue. Because all of that, it is essential to be active on this network, sharing and participating, not just looking at users' comments.

Google+

Google+ allows users to group their contacts into circles; this allows users to regulate who sees what information. Be mindful of what content you make public. Content shared on Google+ shows up in Google search results, so make sure it is positive. Each user has a profile that displays information about them. Normally you can use your Google account to improve this network.

Klout

Klout gives you a way to gauge the effectiveness of your online branding and social media efforts. It provides a score that tells you how influential you are, tells you what areas you are influential in, your personal reach and the breadth of your posts.

CONCLUSION

We have discussed the importance of personal branding and online reputation, not just as an academic but as a person. It is very important to be aware of what are people saying about us and what we want the Internet to show about our work and ourselves. It is not only important for our research reputation and our brand or curriculum and where we publish our work, it is important what people say about our work and history.

Remember to be aware about what the Internet says about you and don't forget that you can influence online sharing value-added content referring to your work on all the network and social media tools we have been talking about. Your personal brand will appreciate it, and your institution will do so as well. Nowadays it is as important for the faculty as for the institution, the more good and valuable information which can be found about the persons in an institution, the more important it will be. And if the information is important in a related theme, this will help the institution to grow in the rankings. The benefits for researchers are many, spreading your work will help you to discover new colleagues, and will make the academic network grow, bringing a lot of new opportunities.

REFERENCES

Heath, R. L., & Ni, L. (2008). *Corporate Social Responsibility*. Retrieved September 2, 2015, from http://www.instituteforpr.org/corporate-social-responsibility

Junco, R. (2012). The relationship between frequency of Facebook use, participation in Facebook activities, and student engagement. *Computers & Education*, *58*(1), 162–171. doi:10.1016/j.compedu.2011.08.004

McCarthy, L., & Lindsay, K. (2014). The 23 Things for Research. *23 Things blog*. Retrieved October 14, 2015, from http://blogs.bodleian.ox.ac.uk/23things/schedule/

Mutom, D. S. (2011). *Social media for researchers and online personal branding*. University of Warwick. Retrieved October 13, 2013, from https://www.academia.edu/2612640/SOCIAL_MEDIA_FOR_RESEARCHERS_and_ONLINE_PERSONAL_BRANDING_e-book_

Peters, T. (1997). *The Brand Called You*. Retrieved December 15, 2015, from http://www.fastcompany.com/28905/brand-called-you

Rasmussen, D. (2012). *Forget Job Security: Build your Marketability!* CreateSpace Independent Publishing Platform.

Chapter 2
We the New Media:
The Disruption of Social Media in Interpersonal and Collective Communication

Miguel del Fresno
Universidad Nacional de Educación a Distancia (UNED), Spain

ABSTRACT

The Internet have led to the emergence of a new communication ecosystem that is not restricted to the online context and in which professional and social media intersect and cohabitate to compete for the attention of audiences. This ecosystem occurs in cyberspace; a placeless space where new forms of power, influence, control, and management of collective perceptions have developed. These emerging macro social platforms are giving rise to a new paradox: people need social media to express their opinions and creativity, while the major platforms tend to view users as products. A new model is presented to describe and explain this new communication ecosystem that has put an end to the mediating exclusivity of professional media and maximizes collective interpersonal communication on one and the same social continuum. It is a new interpersonal and collective communication ecosystem whose deep logic is necessary to understand and in which academics must compete.

INTRODUCTION

Communication can be understood as the act of sharing "meaning through the exchange of information", which is defined "by the technology of communication, the characteristics of the senders and receivers of information, their cultural codes of reference and protocols of communication, and the scope of the communication process" (Castells, 2009, p. 88). Communication is also defined as "a binding force in social relationships without at the same time being visible or having tangible and permanent forms… within a given structure of relationships and have consequences for this structure without being readily open to observation" (McQuail & Windahl, 1993, p. 4). Communication is therefore a property of col-

DOI: 10.4018/978-1-5225-0917-2.ch002

lective interpersonal action rather than of those who engage in it, although anyone may be a potential communicative agent.

The *Era de la información* (the Information Age) (Castells, 2001), or the *Société de l'information* (the Information Society) (Mattelart, 2003), is characterized by the social changes that have occurred since the late twentieth century as a result of the evolution of digital information and communication technologies as well as the emergence of a social network structure that has affected all areas of human activity on a global scale. Social changes brought about by technological innovations have always given rise to "changes in the order of interpersonal relationships" (del-Fresno, 2011a, p. 20). The uninterrupted technological spans have been presented as a gradual extension of our senses. As McLuhan (1964) stated, "we have extended our central nervous system in a global embrace, abolishing both space and time as far as our planet is concerned" (p. 3), thus leading to the growing technologization of reality (Virilio, 1991). This means that new media and technologies are transforming not only the "*how*" of communication, but the "*meaning*" of what is being communicated.

The Internet is one of the most disruptive technological phenomena for communication in history, especially collective interpersonal communication. The Internet, even in the second decade of the twenty-first century, is the result of decisions taken regarding its architecture and code in the early 1990s. Indeed, "the first generation of these architectures was built by a noncommercial sector—researchers and hackers, focused upon building a network" (Lessig, 2009, p. 7). As William Mitchell (1996) put it, the "code is cyberspace's law" (p. 111) or as Joel Reidenberg (1998) called it, "*Lex Informatica*" or "*control of code is power*" (Lessig, 2009, p. 79). As regards the social media in particular, the code is "the unavoidable boundary around which no detour exists in order to participate fully in modern life. It is ubiquitous" (Berry & Pawlik, 2008, p. 58).

The code has a performative function because it is the condition for allowing communication on the Internet, and the means by which the expressive capacities of people are enhanced or restricted, and which facilitates, or not, the diversity of communicative practices in cyberspace. Thanks to the Internet, the previous communication ecosystem has been reabsorbed and a new one comprised of numerous and diverse players (both professionals and non-professionals) has emerged that enables the constant production and exchange of exponential amounts of data and information anytime and anywhere.

Early on, it was anticipated that this phenomenon would be a collective problem in the form of *information overload* (Toffler, 1970; Toffler & Toffler, 1994; Lewis, 1999) or an *accident*: "the damage caused by the explosion of unlimited information" will be "the great accident of the future, the one that comes after the succession of accidents that was specific to the industrial age" (Virilio, 1995, para. 14), where each technology entails its own accident. It is estimated that, thanks to technological development, more information was available in the second half of the twentieth century than ever before in the history of humankind (Feather, 2008) and that the Internet has played a key role in information overload since its entry into society (Swash, 1998).

Following the expansion of the Internet at the global scale, there emerged a "new form of interactive communication, characterized by the capacity of sending messages from many to many, in real time or chosen time, and with the possibility of using point-to-point communication, narrowcasting or broadcasting depending on the purpose" (Castells 2009, p. 55), what Castells referred to as *mass self-communication*. It is, at one and the same time, "mass self-communication because it can potentially reach a global audience" and "self-communication because the production of the message is self-generated, the definition of the potential receiver(s) is self-directed, and the retrieval of specific messages or content… is self-selected" (Castells, 2009, p. 55). This process "has broadened the collective interpersonal communication system, which has had an effect on the reconfiguration of social reality itself' (del-Fresno, 2015, p. 75). Therefore it makes perfect sense to examine the communicative practices arising from the intersection of sociability and communication in the Internet, whose result is understood as an intentional, interpersonal and collective, global and local social action around communication that is not restricted to the digital ecosystem.

These new forms of communication marked a break with the dominant communication model imposed by the mass media prior to the emergence of the Internet. In the mass media, people were relegated to being receptive and more or less passive audiences. In contrast, social media have facilitated the emergence of audiences with the ability and capacity to be senders and receivers, the *micromedia* (del-Fresno, 2011b), thus causing a growing process of disintermediation of professional media. A micromedium can be any individual or entity that, through a simple technological device with Internet access and increasing code, web or mobile applications, can choose at will to change their role as active/passive receivers/senders in the symbolic distribution of content and information to other micromedia or professional media. In the Internet age, where "real time is prevailing over real space" (Virilio, 1996) and "the idea of time can be reduced to a point of view" (Virilio, 1991, p. 105), audiences are no longer stable, homogeneous and passive masses, but rather fragmented, reshaped and (re)presented in interchangeable roles.

This chapter explores how the Internet, as well as its architecture and code, has enabled the emergence of social media—Web 2.0 social platforms (O'Reilly, 2005)—which come and go according to the same market logic as professional media. This has sped up the development of a new communication ecosystem that is not restricted to the online context, and which is characterized by the intersection and cohabitation of professional media and social media, that is, professional communicators and micromedia. It is a new social ecosystem and communicative *placeless space* where there have arisen new forms of power, influence, control and collective management of perceptions on any relevant topic in societies at the local and global scale. As a result, "the communication ecosystem can simultaneously contain without any trace of contradiction noise and signal, truth and lies, virtue and vice, news and rumors, the original and duplicate, voyeurism and exhibitionism—all within the same bold universe of inclusiveness" (del-Fresno, 2015, p. 76).

From the viewpoint of communication, it has always been an easy task to identify senders and receivers. Since the emergence of the Internet and social media, however, these concepts are no longer so obvious or explanatory. The rise of social media communication is "another example of the exceptional ability of people to collectively communicate meaning and ideas" (del-Fresno, 2015, p. 75). We are witnessing a paradigm shift in communication where the media and professional communicators must cohabitate, form part of and compete with social media and micromedia. Following in the path of the traditional models used to study mass communication (McQuail & Windahl, 1993), a new model has arisen to describe and explain this new communication ecosystem that has put an end to the mediating exclusivity of professional media and maximizes collective interpersonal communication on the same social continuum, where the notion of offline and online have lost their traditional explanatory power. In terms of accident, as understood by P. Virilio, in exchange for greater resources and opportunities for global and local, synchronous and asynchronous, and interpersonal and collective communication we must accept that the boundaries between communication and pseudo-communication, between signal and noise, between the relevant and the irrelevant are being blurred as a sign of our times.

This is the social and communicative context in which individuals and academics seek to build their identity and position themselves in the online environment. A proper understanding of this reality is critical for individuals and organizations to successfully undertake projects, as the social media is another example of the extraordinary ability of individuals to generate, disseminate and exchange meanings in a massive network system in real time.

BACKGROUND

Throughout human history, communication has always been tied to a specific technology through drawing, painting, writing, printing, etc. up to Internet and smartphones. In the late nineteenth and early twentieth century, communication occurred as a result of technological changes (sped up by printing and distribution), the dissemination of written press and the break from a dominant localist, elitist and minority type of information with ideological-political, economic-financial and religious intentions. In 1883, Benjamin Day, the editor of the *New York Sun*, decided to lower the price of the daily by reducing printing costs in exchange for advertising to make written press accessible to the working classes, thus founding *The Penny Press*. In the 1990s, the owners of mass media viewed the Internet, above all, as a new advertising medium, but did not fathom the extent to which collective interpersonal communication would be reorganized globally or let alone how their business model and mediation exclusivity would be plunged into crisis.

From N. de Condocert, L. Mumford, H. Innis and M. McLuhan to the cyberpunk authors, there has been a long tradition of receiving newly emerging media in society as the bearers of a new civilization, thus "the utopian, as well as dystopian, terms in which new media have been received have caused several media historians to record a sense of déjà vu, the feeling that we have been here before" (Lister, Dovey, Giddings, Grant, & Kelly, 2009, p. 65)

The sociologist A. Mattelart (2003) associated this sense of déjà vu with the telegraph in as early as the eighteenth century, claiming that it is possible to identify a recurring "prophetic discourse on the democratic virtues of long-distance communication" although such emancipatory expectations "would soon be contradicted by the continuation of the embargo on codes and encrypted language of signs and by the refusal to authorize their use for civilian purposes in the name of domestic security and national defense" (p. 23). He added that each new technology generation "revived the discourse of salvation, the promise of universal concord, decentralized democracy, social justice and general prosperity. Each time, the amnesia regarding earliest technology would be confirmed" (Mattelart, 2003, p. 23). If each technology is presented, at least in its initial stages, as a door to a new era and those expectations are repeated with each new media, it should come as no surprise that this also occurred with the Internet.

In the dawn of the Internet, cyberpunk authors encountered the metaphorical and life-saving hybridization of man and machine in computing and technology. Cyberpunks envisioned different scenarios in which the established social order would breakdown and an ideal community would emerge around a techno-emancipatory cyberculture where human capacities are enhanced by technology or, alternatively, machines are created and evolve to finally overcome the contradictions of society and pave the way to a new one. Some of the most important topics related to the notion of overcoming previous stages of human development include gender contradictions (Haraway, 1989; Franck, 1995), the evolution of our consciousness or identity parallel to technological development (Turkle, 1984, 1994), the broadening of our cognitive capacity to gain a better understanding of our place in the world and access a larger reality (Heim, 1998), or the cyberspace experienced in religious terms as the spiritual realm of the non-physical space, a new *soul space* (Wertheim, 1999). Thus, since the early 1990s, the Internet has been welcomed and defended—in the same spirit of its researcher and hacker founding fathers—as the door to a new era. Although these notions have undergone successive revisions on a par with the rapid development of technology (Sirky, 2009; Bertot, Jaegar, & Grimes, 2010; Li & Bernoff, 2008), to some extent they still prevail today and continue to conceive of the Internet as an exceptional space for social emancipation, frictionless horizontal communication, mass collaboration (Tapscott & Williams, 2006), participatory culture and collective intelligence (Jenkins, 2006). The Internet is so far removed from both the techno-utopian and dystopian visions (Nie & Erbring, 2002) that they have tended to accuse the Internet of being a new form of social isolation where contact with the social environment is lost, even more so than television.

More than two decades after the dawn of the Internet, the analysis of its effects has shown that the Internet has had a greater capacity for eroding or destroying business models than for creating new ones; a promise of *creative destruction* (Aghion & Howitt, 1992, 1997; Foster & Kaplan, 2001; Cowen, 2004) that has been repeatedly postponed and unfulfilled (Lanier, 2011). This phenomenon of erosion or destruction of business models brought about by the Internet (film, music, travel, the hospitality industry, commerce, publishing, etc.) is also evident even with regard to classic mass media that have adapted and evolved since *The Penny Press*.

The triangulation of sociability, communication and technology means that we are immersed in a dense web of social interactions forming an aggregate and vast social network that "connects people, information, events and places, thus facilitating or restricting the flow of information, ideas and perceptions in an instant and massive network communication system" (del-Fresno, 2014, p. 246). The importance of the Internet and its evolving code does not lie so much in its potential as an emancipatory lever or its accidents but in the evidence that it has both reorganized mass communication and deeply modified the conditions that enable collective interpersonal communication.

The Internet has given rise to a triple hybridization: 1) between offline and online sociability where "an ontological and phenomenological distinction between the online and offline world is no longer timely or meaningful" (del-Fresno, 2011a, p. 61) "as part of the same social continuum" (del-Fresno, 2014, p. 247); 2) between professional media and social media in a single and inclusive macromedium with all types of information and content; and 3) between individual and collective communicative practices in one macro communication ecosystem.

The most significant effects of the social media on the professional media forced to cohabit with them are:

1. The introduction in the communication ecosystem of the capacity for feedback and the circularity of social media insofar as audiences have the potential—regardless of whether they use it for good or bad purposes or do not use it at all—to create, modify, share, accept or reject and redistribute their own or others' information and content in an autonomous and immediate way.
2. The loss of the mass media's exclusivity to mediate vertically with audiences, symbolically control meanings and manage the social synchronization of perceptions.
3. The radical reorganization of the advertising market where the battle once waged by professional media to capture attention translated into advertising revenue is now focused on social media, which have become a competitor with a high potential for the cannibalization of revenues.

The result is not a communication model restricted to the digital context, nor is it a utopian or dystopian one from the viewpoint of emancipatory expectations. The information circulating in the new communication ecosystem remains as relevant and/or as trivial as before and suffers from the same problems of veracity as in the previous model, but is of a greater complexity in terms of understanding how information is produced and distributed and how opinions and behaviors are shaped and spread at an unprecedented scale.

Social Media

A definition of social media must incorporate two concepts dating from its origins: the Web 2.0 and UGC (user-generated content). The term "Web 2.0" was coined by Tim O'Reilly at the Web 2.0 Conference in 2004 (O'Reilly, 2005; Graham, 2005). The Web 2.0 is considered the ideological and technological platform for the development of what is now known as social media. From a technical standpoint, the Web 2.0 as a governing social media refers to the group of code applications that exploit the ideological and technological foundations of Internet architecture to facilitate the creation and exchange of UGC (Kaplan & Haenlein, 2010). Web 2.0 research takes a user perspective, the expression of the online identity and reality (Turkle, 1995) that gives full meaning to social media.

However, the communication enabled by social media goes considerably beyond the mechanics of how it is produced. Although SixDegrees (1997) is often claimed to be the first social network (Boyd & Ellison, 2007), the Bulletin Board System (BBS), USENET newsgroups, the first CompuServe and later America Online (AOL) forums could be considered the first social media although they were not referred to as such in the early days of the Internet.

Social media can be organized into different categories which include, although not exclusively, social networking platforms (Facebook, LinkedIn), microblogging (Twitter, Weibo), photography (Flickr, Instagram, Pinterest), video (YouTube, Vimeo, MetaCafe), social news (Menéame, Digg, Reddit), live broadcasts (Livecast, Ustream), social gaming (World of Craft), bookmarking (Delicious, StumbleUpon), blogs (WordPress, Blogger), and many more. Social media have become the source of an enormous amount of data produced ubiquitously, relentlessly and massively known as *big data* (Mayer-Schönberger & Cukier, 2014). This type of data, which is studied using social media mining techniques to uncover meaningful patterns, offers numerous opportunities and challenges for researchers, the most interesting of which is not the sheer amount of the data, but what can be done with them that cannot be done with small amounts.

In the scientific literature (Pérez, Portilla, & Sánchez, 2011), social media are understood as spaces for *mass collaboration* (Tapscott & Williams, 2006) and media convergence, participatory culture and collective intelligence (Jenkins, 2006), which gives way to a growing social component in individual decision-making processes (del-Fresno, 2012 a). Similarly, social media and the new forms of social interaction via technology have changed the way people form groups and how they coexist without the need for traditional organizational structures, which ultimately has profound economic and social effects (Shirky, 2009). Social media enable forms of communication in which the one-way control of information is weakened and reduced, when precisely the rigid model to control the emission and distribution of information has been one of the key functions and policies of mass media (Li & Bernoff, 2008). Social media can also be understood as the age of instant communication, transparency, narcissism and participation, calling into question organizations' one-way communication towards people (Qualman, 2009). Moreover, the potential of social media is recognized as a means of communication and expression to promote openness, transparency, combat corruption and improve cultural attitudes about government transparency (Bertot et al., 2010). From a sociopolitical perspective, it is a space where new battles for power and counter-power are waged and "shows the direct link between politics, media politics, the politics of scandal, and the crisis of political legitimacy in a global perspective" (Castells, 2007, p. 238). In short, from the viewpoint of collective communication, *social media* competitively cohabitate with traditional media in terms of immediacy, coverage, scope, visibility and relevance due to their circularity, marketability and open communication practices.

As subjects of social media, micromedia have the "dual property of being both part of the medium as well as part of the message" (del-Fresno, 2015, p. 76) insofar as other peers grant varying degrees of credibility to their individual (one to one) or social (one to many and many to many) communications in a synchronous or asynchronous manner. Thanks to the code applications of social media, micromedia are able to create, modify, share and discuss or reject all kinds of information from any source, while at the same time they can provide late-breaking or redundant information live, thus resulting in a collective and informative narrative that is not subject to an agenda of unilateral interests and generating relevant and/or trivial information and content with the same problems of veracity that already occurred with mass media.

The originality of the new communication ecosystem following the emergence of social media lies in:

1. **Who Sends:** The hegemony and mediation exclusivity of professional media with audiences has been broken; they are now forced to cohabit with social media, which entails the growing deinstitutionalization of mass media.
2. **What is Sent:** The volume of data and information in circulation—signal and noise on an unprecedented scale—requires new ways of capturing, storing, analyzing and producing knowledge that has led to the emergence of social media techniques such as mining.
3. **In Which Channel:** The control of channels by major media groups undergoing constant merger processes continues to exist on a par with the explosion of social media channels and the multiplication of individuals as potential micromedia broadcasters or narrowcasters.
4. **To Whom:** The media are based on a one-way, vertical communication from professional media owners—and their agenda setting—towards audiences understood almost as a property of professional media, thus commercializing attention in the form of advertising. This aspect must also coexist with the new, major social media players that are capturing increasing shares of the advertising market.

The definition of social media proposed here is therefore: media resulting from collective and interpersonal communication thanks to Internet architecture and the creation of web or mobile code applications that enable synchronous and/or asynchronous communication by (multi)decentralized, (de)localized, (co)generative micromedia with the capacity to provide constant feedback to audiences in an ongoing process of reconfiguration, without a predefined geographical scope of a non-recurring, non-normalized and heterogeneous set of symbolic content or messages (their own or of others) that is not subject to a performative agenda.

Thus, it is not a question of somehow reformulating the theory of meaning such as that derived from the sentence *the medium is the message* (McLuhan, 1964). The novelty stems from the fact that the new communication ecosystem has emerged as a result of both the technology that enables the expansion of innovative communication practices and that, for the first time in the history of communication, people as micromedia play the simultaneous and interchangeable role of audience (passive or active) and sender (narrowcaster and broadcaster) with the capacity of response; a phenomenon that has led to profound

Figure 1. Professional and social media: New communication ecosystem
Source: Author

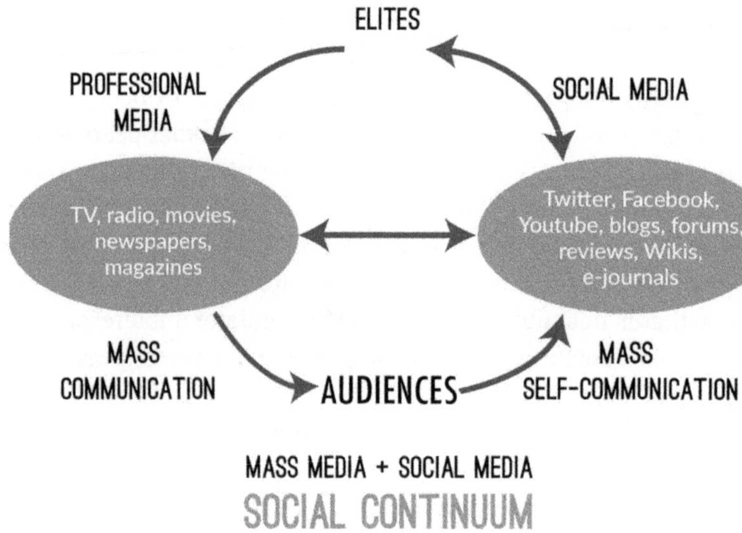

changes in the democratization of communication (Kietzmann, Hermkens, McCarthy, & Silvestre, 2011). Indeed, this is how the shift from "other" the media to "we" the media has occurred.

The emergence of social media has resulted in the reorganization of the ecosystem of influence and collective management of perceptions once monopolized by the mass media, particularly large global communication groups. The traditional system of influence oriented towards the massive-scale synchronization of perceptions has been disrupted by social media and the mass self-communication of micromedia, giving rise to a new media ecosystem (Figure 1).

The mass media's ecosystem of vertical, one-way influence survives embedded in the emerging system, forcing professional media to take part in an uncomfortable—and undesired—cohabitation with social media. Social media and thousands or millions of micromedia have taken a leading role in the battle for the management of social perceptions by influencing beliefs and the dissemination of opinions and behaviors, thus unifying interpersonal and collective practices on one and the same continuum. This "hybrid, all-media space forms a labyrinth linking people who share information, news, perceptions, beliefs, and rumors in a real-time, immense, networked communication system" (del-Fresno, 2015, p. 77).

COMMUNICATION MODELS

Models for Studying Collective Communication

With the emergence of mass consumption in the 1920s, the first "notions of opinion management and engineered consent began to circulate, pointing to an infatuation with the new method of governing mass democracy, involving a carefully calculated combination of information and censorship" (Mattelart, 2003, p. 38) at the hands of pioneering researchers such as W. Lippmann (1922) and H. Lasswell (1938). Coinciding with the pre-war and wartime evolution of World War II in the 1940s, the written press and other mass media, such as radio, film and later television, showed increasing interest in the systematic research of figures like P. Lazarsfeld and R.K. Merton (1940 and 1943). At that time, there came a need for concise models to explain the phenomenon of mass communication in a historical context and how communication was changing on a par with the social and technological changes of the time. Models are useful as tools depending on how effective they are in simplifying complex, abstract or ambivalent phenomena and help to reflect on changes in communication practices. The speed with which social changes are occurring requires simple communication models that provide maximum explanatory power.

According to McQuail and Windahl (1993), a model of collective communication is always "a consciously simplified description in graphic form of a piece of reality [that] seeks to show the main elements of any structure or process and the relationships between these elements" (p. 30). However, because mass communication is an abstract and dynamic process, there is no multipurpose model that can give a definitive analytical, heuristic or epistemological answer.

For McQuail and Windahl (1993), a model for studying collective communication is useful if it provides:

1. An *organizing function* and presents an overall picture to explain the whole phenomenon of communication at a given moment;
2. A *heuristic function* that allows communication researchers to focus on key aspects of communicative processes and practices;
3. A *predictive function* of how results can be achieved and which facilitates the presentation of research hypotheses; and
4. A *descriptive function* if communicative practices can be explained in terms of "forces and their direction, the relations between parts and the influence of one part on another" (p. 3).

The most relevant models are those which have the ability to be explanatory and address the key conceptual core of collective communication.

Harold D. Lasswell was the first to propose a model for the study of mass communication, which was based on the effects of propaganda on World War II fighters (Lasswell, 1948). Lasswell's classic model sums up communication as: *Who–says what–in which channel–to whom–with what effect*. The most significant aspect of Lasswell's model is that it revealed the keys of mass media early on: one-way communication, the highly passive assumption—especially from the advertising point of view—of receiving the message and the absence of feedback from audiences. All the models developed after Lasswell's have a common purpose: to show how a privileged sender tries to influence a large audience in a one-way and intentional manner.

Theodore M. Newcomb's ABX model offered a different approach in assuming that "communication among humans performs the essential function of enabling two or more individuals to maintain simultaneous orientation toward one another as communicators *and* toward objects of communication" (Newcomb, 1953, p. 393). However, the model was naïve in two ways:

1. It assumed a natural tendency for consensus in communication, and
2. Extrapolated communicative processes between individuals or small groups to society as a whole.

In their co-orientation model, J. McLeod and S. Chaffee (1973) incorporated the concept of conflict as a variable of mass communication together with those of accuracy, congruency and the search for informative support. The co-orientation model exerted considerable influence in the twentieth century as it provided strong heuristic and descriptive power in mass media by incorporating a new technology: television. Since then, the most widely accepted definition of mass communication is:

The transmission from a single or centralized (and organized) sender to all or most of a population of a recurring and standardized set of messages (news, information, fiction, entertainment and spectacle), without there being much possibility of responding or answering back...Mass communication enabled the symbolic and informational dominance of a whole society by those with control of the means of dissemination. (McQuail, 2002, p. 5)

Countless models for the study of collective communication in relation to the media have been proposed, but there is a gap in the literature (Lister et al., 2009) as models that include the collective interpersonal communication of the Internet, especially social media, are lacking.

Model for Studying Collective Interpersonal Communication in the Internet Age

We have seen how mass media channels and messages were the only game in town and they owned both the message and the means of communicating that message. Within a mass self-communication system, micromedia are able to extend messages through social media from and to others (peers or professional media) via a multidirectional communication system. "This system reflects individual (one to one) or social (one to many and many to many) interactions without the imposition of any agenda from the larger professional media" (del-Fresno, 2015, p. 76). The model proposed here (Figure 2) (del Fresno, 2012 b) to study collective interpersonal communication incorporates elements of mass media collective communication models together with new post-Internet elements.

The lines connecting the concepts in the figure indicate one-way or two-way relationships (direct or indirect), attitudes, perceptions and a feedback process. The traditional one-way communication of mass media is indicated by a continuous line on the left; while the dotted line on the right represents social media communication.

The *ELITES* are interest groups that may be in conflict or in competition with each other (political parties, corporations, lobbies, governments, brands, etc.) for the symbolic and informational dominance of the largest possible set of audiences. These groups defend their interests by exerting their influence (ideological or economic pressure as advertisers) on the *PROFESSIONAL MEDIA* or through the direct ownership of them. The professional media have lost their classic mediating exclusivity and have also been forced to a renounce a key aspect of their history: to be the first in breaking the news. Indeed, there are countless cases where the micromedia offer exclusive or breaking information and news before the mass media (Ibáñez & Baraybar, 2011; Weaver, Johnson, Sweetser, & Howes, 2009).

Since the elites have recognized *SOCIAL MEDIA* as significant collective interpersonal channels of communication, they have begun to deploy ownership strategies (by purchasing stocks), thus incorporating the social media corporate layer into their communication and public relations strategies in a direct (Verhoeven, Tench, Zerfass, Moreno, & Verčič, 2012; Bajkiewicz, Kraus, & Yeon Hong, 2011; Eyrich, Padman, & Sweetser, 2008) or indirect manner (del-Fresno & López-Peláez, 2014) in addition to investing in advertising in major social media platforms.

Topics on any matter that are susceptible to social exchange in both professional media and social media circulate in the communication ecosystem. *AUDIENCES* interested in or related to any of the topics circulating in the informational ecosystem are the target group to be impacted upon.

Figure 2. Collective interpersonal communication model
Source: Author

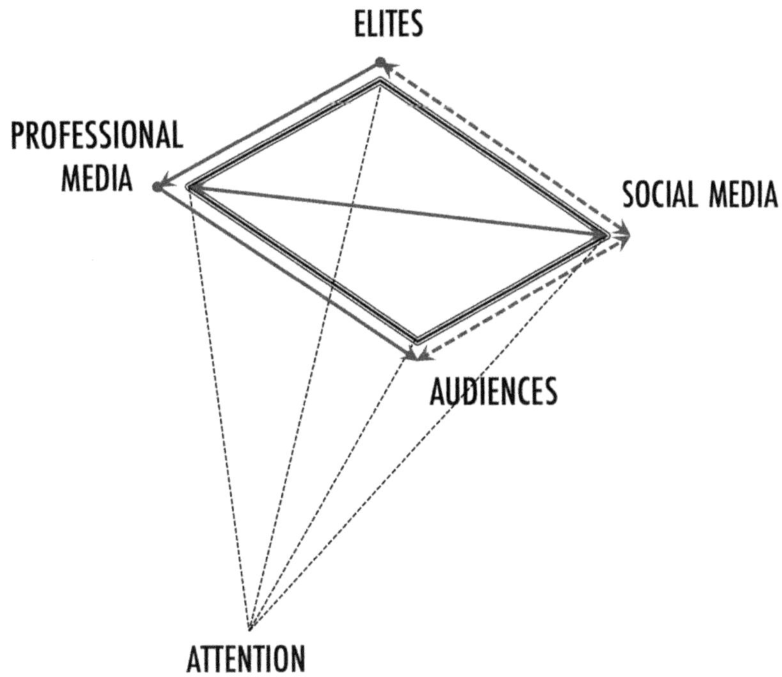

The relationship between professional media and social media is two-way since the former draw on information from the latter as sources to (re)elaborate the content and information (Hong, 2012; Canavilhas & Ibars, 2012; Weaver et al., 2009). Information considered interesting or of value is filtered from social media and selected by the micromedia (blogs, forums, information and opinion aggregators, etc.) of professional media to be (re)sent and spread virally among (micro)audiences in constant reconfiguration.

Cohabitation between the professional media and the social media is therefore 1) a feedback process related to the flow of data and information circulating in the ecosystem and 2) a competitive cohabitation in the battle to capture audiences' *ATTENTION*. As information grows exponentially, the audience's *attention* becomes an increasingly scarce resource (Goldhaber, 1997; Lanham, 2007). The purpose of all media is to attract and retain the attention of the audience, since it has both a symbolic and economic value. Unlike mass media, social media attention is not merely passive, but can be monetized in the market in the form of both advertising or e-commerce through online behavior data, private and personal data, private aggregated metadata, etc.

While media, brands, companies, organizations and others compete for the attention of citizens and consumers, other individuals tend to develop cognitive mechanisms and strategies of defense against the excessive demand for attention (Datchary, 2005; Pashler, 1998). The ability to reconcile these two aspects is a challenge for the media and a competitive advantage in the digital economy (Kessous, Mellet, & Zouinar, 2010).

From the cognitive point of view, the novel contribution of social media is the new emphasis on sensory impressions integrated into a new perception regime (Styhre, 2009). This has led to new forms of perception and attention-action due to the multimedia integration of text, sound, images or tactile impressions characteristic of virtual media called *haptic sense-impressions* (Prentice, 2005).

The relevance of introducing attention as a scarce resource within the collective communication model presented here is how the tension between two logics are depicted: protecting individuals' attention against the risk of cognitive overload versus the *market value* of attention that all media tend to saturate with advertising and calls-to-action.

The value of *attention* in the proposed communication model is manifold as it is presented as:

1. A scarce resource;
2. As something to capture, win, retain and capable of being monetized;
3. The variable ensuring strong competition between media; and
4. The cornerstone of the economic viability of both professional media and large social media platforms based on a market logic with their IPOs (Facebook, Twitter, LinkedIn, Yelp, Zinga, etc.).

Finally, the model includes social change in its heuristic. This is because while collective communication in the mass media occurred through information flows *towards* people (as in all classic models of collective communication), by including social change, the model captures the quality of the communication and extends it to the interpersonal collective scale, thus reflecting communication as a flow *of* people as micromedia at an unprecedented scale.

In short, the model also reflects the difficulty of drawing the boundaries between media. Thus, as McQuail and Windahl (1993) predicted, "we anticipate a future in which existing boundaries will become even less clear than they are at present and when communication technology and new expressions of communication needs will produce different structures and relationships" (p. 2).

Both individually and collectively, we are facing a new context of collective interpersonal communication at the local and global scale as a result of the emergence of the Internet; the most disruptive technological phenomenon in recent decades, which has led to a radical reorganization of communicative practices. In this chapter, a definition of social media has been proposed that contrasts with the classic notion of media, as well as a new model to study the collective interpersonal communication of our time, where all media and micromedia are exposed to new technological developments, new competitors, new options for audiences and new challenges that also define them as they unfold.

The proposed communication model unifies and explains the cohabitation of the mass media and social media, thus uniting the vertical, one-way communication typical of mass media with the multidirectional communication of social media. An additional key contribution is the incorporation of the variable *Elites*' fight for the *Attention* of local and global audiences in the model, where elites are understood as interest groups that may be in conflict or in competition with each other for the symbolic and informational dominance of the largest possible set of audiences through the direct or indirect control of classic and competing broadcasting channels or new global digital corporations at the local or global scale such as Facebook, Google, Apple, Amazon, Twitter, and others.

Micromedia can take an active part in the whole communication process by creating, distributing and disseminating content and information through multidirectional interaction. The tension between senders and audiences is reflected in two distinct logics: the protection of individual privacy versus the *market value* of their attention.

FUTURE RESEARCH DIRECTIONS

As a result of increasing computational capacity and the development of ad hoc technologies and Internet architecture, it is possible to capture, represent and analyze online relational data in order to broaden our knowledge. The new communication ecosystem presented via a model entails new challenges for researchers who must deal with large amounts of information. What is of interest, however, is not only the amount of data generated, but what can be done with such large amounts that cannot be done with small ones. In this regard, the two major challenges for researchers are:

1. The *challenge of complexity* or how to capture and consistently aggregate multidimensional data that are not homogeneous, unstructured and massive and produced *ad infinitum* in any time or place, as well as heterogeneous and unstable media (which may come and go), while maintaining the central objective of seeking out and identifying significant patterns.
2. The *challenge of n=all* or how to develop methods to work with all the data that is produced, that is, how to investigate with complete universes which are now accessible thanks to big data and data mining. Sampling is a technique that was developed two centuries ago to respond methodologically to times of scarce information, something which no longer seems to be the problem.

CONCLUSION

Online sociability is defined by the code, which according to Berry and Pawlik (2008), "…harmonised with the language of machines, our life history, tastes, preferences and personal details become profiles, mailing lists, data and ultimately markets. Societies of control regulate their population by ensuring their knowing and unknowing participation in the marketplace through enforced compatibility with code" (p. 58). Thus, the code of major social platforms like Facebook imposes some very specific forms of normative sociality and tends to hide a "logic of connectivity imbued in the commercial drives and coercive formats" (Van Dijck, 2013, p. 155). This has given rise to tensions and, to a large degree, the paradox between the original meaning of the social media's communicative community and the market-oriented connectivity of large platforms aimed at gaining revenues.

As has occurred throughout the history of the media, much of the global market share (attention, audience, etc.) is in the hands of a small number of social media platforms. These large platforms not only monetize attention via advertising, but also through the commercialization of the aggregate metadata derived from audience activity and the personal data that they capture or are provided by users (socio-demographic data, friendships, likes, images, videos, etc.). In this way, privacy becomes a monetized product. If platforms like Facebook are free it is because people are invariably understood as a product due to the capturing of both attention—which generates advertising revenues—and individual and aggregate data and metadata on audience participation and activity. The fact that social media fall under the *freemium* umbrella does not mean that users have to pay nothing: through the transfer of private data they become part of the product, thus opening up a new mass commercial market.

Whether media users have control over their data or not (private, behavioral, etc.) or are the property of major media appears to be a battle dominated by corporate forces, which take advantage of their global presence and local laws to impose privacy policies or, more commonly, unilateral and ever-changing terms of service (ToS) at their whim, which is practically the same as saying that de facto privacy policies do not exist.

As mentioned above, an additional effect of major social media social platforms is that they frequently impose a particular form of online sociality through the code. One example is the Facebook timeline changes that allow the users themselves to enrich (enrichment) databases (basic demographic data, friendships, preferences, complete biographies, photos, interests, etc.). In this regard, the cover of the *MIT Technology Review* (2012) stated that Facebook "has collected more personal data than any other organization in human history. What will it do with that information?" and that "the company's social scientists are hunting for insights about human behavior. What they find could give Facebook new ways to cash in on our data—and remake our view of society" (Simonite, 2012, p. 42). What the big social media platforms are pursuing is "to lure and lock users into their chairs of platforms" (Van Dijck, 2013, p. 156) and "discipline their users into particular roles and modes of behavior" (Van Dijck, 2013, p. 157) by restricting to the largest possible degree any attempt to opt-out of them. This implies a clear ideological vision of the mainstream platforms based on a corporate market logic as opposed to the original Web 2.0 ideas based on the sharing, community, freedom and egalitarian interaction of liberal democratic ideology.

The accident of social media, in P. Virilio's terms, is not only a question of contributing vast amounts of new data, but an increasing loss of privacy, the gilded cages of lock in and opt-out restrictions and what is possible or not for the people restricted by the code; a code that was not developed by the successors of those scientists and hackers in the dawning of the Internet but by brilliant corporate programmers. A growing number of users and some governments are beginning to see the risks of a few big social media corporations engaging in the privacy trade as a threat to public and private democratic principles.

We are now faced with technologies of the self, which for M. Foucault had a positive rather than a negative meaning, as the development of such technologies entailed the improvement of the self. That is, they are techniques that allow individuals to construct and change themselves in an effort to develop their own potential "in order to attain a certain state of happiness, purity, wisdom, perfection or immortality" (Foucault, 1988:18). Like all technology, however, the new technologies of the self cause their own accidents. Among the negative effects technologies of the self have on redefining the interpersonal and mass communication ecosystem is the need for greater personal branding, which entails the risk of progressively reducing researchers to a mere market product or service through the social marketing of the self. Yet communicating and promoting oneself to gain visibility is not the same as being a brand or explicitly waiving growing areas of privacy. The private has been made public in the form of a social conquest and can be viewed as a new step towards the expropriation of privacy, which due to increasing corporate concentration, reveals the potential of the technological accidents arising from the new technologies of the self.

The paradox becomes increasingly evident: people need social media to express their opinions and creativity, while the major platforms tend to lock-in and understand users as products to generate revenues, which are two opposing ideological premises. Nonetheless, the dominant position of some mainstream platforms are more precarious than would seem, as evidenced by the rise and fall of other online giants (AOL, Yahoo, Second Life, MySpace, etc.). Just as Google defended itself against antitrust charges, there can be no monopoly if the competition is just a click away. One click today can be a great act of dissent because, ultimately, we are the new media.

REFERENCES

Aghion, P., & Howitt, P. (1992). A Model of growth through Creative Destruction. *Econometrica*, *60*(2), 323–351. doi:10.2307/2951599

Aghion, P., & Howitt, P. (1997). *Endogenous Growth Theory*. Cambridge, MA: MIT Press.

Anderson, C. (2008). The end of theory: the data deluge makes the scientific method obsolete. *Wired*, *16*(7). Retrieved June 15, 2015, from http://archive.wired.com/science/discoveries/magazine/16-07/pb_theory

Bajkiewicz, T. E., Kraus, J. J., & Yeon Hong, S. (2011). The impact of newsroom changes and the rise of social media on the practice of media relations. *Public Relations Review*, *37*(3), 329–333. doi:10.1016/j.pubrev.2011.05.001

Berry, D. M., & Pawlik, J. (2008). What is code: A conversation with Deleuze, Guattari and code. In *Libre culture. Meditations on free culture*. Winnipeg: Pygmalion Books. Retrieved June 15, 2015, from http://sro.sussex.ac.uk/46287/1/BERRYMOSS-LibreCulture2008.pdf

Bertot, J., Jaeger, P., & Grimes, J. (2010). Using ICTs to create a culture of transparency: E-government and social media as openness and anti-corruption tools for societies. *Government Information Quarterly*, *27*(3), 264–271. doi:10.1016/j.giq.2010.03.001

Boyd, D., & Ellison, N. (2007). Social network sites: Definition, history, and scholarship. *Journal of Computer-Mediated Communication*, *13*(1), 210–230. doi:10.1111/j.1083-6101.2007.00393.x

Canavilhas, J., & Ivars, B. (2012). Uso y credibilidad de fuentes periodísticas 2.0 en Portugal y España. *El profesional de la información*, *21*(1), 63-69.

Castells, M. (2001). *La era de la información: economía, sociedad y cultura*. Madrid: Alianza editorial.

Castells, M. (2007). Communication, Power and Counter-power in the Network Society. *International Journal of Communication*, *1*, 238–266.

Castells, M. (2009). *Communication and power*. Oxford, UK: Oxford University Press.

Cowen, T. (2004). *Creative Destruction: How Globalization Is Changing the World's Cultures*. Princeton, NJ: Princeton University Press.

Datchary, C. (2005). Se disperser avec les tics une nouvelle compétence? In *Le travail avec les technologies de l'information* (pp. 157-173). Paris: Hermès-Lavoisier. Retrieved June 15, 2015, from http://hal.archives-ouvertes.fr/docs/00/33/17/82/PDF/_Datchary.pdf

Del Fresno, M. (2011a). *Netnografía*. Barcelona: UOC.

Del Fresno, M. (2011b). Infosociabilidad: monitorización e investigación en la web 2.0 para la toma de decisiones. *El profesional de la información*, *20*(5), 548-554.

Del Fresno, M. (2012 a). *El consumidor social. Reputación online y social media*. Barcelona: UOC.

Del Fresno, M. (2012 b). Comprendiendo los social media y mass media un modelo para el estudio de la comunicación interpersonal colectiva en tiempos de Internet. *Derecom*, *11*, 99–109.

Del Fresno, M. (2014). Haciendo visible lo invisible: Visualización de la estructura de las relaciones en red en Twitter por medio del Análisis de Redes Sociales. *El profesional de la información*, *23*(3), 246-252.

Del Fresno, M. (2015). We are the media: the ongoing disruption of social space via social media. In *#Commoncore Project. How social media is changing the politics in education*. Consortium for Policy Research in Education Pennsylvania. Retrieved June 15, 2015, from http://www.hashtagcommoncore.com

Del Fresno, M., & López-Peláez, A. (2014). Social work and Netnography: The case of Spain and generic drugs. *Qualitative Social Work: Research and Practice*, *13*(1), 85–107. doi:10.1177/1473325013507736

Eyrich, N., Padman, M., & Sweetser, K. (2008). PR practitioners' use of social media tools and communication technology. *Public Relations Review*, *34*(4), 412–414. doi:10.1016/j.pubrev.2008.09.010

Feather, J. (2008). *In The information society: A study of continuity and change*. London: Library Association.

Foster, R., & Kaplan, S. (2001). *Creative Destruction: Why Companies that are Built to Last Underperform the Market – And how to Successfully Transform Them*. New York: Crown Business.

Foucault, M. (1998). *Technologies of the Self. A seminar with Micher Foucault*. The University of Massachussets Press.

Franck, K. (1995). When I Enter Virtual Reality, What Body Will I Leave Behind? *Architectural Design*, *65*(118), 20-22. Retrieved June 15, 2015, from digitalarts.bgsu.edu/faculty/wmadsen/.../in_VR_Where_is_my_body.pdf

Goldhaber, M. (1997). The attention economy on the net. *First Monday*, *2*(4). doi:10.5210/fm.v2i4.519

Graham, P. (2005). *Web 2.0 Want to start a startup?* Retrieved June 15, 2015, from http://www.paulgraham.com/web20.html

Haraway, D. (1989). *The Cyborg Manifesto*. Retrieved June 15, 2015, from http://faculty.georgetown.edu/irvinem/theory/Haraway-CyborgManifesto-1.pdf

Heim, M. (1998). *The virtual reality of the tea ceremony*. Retrieved June 15, 2015, from http://www.mheim.com/files/vrtea.pdf

Hong, S. (2012). Online news on Twitter: Newspapers' social media adoption and their online readership. *Information Economics and Policy*, *24*(1), 69–74. doi:10.1016/j.infoecopol.2012.01.004

Ibáñez, J.A., & Baraybar, A. (2011). Fuentes 2.0 y periodistas. Transformaciones en la comunicación museística. *El profesional de la información*, *20*(6), 634-638.

Jenkins, H. (2006). *Convergence Culture: Where Old and New Media Collide*. New York: New York University Press.

Kaplan, A., & Haenlein, M. (2010). Users of the world, unite! The challenges and opportunities of Social Media. *Business Horizons*, *53*(1), 59–68. doi:10.1016/j.bushor.2009.09.003

Kessous, E., Mellet, K., & Zouinar, M. (2010). L'économie de l'attention: Entre protection des ressources cognitives et extraction de la valeur. *Sociologie du Travail*, *52*(3), 359–373. doi:10.1016/j.soctra.2010.06.009

Kietzmann, J., Hermkens, K., McCarthy, I., & Silvestre, B. (2011). Social media? Get serious! Understanding the functional building blocks of social media. *Business Horizons*, *54*(3), 241–251. doi:10.1016/j.bushor.2011.01.005

Lanham, R. (2007). *The Economics of Attention: Style and Substance in the Age of Information*. Chicago: University Of Chicago Press.

Lanier, J. (2010). *You are not a gadget: A manifesto*. New York: Vintage.

Lasswell, H. D. (1938). *Propaganda techniques in the world war*. New York: Peter Smith. Retrieved June 15, 2015, from http://babel.hathitrust.org/cgi/pt?id=mdp.39015000379902,view=1up,seq=9

Lasswell, H. D. (1948). The structure and function of communication in society. In *The communication of ideas*. New York: Harper. Retrieved June 15, 2015, from www.irfanerdogan.com/dergiweb2008/24/12.pdf

Lazarsfeld, P. (1940) *Radio and the Printed Page: An Introduction to the Study of Radio and Its Role in the Communication of Ideas*. New York: Duell, Sloan and Pearce. Retrieved June 15, 2015, from https://archive.org/stream/radiotheprintedp00lazarich/radiotheprintedp00lazarich_djvu.txt

Lazarsfeld, P., & Merton, K. (1943). *Studies in radio and film propaganda*. Retrieved June 15, 2015, from http://onlinelibrary.wiley.com/doi/10.1111/j.2164-0947.1943.tb00897.x/abstract

Lessig, L. (2009). *Code: versión 2.0*. Retrieved June 15, 2015, from http://codev2.cc/download+remix/Lessig-Codev2.pdf

Lewis, D. (1999). *Information Overload*. London: Penguin.

Li, C., & Bernoff, J. (2008). *Groundswell: Winning in a World Transformed by Social Technologies*. Boston: Harvard Business Press.

Lippmann, W. (1922). *Public opinión*. London: Allen and Unwin. Retrieved June 15, 2015, from http://monoskop.org/images/b/bf/Lippman_Walter_Public_Opinion.pdf

Lister, M., Dovey, J., Giddings, S., Grant, I., & Kelly, K. (2009). *New Media: A Critical Introduction*. London: Routledge.

Lycett, M. (2013). Datafication: Making Sense of (Big) Data in a Complex World. *European Journal of Information Systems*, *22*(4), 381–386. doi:10.1057/ejis.2013.10

Mattelart, A. (2003). *The Information Society: An Introduction*. London: Sage.

Mayer-Schönberger, V., & Cukier, K. (2014). *Big Data: A Revolution That Will Transform How We Live, Work, and Think*. London: John Murray Publishers.

McLeod, J., & Chaffee, S. (1973). Interpersonal approaches to communication research. *The American Behavioral Scientist*, *16*(4), 469–500. doi:10.1177/000276427301600402

McLuhan, M. (1964). *Understanding media: The extensions of man*. Cambridge, MA: MIT Press.

McQuail, D. (Ed.). (2002). *McQuail's reader in mass communication theory*. London: Sage.

McQuail, D., & Windahl, S. (1993). *Communication Models for the Study of Mass Communications*. New York: Longman.

Mitchell, W. (1996). *City of Bits: Space, Place, and the Infobahn*. Cambridge, MA: MIT Press.

Newcomb, T. (1953). An Approach to the Study of Communicative Acts. *Psychological Review*, *60*(6), 393–404. doi:10.1037/h0063098 PMID:13112341

Nie, N., & Erbring, L. (2002). Internet and Society: A Preliminary Report. *IT & Society*, 275-283. Retrieved June 15, 2015, from www.stanford.edu/group/siqss/itandsociety/v01i01/v01i01a18.pdf

O'Reilly, T. (2005). *What Is Web 2.0*. Retrieved June 15, 2015, from http://oreilly.com/web2/archive/what-is-web-20.html

Pashler, H. (1998). *The Psychology of Attention*. Cambridge, MA: MIT Press.

Pérez, F. J., Portilla, I., & Sánchez, C. (2011). Social Networks, Media and Audiences: A Literature Review. *Comunicación y Sociedad*, *24*(1), 63–74.

Prentice, R. (2005). The anatomy of surgical simulations: The mutual articulation of bodies in and through the machine. *Social Studies of Science*, *35*(6), 837–866. doi:10.1177/0306312705053351

Qualman, E. (2009). *Socialnomics: How Social Media Transforms the Way We Live and Do Business*. New York: John Wiley.

Reidenberg, J. (1998). Lex Informatica: The Formulation of Information Policy Rules Through Technology. *Texas Law Review*, *76*(3), 553–593.

Shirky, C. (2009). *Here Comes Everybody: The Power of Organizing Without Organizations*. New York: Penguin.

Silver, N. (2013). *The signal and the noise: why so many predictions fail — but some don't*. New York: Penguin Books.

Simonite, T. (2012). What Facebook knows. *Technology Review*, (July-August), 42–49.

Styhre, A. (2009). The cinematic mode of organizing: Media and the problem of attention in organization theory. *Information and Organization*, *19*(1), 47–58. doi:10.1016/j.infoandorg.2008.06.001

Swash, G. (1998). UK business information on the Internet. *New Library World*, *99*(1144), 238–242. doi:10.1108/03074809810236793

Tapscott, D., & Williams, A. (2006). *Wikinomics: How Mass Collaboration Changes Everything*. New York: Portfolio.

Toffler, A. (1970). *Future shock*. New York: Bantam Books. Retrieved June 15, 2015, from http://goo.gl/KDGWwn

Toffler, A., & Toffler, H. (1994). Surfing the third wave. *Microtimes*, *118*. Retrieved June 15, 2015, from http://www.yoyow.com/marye/tofflers94.html

Turkle, S. (1984). *The Second Self. Computers and the human spirit.* Cambridge, MA: MIT Press. Retrieved June 15, 2015, from http://mitpress.mit.edu/books/chapters/0262701111intro1.pdf

Turkle, S. (1994). *Constructions and reconstructions of the Self in Virtual Reality.* Cambridge, MA: MIT Press. Retrieved June 15, 2015, from http://web.mit.edu/sturkle/www/constructions.html

Turkle, S. (1995). *Life on the Screen: Identity in the Age of the Internet.* New York: Simon and Schuster.

Van Dijck, J. (2013). *The Culture of Connectivity: A Critical History of Social Media.* New York: Oxford University Press. doi:10.1093/acprof:oso/9780199970773.001.0001

Verhoeven, P., Tench, R., Zerfass, A., Moreno, A., & Verčič, D. (2012). How European PR practitioners handle digital and social media. *Public Relations Review, 38*(1), 162–164. doi:10.1016/j.pubrev.2011.08.015

Virilio, P. (1991). The aesthetics of disappearance. New York: Semiotext(e).

Virilio, P. (1995). *Speed and Information: Cyberspace Alarm!* Retrieved June 15, 2015, from http://www.ctheory.net/printer.asp?id=72

Virilio, P. (1996). Speed Pollution. *Wired.* Retrieved June 15, 2015, from http://archive.wired.com/wired/archive/4.05/virilio.html

Weaver, R., Johnson, E., Sweetser, K., & Howes, P. (2009). An examination of the role of online social media in journalists' source mix. *Public Relations Review, 35*(3), 314–316. doi:10.1016/j.pubrev.2009.05.008

Wertheim, M. (1999). *The pearly gates of cyberspace. In Cyber_reader. Critical writtings for the digital era.* New York: Phaidon Press.

KEY TERMS AND DEFINITIONS

Communication: To share meanings through the exchange of information where the technology used, the characteristics of the senders and receivers, the cultural codes of reference, the protocols used and the scope of the process are significant.

Information Society: The era characterized by social changes occurring since the late twentieth century as a result of the evolution of digital information and communication technologies alongside the emergence of a social network structure, which has affected all areas of human activity on a global scale.

Micromedia: Any individual or entity which, through a simple technological device with Internet access and increasing web or mobile code applications, can choose to change their role as active/passive receivers/senders in the symbolic distribution of content and information with other micromedia or professional media.

Social Media: The media resulting from Internet architecture and the creation of code-based web or mobile applications which enable synchronous and/or asynchronous communication by people who, as micromedia, acquire the ability to provide constant feedback to audiences in constant reconfiguration and without a predefined geographical scope of a non-recurring, non-normalized and heterogeneous set of symbolic content or messages (their own or of others) not subject to a performative agenda.

Social Media Mining: The process of extracting, storing, representing, visualizing, and analyzing massive amounts of user-generated data in order to uncover meaningful patterns from interactions within social media.

Web 2.0: The group of code applications that exploit the ideological and technological foundations of the original architecture of the Internet to facilitate the creation and exchange of user-generated content.

Chapter 3
Building Academic Branding:
The Digital Branding as Academic Footprint

Ruth Matovelle Villamar
Escuela Superior Politécnica del Litoral, Ecuador

ABSTRACT

Nowadays, many professionals of different areas of expertise have found themselves at the needs to incorporate in the digital world, their personal digital branding as a result of entrepreneurship projects, or as a crisis of business sectors. A different phenomenon occurs in the academic field; the researchers are forced to incorporate themselves in the digital field. The research professors have a responsibility to society, popularizing science, this represents to value their own work and contribute to its visibility, and the starting point is to have "digital academic Branding. This chapter explains step-by-step techniques that develop a digital academic branding, starting from an interior analysis of the academic up to the formulation of the visual and conceptual brand; it is important is to make in a sequential manner every one of the activities described here.

INTRODUCTION

This chapter is the result of the synthesis and reorder of ideas, from different authors, plus the academic experience of the author. This model of brand building does not intend to create an industrialized model or series making, on the contrary its objective is to enhance their individuality. It seeks to show clearly the steps to follow to discover and create your personal brand, using basic indications, methodologies and theories used in the business, psychological and communicational field.

Starting from an interior analysis of the academic up to the formulation of the visual and conceptual brand. The importance of this chapter is to make in a sequential manner every one of the activities described here, rest assured that the result would be entirely satisfactory. Welcome to the world of personal branding.

DOI: 10.4018/978-1-5225-0917-2.ch003

BACKGROUND

Nowadays, colleges and universities have a fundamental roll in the human and social development, channeled primarily through the investigations made by their teacher-researchers. In recent years there has been an increase in the number of arguments supporting reforms to create socially responsible scientists, considering the need to democratize knowledge and provide expert advice and plural to the democratic institutions and the general public, thus increasing the capacity to discuss and finally meet the expectations of citizens. (Vessuri, 2010).

This leads us to think then that the researcher not only has to develop a task of research, but is also relevant his task as a communicator of: the scientific community interested in his field of work; to the media and citizens.

New technologies have contributed to the awareness of many communicators that previously limited themselves to encourage scientific knowledge from more minority plots (research department). This is generating real networks of scientific communication and research teams are increasingly aware of the need to explain what they do to their peers, but also the general public. (Sapiña, 2013).

The 2.0 environment offers a variety of platforms and scientific networks broadcasting scientific work and publications, the most important is to know and select the appropriate ones for each particular researcher. But if the center of everything is the disclosure, what role does personal branding do in this? Well a fundamental role is the starting point for disclosure.

Having a digital academic branding allows the academic to be identified by its target, it places you as an expert in a specific field, facilitates the search for other researchers to disseminate the work done, in order to establish new partnerships and attract funding.

The academic success of their digital branding, will contribute greatly to the success of your academic institution, which will be reflected in the positions in rankings that measure the quality of the IES1.

HOW TO CREATE A PERSONAL DIGITAL BRANDING

The author Marcos (2009), assures that the creation of a personal brand, "has to do with the creation and keeping up the credibility" and reassures that the best personal brands are the ones that are authentic, under any aspect of what they are and do strengthens the set, these "have a visibility based on placement, contact consistently regardless of the medium used" (Marcos 2009). A brand built from ideologies, placing dreams instead of real values can be compared as building a fictional character in a play, which could eventually be found out.

Since 1996, different authors have written about how to construct a Brand, there are many books dedicated to this subject. Nowadays, this can be seen from a simple point of view, focus the principles of branding to the people, but the current job goes beyond that, this chapter addresses the issue from a deeper and personal vision aimed to the academic researchers. I have developed a scheme to work on brand building, made of 4 phases: Know yourself; your personality, place it on your brand; express your thoughts; express your emotions and finally build your academic branding.

Figure 1.
Source: Author, 2015

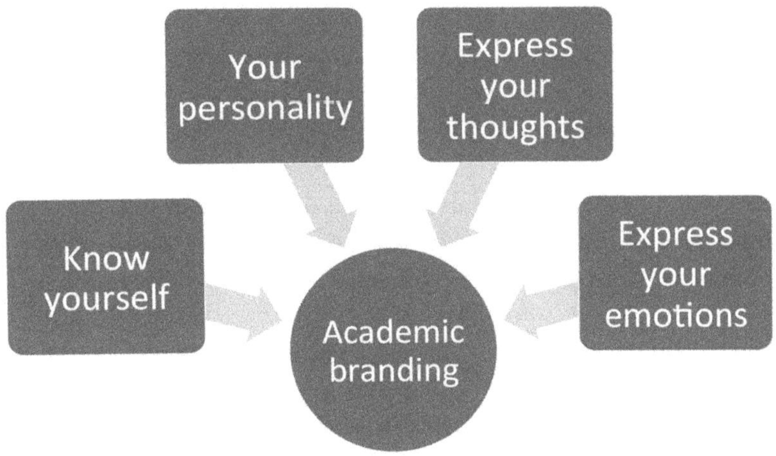

Know Yourself

The first step to knowing yourself, is to perform the technique Swot(Strengths, Weaknesses, Opportunities and Threats). This technique currently has a high degree of applicability in the business field, but can also be used in the personal field, that is called Personal SWOT specifically focused on the human being. This will evaluate the strengths and opportunities of a person just as their weaknesses and threats that revolve their environment. Originated by Albert S Humphrey in the 1960s, SWOT Analysis is as useful now as it was then. You can use it in two ways - as a simple icebreaker helping people get together to "kick off" strategy formulation, or in a more sophisticated way as a serious strategy tool (Swot Analysis, 2011).

This will be a self conscience exercise, one should be fair enough to yourself, and place the corresponding information no matter what it is, if the reader finds it very difficult, he or she can ask help from someone of their trust. The focus will be specifically on the academic role, it is required to have at hand a lot of information: a list of research conducted throughout his career, published articles, training events, debates, all the congresses where he or she has participated in, a list of places where they usually go on vacation, list tasks often performed on weekends, etc.

Now take a pencil and a blank sheet of paper, divide it into four sections and place the following titles: Strengths, Weaknesses, Opportunities and Threats, next do a breathing exercise and fill in each division the information required:

- **Strengths:** Identify the best you do as an academic, example: write, speak in public, participate in debates, research, etc., if you have any difficulty you can help yourself answering the following questions: ¿What do you do better than anyone else?; ¿What was the reason you were asked to participate in a project?; ¿What are your strengths recorded in the evaluations that the institution where you work made to you?.
- **Weaknesses:** Identify in what specific activity your are no doing to well, how would you like, maybe you do not like to participate in conferences with a large public, or debates, tv, radio or other interviews, if you have any difficulties you can help yourself answering the following ques-

tions: ¿In which of my professional activities I feel uncomfortable?; ¿Which are the factors where you have the lowest scores in your professional evaluations?.

- **Opportunities:** Identify what you consider may arise along the way and you can take advantage of: a scholarship, a postdoctoral study, a congress where you can participate, etc., if you have trouble you can help yourself answering the following questions: ¿What benefits can come if I study to work for an innovating project?;¿What mentions or awards can I apply in the university that helps me create new partnerships?.
- **Threats:** Identify what could affect your academic development that can come from an external agent. If you have trouble you can help yourself answering the following questions? ; Which obstacles do you confront?;Can any of your weaknesses put in serious jeopardy your job?.

Once you have finished filling out the four variables, the reader will already have his first instrument.

Identify Relevant and Sustainable Differential Values

A valuable person is a person that has interior values and that lives by it. Therefore a man is worth what his values are worth and the way he lives them (Ardilla & Orozco)

The "*El libro de los valores*", by the authors Ardilla & Orozco mentions that the values are fundamental beliefs that help us decide, appreciate and choose some things over others, we acquire them during the process of growing and are influenced by the environment that surrounds us and mostly by the family. They are a guideline for formulating goals and personal or collective purposes and reflect our interests, feelings and most important convictions.

Several authors have written about this exciting topic from different perspectives and areas of knowledge, but a general acceptance is that these are classified into seven values: biological, sensitive, aesthetic, intellectual, economic, moral and religious.

The quality of our lives depends much on the values we have, what makes us unique and different. With this in mind, it is necessary to make a self-analysis and identify your values, just as in the previous section be sincere and objective, to make this process easier below is a chart with a list of values (both work and personal) (Metcalf & Palmer, 2011), select the ten that are most important to you—as guides for how to behave, or as components of a valued way of life. Feel free to add any values of your own to this list.

Upon completion of this activity, the reader now has his second instrument of the process.

Establish Where You're Headed and Your Role in the Future

The best way to do this activity is to do it in an atmosphere of meditation in a personal space, without interruptions.

The mission is about our reason of being in life, it is about the human need of identifying and expressing our purpose in life.

Understand that the personal mission is about answering the following questions: * who am I?* who do I want to be? Who do I want to be in the future?* Why am I here?* Am I successful? Why or why not?

For many of the readers it may seem unnecessary the elaboration of the mission, because of the day by day activities it helps the fulfillment, but it results very interesting to make this self-analysis by an-

Figure 2.
Source: http://www.metcalf-associates.com/

Personal Values Checklist		
▪ Achievement	▪ Fast-paced work	▪ Pleasure
▪ Advancement and promotion	▪ Financial gain	▪ Power and authority
▪ Adventure	▪ Freedom	▪ Privacy
▪ Arts	▪ Friendships	▪ Public service
▪ Autonomy	▪ Having a family	▪ Recognition
▪ Challenge	▪ Health	▪ Relationships
▪ Change and variety	▪ Helping other people	▪ Religion
▪ Community	▪ Honesty	▪ Reputation
▪ Compassion	▪ Independence	▪ Security
▪ Competence	▪ Influencing others	▪ Self-respect
▪ Competition	▪ Inner harmony	▪ Serenity
▪ Cooperation	▪ Integrity	▪ Sophistication
▪ Creativity	▪ Intellectual status	▪ Spirituality
▪ Decisiveness	▪ Leadership	▪ Stability
▪ Democracy	▪ Location	▪ Status
▪ Economic security	▪ Love	▪ Time away from work
▪ Environmental stewardship	▪ Loyalty	▪ Trust
▪ Effectiveness	▪ Meaningful work	▪ Truth
▪ Efficiency	▪ Money	▪ Volunteering
▪ Ethical living	▪ Nature	▪ Wealth
▪ Excellence	▪ Openness and honesty	▪ Wisdom
▪ Expertise	▪ Order (tranquility/stability)	▪ Work quality
▪ Fame	▪ Peace	▪ Work under pressure
▪ Fast living	▪ Personal development/learning	▪ Other: __________

swering all the questions in the previous paragraph, and with this make the mission. The result of this exercise may bring many surprises, it may come up that your daily work does not contribute in anything your mission, or on the contrary it rectifies it.

The vision is about what we want to achieve in the long term, the place where we want to be after several years, it is the mark the academic wants to leave as his legacy, it is important to establish it in a maximum of two lines projecting yourself in a period of at least 15 years.

The mission and the vision are closely related and to write a personal vision, we must consider the mission.

By finishing this activity, the reader now has his third instrument of the process.

Your Personality Should be Shown on Your Brand

An identity is a self-definition, implicitly, that a human being should be able to develop in the course of his conversion to adulthood and continue redefining throughout his life (Erikson, 1950)

To build an identity three dimensions should be considered: The personality the brand builds while it communicates, the message of all you have to say and the expressions to be used so they can be understood.

Brand personality is defined formally here as "The set of human characteristics associated with a brand " (Aaker, 1997). In contrast to "product-relate attributes" which tend a serve a utilitarian function for consumers, brand personality tends to serve a symbolic or self-expressive function (Keller, 1993).

Conducting a review of authors on the construct Brand, (Bailey II & Schechter, 1994,) and (Grossman, 1994), refer to the brand only for it's tangible elements, defining it by it's most basic expression, on the other hand is the point of view, (Kapferer, 1992; Fournier, 1998; Aaker D., 1996; Sweeney & Brandon, 2006), in which the emotional, symbolic and intangible elements of the Brand are predominant. In these symbolic elements the personality of the Brand, the way the Brand reinforces the identity of the consumers and the ability of the brands to represent the consumers before others are included (Schlesinger Díaz & Cervera, 2008).

These authors conclude that the brand has a personality that is perceived by the consumers, whether it represents a product a tangible or intangible commodity, since the topic of this book is centered around personal academic branding, the Brand that we should define belongs to a human being, whom already has an established personality, following is a list of elements that help establish the personality of the Brand:

1. Country of origin.
2. Sector of activity.
3. Style of communication.
4. Image of spokesman.
5. Image of the users.

Country of Origin

The country of origin of the academic will carry to the Brand those prototypical features of the image representing the citizen of that country, just like Roobert Govers explains it in his book titled "Place branding: Glocal, virtual and physical identities, constructed, imagined and experienced" where he affirms:

The identities of the places are constructed by historical, political, religious and cultural discourses, also through local knowledge and the inevitable interference of power struggles. The real identity of places is managed by the realization of the group of unique features and / or group of meanings existing in a place and its culture at a particular moment in time (Govers & Go, 2009)

The reader must now identify the values and positive traits that transmits the Brand of his country, and place them in a contribution under this name.

Sector of Activity

The organizations working in the same sector must necessarily share values because the category requires it, the same way and for the same reasons some general personality traits are shared (Avalos, 2013).

The claim of Avalos is quite successful at present time by the way the global markets behave, this is the fundamental reason why sectorial brands are made, their creation is focused primarily towards the industrial area, this solidifies the placement of the industry mainly to search, maintain and strengthen its presence in international markets.

But, is this the same behavior in the field of higher education? An immediate answer would be to think that this field is appropriately represented by its sectorial Brand bonded with the country Brand, and the combination of both elements will help us see certain glimpses of the sector's quality.

The other answer would be no, a sectorial Brand can not represent the sector, since it has special particular characteristics and as a result it is necessary to implement a different methodology. Precisely this has been the worldwide way and today in the academic field what prevails in first instance is the reputation earned by the school's Brand in particular.

The publications showing the lists of institutions of higher education or programs supposedly in order of their quality are part of the current landscape of higher education systems in many countries. As in many cases, this phenomenon has appeared with a lot of anticipation in the United States but, at least from the decade of 1990 it has expanded all over the world (Martínez Rizo, 2011).

Currently there are many rankings that measure the academic quality of Universities all over the world, some of these are:

The Academic Ranking of World Universities (ARWU) was first published in June 2003 by the Center for World-Class Universities (CWCU), Graduate School of Education (formerly the Institute of Higher Education) of Shanghai Jiao Tong University, China, and updated on an annual (Academic Ranking Of World Universities, s.f.).

The Times Higher Education World University Rankings 2014-2015 list the best global universities and are the only international university performance tables to judge world-class universities across all of their core missions - teaching, research, knowledge transfer and international outlook. (The World University Rankings, s.f.).

Nowadays the Shanghai Ranking is the most widely accepted ranking among academics, as it is stated in Salinas Torres and Docampo in their article entitled "The new Highly cited researchers list from Thomson Reuters and the Shanghai: Spain's position and university map", in which it is mentioned that this ranking has currently become a "performance measurement tool of national and worldwide university", joining the competition among institutions to climb positions, are causing universities to pay more attention to the variables that make up the different scores of the ranking (Torres & Docampo, 2013).

The measurements have evolved, now it is not enough to only measure the quality of the universities, works are now being done at a more detailed level, It aims to identify the potential of researchers who work in these institutions, we will discuss more about this topic in the following chapters, for now we will mention the Highly Cited Researcher (HCR), made by Thomson-Reuters, they publish an annual list of heighted cited researchers openly available on the site http://highlycited.com.

The importance of the HCR is that from 2012 Thomson-Reuters incorporates this as a component of the Academic Ranking of World Universities (ARWU)- Ranking of Shanghái, which has turned the highly cited researchers into a "solid asset for their universities and has generated, just like in the style of great sport leagues, recruitments with clear intentions to quickly scale positions in the ranking" (Torres salinas & Docampo, 2013).

After the broad explanation of how the sector transmits and shares values and personality traits towards academic branding, the reader should identify in an impartial way the values that the identity of the institution of higher education transmits and place them in a section entitled "values of the sector", as an additional exercise the reader can idealize what are the values they wish to reach and place them here, keeping in mind "my goal institution" at where they would like to work at a near time.

Figure 3.
Source: http://highlycited.com/

Establish Your Style of Communication

Each person has their own style of communication, the ones that let us establish relationships with other people creating a link with other persons, this process flows through the components of social abilities (behavioral, cognitive, and physiological).

Monjas (1993) defines social skills as "specific behaviors or social skills required to competently perform a task of interpersonal nature, involving a set of learned and acquired behaviors and not a personality trait. They are a group of complex interpersonal behaviors that come into interaction with other people".

Rich and Schroeder (1976) state that this set of behaviors are closely linked to the physical and psychological characteristics of the individual, and the degree of awareness that has the communication process. While more aware the person is about the process and their own social skills, he will be able to control, change or improve those weak components, avoiding they act against him.

The reader should ask himself or herself now what are my social abilities? What are my weak components? to answer these questions it is necessary to look at the social abilities and their components, however I must anticipate that there are many authors who have written about this concept, so it has a lot of definitions with different approaches, shown here is the theory presented by (Caballo, 1998)

The components that constitute the social abilities are three:

The behavioral components classify themselves into:

- Non verbal
  - Within the Kinesis[2] we can distinguish the body posture, eye contact and gestures.
  - Proxemics[3] concerns the appropriate use of space.
  - In the paralinguistic we can differentiate the volume and tone of voice, inflection, speed and fluency in speech and silence.
- Verbal
  - Empathy. The skilled person transfers his empathy with the expressions used.

Figure 4.
Source: Author, 2015.

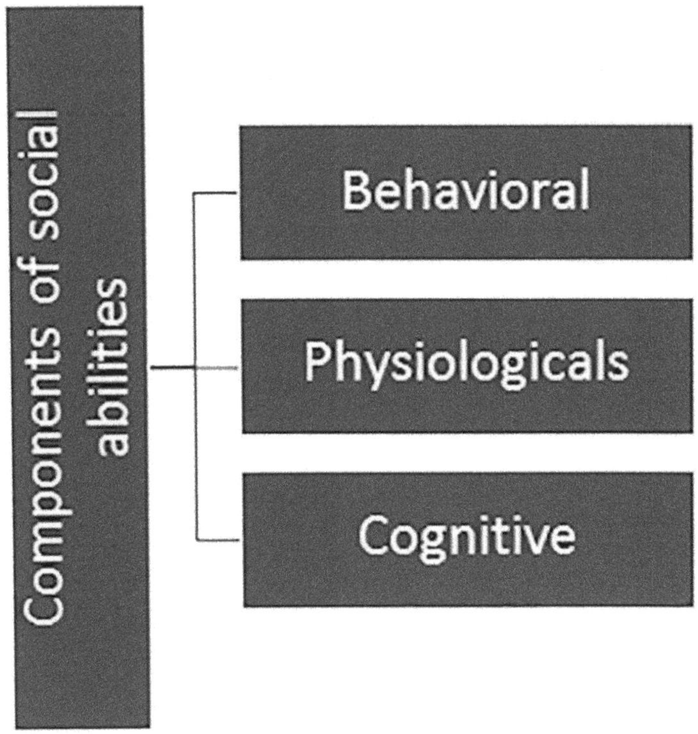

Figure 5.
Source: Author, 2015.

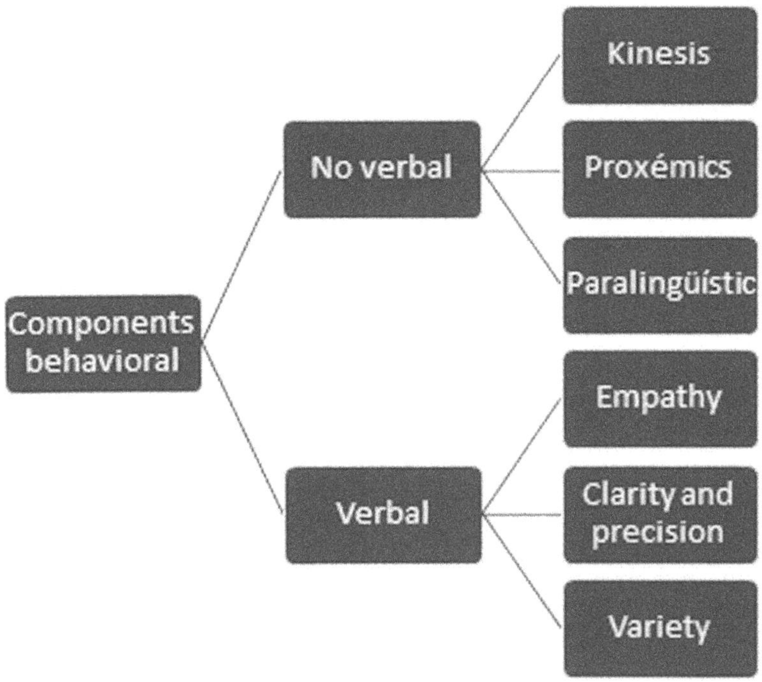

- Clarity and precision.
- Variety. The unskilled people have a smaller variety of subjects for use in the conversation, minor expressions of interest to the interlocutor, the more utterances of interest to oneself and excessive self-revelation.

The physiological components refer to:

- The pulse.
- Blood pressure and blood flow.
- The electrodermal responses (sweating).
- The electromyographic responses (muscle contraction).
- Breathing (depth and frequency).

In practice this is what is identified as the degree of anxiety that occurs at the time the person is interacting.

The Cognitive components may be the cause why a person can fail when displaying a socially responsible behavior, they are:

- Lack of acquisition or learning the skill: it is possible that a certain type of skill is never acquired, or has been inadequately learned.
- Failure to use the ability in certain situations: sometimes a social skill is shown in some situations and not in others.
- Influence of situational variables: people learn what social skills should be strengthened in certain situations and adapt their behavior accordingly.

Communication styles could then be described in terms of the components of social skills used by the individual. This gives us the sum of three main styles (Vander Hofstadt, 2003):

- **Assertive Communication:** It is the most natural, straightforward style. It is used by people with self-esteem and security in them, that through communication seek to raise issues that are satisfactory for all, without resorting to manipulation or pretense.
- **Aggressive Communication:** It is the style of who seek to achieve its objectives without worrying about the satisfaction of the other. Often they use strategies like the feeling of guilt, intimidation or anger.
- **Passive Communication:** It is the style used by people who avoid confrontation and attention. To do so they respond passively, without getting involved in the issue or showing conformity with whatever arises.

Features that identify each style of communication.

The reader now should identify from the given list their social skills and communication style. This is a really easy exercise if you really know your "inner self", if instead you have got confused with the information, take your time and read again in detail the information provided in this section, than look closely your daily actions one by one at the moment of communicating and interacting with your environment,

Figure 6.
Source: http://www.slideshare.net/ ; Steven M. Gerson & Sharon J. Gerson.

Characteristics of Communication Styles
(adapted from Sherman, 2001)

Characteristics	Aggressive	Passive	Assertive
Mottoes and Beliefs	"Everyone should be like me." "I am never wrong."	"Don't make waves." "Don't disagree."	"I have rights, and so do others."
Communication Style	Close minded Poor listener	Always agrees Hesitant	Active listener States ideas directly
Characteristics	Domineering Bullying	Apologetic Does not express self	Confident Non-judgmental
Behavior	Puts others down Bossy	Asks permission unnecessarily A fence sitter	Firm Action oriented
Non-verbal Cues	Points with jabbing finger Frowns	Looks down Slouches	Direct eye contact Confident and relaxed posture
Verbal Cues	"Don't ask why. Just do it." Verbal abuse	"I'll try to do what you suggest. . . ."	"I choose to . . . ." "What are my options?"
Confrontation and Problem Solving	Must win arguments Threatens	Avoids or postpones Needs supervision	Negotiates Confronts problems when they occur
Feelings Felt	Anger Hostility	Powerlessness	Enthusiasm Well being
Effects	Wastes time and energy micromanaging Provokes others	Loses self esteem Builds dependent relationships	Achieves self esteem and confidence

you can ask for help from a friend but sincere in the information he will provide you. When you have the required information place it in a section under the name "Social abilities and communication styles".

After this broad explanation the reader with full security can answer the questions posed at the beginning of this subject and enter them in the corresponding section.

- What are your social skills?
- What is your communication style?

Image of Spoke: Man

The use of what is technically called spoke-man or spokesman for the Brand consists of choosing a specific person with certain associated values, to represent the brand and transfer its aesthetics, values and attitudes to this, it is the common procedure that is used in the business world, but how does it work in the academic branding?, the answer is very simple, you are the Brand and the only authorized spokesman to represent it.

Therefore, nothing has to be done in this section, just place here the values that you identified in section 2 of this chapter.

Image of Brand Users

The image we have of brand users is perhaps one of the strongest determinants of personality because it is mediated; the experience is personal and direct, proving to be one of the basic techniques in brand management (Avalos, 2013). You can also experience a backlash towards the expected, you can rule out a brand when you see that a person we do not like is using it.

Turning to the family level, which plays a large component in the formation of the personality of the individual in the growth stage, some have to feel familiar with some proverbs, which are proverbial[4] sayings or phrases used in Spanish since the Middle Ages and have come down to us through oral tradition, the well known phrase "Tell me who you walk with and I'll tell you who you are." You can guess the likes and hobbies of someone by their friends and environments they frequent. Similarly, this proverb warns of the great influence on the behavior or customs on someone the companies of the rest, whether good or bad.

The statements from the business and family point of view about the users of the brand, gives us a space to think about who are our colleagues, that we frequent, and therefore identify us. What values am I achieving to transmit through my fellow researchers, the circle that surrounds me as a professional, is the circle that I really want to be in?

After these reflections, which certainly could be considered more philosophical, we can clearly identify the values that identify the users of my brand, and place them in a section entitled "values of my users"

This activity certainly has resulted strenuous, little by little enter each of the elements that conform the personality, which is the result of a combination of situational conditions moderated by hereditary and environmental factors, which often goes unnoticed by the individual.

Now you have the fourth element of this process, the same that has sub elements represented in the Figure 7.

Express Your Thoughts

The speech is a linguistic message in the broadest and deepest sense, it brings to light all the values and brand promises. All speeches converge towards the establishment and operation of the brand universe. The expression of values through contents (meanings) and forms (significant) ensures cohesion and coherence, or must, at all times.

The speech of the brand is made up by: taglines or slogans, stories, anecdotes and short stories and voice tone.

Figure 7.
Source: Author.

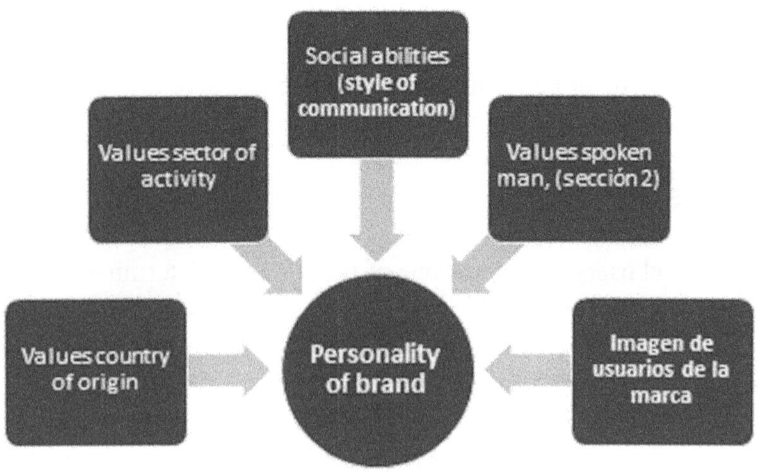

Taglines of Slogans

It is a phrase that immediately follows the logo, generally it synthesizes the concept or essence of the brand promise, this should consider the following features:

1. Be short.
2. Separate itself from the competition.
3. Be unique and original.
4. Capture the essence of the brand and it's placement.
5. Have an easy pronunciation and easy to remember.
6. Motivate an emotional response.

They can be imperative, descriptive, superlative, provocative, specific, here are some examples, from renowned scientists for their presence in the social networks:

Example 1: The scientist Ben Goldacre, cited by Science magazine as a "star" scientist in Twitter with about 409.196 followers (July 10/2015), British physician, academic and science writer. As of March 2015, he is a Senior Clinical Research Fellow at the Centre for Evidence-Based Medicine based at Nuffield Department of Primary Care Health Science, University of Oxford, has the slogan "Bad Science". (Figure 8)

Example 2: Brian Edward Cox, cited by Science magazine as a "star" scientist in Twitter with about 1'611.000 of followers (July 10/2015), is an English physicist, and professor of particle physics in the School of Physics and Astronomy at the University of Manchester, has a slogan made up of questions that clearly conveys the essence of his brand. (Figure 9)

Surely right now you are wondering what is the essence of the brand? Or how do I build my slogan?, well now is the time to use the material produced in section 3 of this chapter, read what you wrote and try to consolidate all of this in a slogan, or multiple slogans, which clearly conveys what you want to

Figure 8.
Source: https://www.facebook.com

Figure 9.
Source: https://www.facebook.com

achieve in life, the reason for your existence, how do you want to be remembered, in other words the mark you want to leave imprinted.

Tone of Voice

A tone of voice both embodies and expresses the brand's personality and set of values. It's about the people that make up the brand – the things that drive them, their loves and hates, and what they want to share with the world. (Cummings, 2015)

A brand's tone of voice should be distinctive, recognizable and unique. This may seem like a tall order until we consider the use of our own language in everyday life. We all employ language - both written and spoken - in our own way. Of course, culture and dialect are the most significant factors dictating our approach to words. But within these, we each have our own idiosyncrasies, favorite expressions, inflections, pace and so forth.

Therefore you should only be "yourself", when you answer questions or write messages spreading your academic work.

Now you can prepare the elements of section five, the "Speech", slogans, at least two expressions that determine your tone of voice.

Express Your Emotions

"Expression" according to the Royal Academy of Language is the representation with words or with other external symbols, of a thought, an idea, or a feeling. In his book Brand Sense, Lindstrom defines the term as:

We store our values and emotions in memory banks for each sense — image, sound, smell, taste, and touch. Events, moods, feelings, and even products in our lives are continuously imprinted in our memories, from the second we wake to the moment we sleep. However, most advertising messages that we're exposed to on a daily basis come to us through only two of the five senses: sight and sound (Lindstrom, 2005)

In this section we shall especially work with these two senses: sight and sound, since the personal brand principally develops online.The sight is the most powerful of our senses and comparatively has the ability of a wider range of subtlety. The elements that could be grouped under this item are the following:

- **Graphic:** Look for a graphic where you can permeate your professional vision (use the instrument prepared in subparagraph 3), and represent you appropriately, consider every detail, if there isn't a graphic that meets your expectations, look for a graphic designer and ask him to design one for you.
- **Objects:** Each object used in your visual communication, whether in imagery as in the information being shared should carefully support your ethical values.
- **Actors and Clothing:** The outfit faithfully reflects your personality, this clearly conveys the values of your country of origin, the values of the institution you represent and your personal tastes for colors and designs.

Example 1: Let's review again the scientist Brian Edward Cox, but now analyzing the existing visual elements on his Twitter profile image, carefully look at Figure 10:

Figure 10.
Source: https://twitter.com/profbriancox

His facial expression detonates strength, force, his fist conveys power, aggressiveness willing to discover new challenges, and he fears nothing. The objects that complement the scene: the helmet, uniform and motorcycle harmoniously complement each other to convey that: adventure, courage, ability.

Example 2: Neil deGrasse Tyson, cited by Science magazine as a "star" scientist on Twitter with about 3'952.599 of followers (July 10/2015), born October 5, 1958 is an American astrophysicist, cosmologist, author, and science communicator. Since 1996, he has been the Frederick P. Rose Director of the Hayden Planetarium at the Rose Center for Earth and Space in New York City., watch carefully each of the elements involved in the picture captured from his Twitter account shown in Figure 11:

What first stands out is his look, it's a happy look, denotes comfort, friendliness, which articulated with his face gesture invites to an open and emotional dialogue. His clothing gives a space of formality, who likes the nightlife, celebrate with friends, the background photo complements this interpretation, it's like you want to come see the sunrise? Do not forget his passion is being a cosmologist.

Example 3: Richard Dawkins, cited by Science magazine as a "star" scientist on Twitter with about 1'211.218 followers (July 10/2015), born March 26, 1941, is an English ethologist, evolutionary biologist,[3] and writer. He is an emeritus fellow of New College, Oxford,[4] and was the University of Oxford's Professor for Public Understanding of Science from 1995 until 2008 (Figure 12)

The image speaks of nature, quickly my mind takes me to the habitat, I can hear the birds and feel the warmth of the sun, transmitted by the cool clothes worn by the protagonist of the image. The word "RELIGION" is strongly emphasized as a label on the shirt, this one word evokes the religious values this scientist surely has, on the other hand his gaze triggers that he wants to see beyond the obvious.

Figure 11.
Source: https://twitter.com/neiltyson

Figure 12.
Source: https://twitter.com/richarddawkins

These three examples to which I have applied a simple visual analysis, helps to convey to the reader everything what an image can say, each element combines an important role in this. There may also be different interpretations of a same image, depending on the degree of experiences, maturity and perspective of the person who makes it, what is difficult is to change the visual cues that establish a context in the image.

In the hearing aspect we can group: sounds, music and broadcasting voice in the online environment, we are talking about the background music we might have in the background of our resources, or some video that we may have made to present them in several selected informative formats.

To this section a little more time has to be dedicated in taking care of every element that will make up your profile, it would be good to ask for help from a photographer friend to make some good shots and we have the sixth instrument.

BUILD YOUR ACADEMIC BRANDING

In the previous sections, the material has focused on getting the essence of the brand and its value proposition that carries out your personality as a valuable human being with whom it is worth sharing, focusing mainly on your professional talent. Now what's left is to develop the visual brand, which shall transmit all this at a glance, it is important to remember the words of the author Schwabel on personal branding, to be psychologically ready in preparing the remaining steps:

The personal brand describes the process by which the person and the businessmen differentiate and stand out among a multitude, identifying and expressing their proposal of unique value, whether it be professional, that later promote on different platforms, with a message and a consistent image that let's them reach a specific goal. This way, people can be recognized more and more as experts in their field, build up a reputation and credibility, further his career and improve your self-confidence. (Schwabel, 2009)

The logo is your own name, which has identified you since you where born, as simple as that, in sight that a personal branding is being formulated. The most appropriate now is to make a broad search on the internet of your name, with the intention to find homonyms, if that were the case write down on a sheet of paper these repetitions, to have a clear position of the situation.

If they are many repetitions, than you would have to add to your logo differential elements, for example: for the field "Names" use the name that less repeats itself, and the most repeated abbreviate it or omit it, in many contexts in a world wide level it is indifferent if a person uses his first or second name; for the field of "last names" it is obligatory to place the first one, if you have many repetitions of this one, in the conjunction add your middle name as a differentiator, on the other hand if your first last name is not too common, forget about placing your second last name because it will visually overload your brand. If you are one of the lucky persons in the world without any homonyms, congratulations, you automatically already have added a unique value and differentiator.

Refining the name, now it's time to decide about the typography, about this topic there are many scientific definitions that could be of great help at the moment of selecting the ideal letter type, but I do not wish to not mention Massimo Vignelli "was, is, the most important and influential designer of the last third of the 20th century and beginning of the 21st, the golden age which the classic design culminates" (Corazón, 2014), with one of his most famous phrases.

In the new computer age the proliferation of typefaces and type manipulations represents a new level of visual pollution threatening our culture. Out of thousands of typefaces, all we need are a few basic ones, and trash the rest. So come and see a few basic typefaces (Vignelli, 2010)

Select the typography with your best criteria for your logo, once this has been done you now have your logo.After finishing this chapter, you must have a total of 7 inputs, allowing you to create any profile on the selected digital formats:

1. A unique name, with a letter type that represents your values and homonym free.
2. A list of professional strengths (personal SWOT).
3. 10 values that identify you as an individual.
4. Mission and Vision.
5. Personality (country values, sector values, social abilities, personal values and finally an image of the users of the brand).
6. The speech, represented by a slogan that transmits your mission and vision.
7. Photographs creating in context, in which the brand is developed, in which each element has a meaning that represents the personality of the brand: expression, environment, position of the hands, eye sight, clothing, complementary elements, etc.

With this material you are ready to open your profile and present your personal academic digital branding.

FUTURE RESEARCH DIRECTIONS

This problem, about the imminent need that the academic researchers should have a digital presence with a unique value, makes space to see future research directions, that allow to correctly direct the actions of the researchers. One of the questions to answer would be: does the academic require a previous communicational work to the present surroundings to strengthen his reputations before formulating the academic branding online? Or failing that, they get better results for example, for his acceptance in competitive opportunities, the academic start their way to become brands versus those who hasn't.

CONCLUSION

This chapter enfasized that to create your academic branding, it is neccessary first to know oneself, which will let you quickly identify your value proposition. Also, you should know which is your current role and future one, in other words, dream, establishing the goals you wish to reach, without forgetting to transmit the values, personality and feelings through expressions and speech, elements that should be reflected by the visual Brand, therefore building an authentic personal Brand.

Personal branding does not focus on the outer side of the person, it is not an image consulting, it is not the way we dress or our appearance. Personal Branding is the set of elements that goes from the inside of the person and how he transmits it, these elements are the ones that will make us be perceived as valuable and reliable professionals, which automatically transforms in their footprint, as explained by Andrés Pérez, "personal branding is about leaving a mark in other people. If that mark you left is well defined and you communicate it effectively, you will influence on their emotions" (Perez, 2014).

Los Academics researches have the responsibility to leave an academic mark, which contributes to the recognizing of the institution of higher education they represent and a commitment to society in disseminating the science. They should also function successfully at virtual organizations, in highly integrated social networks, virtual (professional) communities and virtual laboratories / e-science, resulted from research, so it is necessary to have one academic branding.

It is important to point out that to obtain and academic mark it demands a lot of work, time, dedication, patience and persistence, you cannot reach it from one day to another; it is a process that it's results will be shown on a long term. Therefore a good communicational strategy is required and a plan to acquire visibility of the brand that complements your career as a researcher. There is much to do to make an academic footprint, the following chapters are the guide to the management and fluidity of online academic reputation. Good luck!

REFERENCES

Aaker, D. (1996). *Building Strong Brands*. New York: The free press.

Aaker, J. (1997). Dimensions of brand personality. *JMR, Journal of Marketing Research*, *34*(3), 347. doi:10.2307/3151897

Ardilla, & Orozco. (n.d.). *El libro de los valores*. El tiempo.

Avalos, C. (2013). *La marca identidad y estrategia*. Argentina: La Crujía.

Bailey, I. I. I., & Schechter, A. (1994). The corporation as brand: An identity dilemma. *Chief Executive*, *98*, 42–45.

Caballo, V. (1998). *Manual de evaluacion y entrenamiento de las habilidades sociales*. Madrid: Siglo XXI de España.

Corazón, A. (2014). *El País*. Retrieved from http://cultura.elpais.com/

Cummings, H. (2015). *Finding Your Brand's voice*. Retrieved from https://www.distilled.net/t

Erikson, E. (1950). *Childhood and society*. New York: Norton.

Fournier, S. (1998). Consumers and their brands: Developing Relationship theory in consumer research. *The Journal of Consumer Research*, *24*(4), 343–373. doi:10.1086/209515

Govers, R., & Go, F. (2009). *Place branding: Glocal, virtual and physical identities, constructed, imagined and experienced*. Hampshire: Palgrave Macmillan. doi:10.1057/9780230247024

Grossman, G. (1994). Carefully crafted identity can build brand equity. *The Public Relations Journal*, *50*, 18–21.

Kapferer, J. (1992). *Startegic Brand Management*. Kogan.

Keller, K. (1993). Conceptualizing, measuring, and managing customer- based brand equity. *Journal of Marketing*, *57*(1), 1–22. doi:10.2307/1252054

Lindstrom, M. (2005). *Brands Sense. Build Powerful Brands through Touch, Taste,Smell, Sight and Sound*. New York: The Free Press.

Marcos, T. (2009). *Como crear y mantener una marca personal*. Madrid.

Martínez Rizo, F. (2011). Los rankings de universidades: una visión crítica. *Revista de la educación superior*.

Metcalf, M., & Palmer, M. (2011). *Innovative Leadership Fieldbook*. Tucson, AZ: Integral Publishers.

Monjas, M. (1993). *Programa de enseñanza de habilidades de interacción social. Para niños y niñas en edad escolar*. Valladolid: MIMC.

Perez, A. (2014). *Marca Pesonal para Dummies*. Barcelona: Wiley Publishing.

Rich, A., & Schroeder, H. (1976). *Research issues in assertiveness training*. Academic Press.

Sapiña, L. (2013). *Aprender a divulgar la ciencia*. Metode.

Schlesinger Díaz, M., & Cervera, A. (2008). Estudio comparativo entre personalidad de marca ideal vs. percebida. *Innovar*, 31.

Schwabel, D. (2009). *Me 2.0: A Powerful way to Achieve Brand Success*. New York: Kaplan Publishers.

Sweeney, J., & Brandon, C. (2006). Brand personality: Exploring the potential to move from factor analytical to circumplex models. *Psychology and Marketing*, *23*(8), 639–663. doi:10.1002/mar.20122

Swot Analysis. (2011). Retrieved from http://www.washington.edu/

The World University Rankings. (n.d.). Retrieved from www.timeshighereducation.co.uk

Torres Salinas, D., & Docampo, D. (2013). The new Highly cited researchers list from Thomson Reuters and the Shanghai: Spain´s position and university map. *El profesional de la Información*, 265.

Vander Hofstadt, C. (2003). *El libro de las habilidades de comunicación*. Academic Press.

Vessuri, H. (2010). *El rol de la investigación en la Educación Superior*. Academic Press..

Vignelli, M. (2010). The Vignelli canon. New York: Academic Press.

ENDNOTES

1 Higher Education Institution

2 Kenesis, the movement of an organism in response to a stimulus, as light

3 Proxemics, Sociology, Psychology. the study of the spatial requirements of humans and animals and the effects of population density on behavior, communication, and social interaction.

4 Proverbial, it is a short, pithy and witty statement that conveys an instructive message, prompting the intellectual and moral reflection.

Chapter 4
Flow My Bits, the Professor Screened:
Netnography, Academic Micro-Celebrity, and Personal Branding

Robert V. Kozinets
University of Southern California, USA

ABSTRACT

Academics have always been in the business of producing and promoting ideas. The configuration of their identities or brands can play a central role in the successful realization of this ambition. This chapter uses the participative stance in netnography to discuss academic personal branding, but also interweaves advice, core principles, and a series of short inductive case studies about personal academic brands and fields. Featuring cases like those of Plato, Timothy Leary, John Lilly, and others, the chapter outlines a theory of personal academic branding, relating core identities to academic branding practices and purposes. In particular, this chapter suggests the relationship between thinking innovative thoughts, representing them to various audiences, and having one's idea shared and built upon by a network of interested others. Core to the successful realization of this endeavor are the following nine elements: unconventionality, creativity, internal and external audiences, segmentation, core constituents, co-branding, simplicity, uniqueness, and catchiness.

INTRODUCTION

From time immemorial, learned speakers and writers have plied their trade by forging unique identities that, today, we would consider to be brands. We know the stories of Socrates, and how, while teaching in the market, he inspired Plato to write and teach. Socrates was a man of many questions; one of the major things he espoused was to question authority, and to speak the truth to power. Emboldened by the man, Plato founded the Academy, north of Athens, the world's first formal university. From Socrates forward, learned scholars thought, taught, were venerated, fear, persecuted, and had impact upon one another and upon the world. Their academic identities interact, intersect, and interweave, reticulating

DOI: 10.4018/978-1-5225-0917-2.ch004

with various networks and alliances to reach different audiences to different effects. From academic journals and conferences to students, government, the press, and various publics, academics have always been in the business of devising, promoting and communicating ideas. Like Socrates and Plato, they seek to have an impact upon one another and the world. The role of their identities—their brands, if you will—can play a central role in realizing of this ambition.

This chapter will focus on the role of netnography in academic personal branding, a topic which draws upon and extends my writing in Kozinets (2015). However, it also seeks to do more than this. Interweaving research, advice, and a short series of inductive case studies about personal academic brands I have known and admired, it tries to collect some emerging wisdom about this very new practice of personal academic branding. Through featuring the faces of these talented scholars, I begin to sketch out a theory of personal academic branding, relating core identities to practices and purposes for academics not only in academia but also in society. In particular, this chapter seeks to clear some initial conceptual ground regarding the relationship between thinking innovative thoughts, representing them to various audiences, acting in the world, seeking to cause others to act upon it for social change, and having one's idea shared and built upon by a variety of interested others. The chapter does this through a focusing on netnography, a fluid concept made even more fluid by its recent redefinition and the new conceptual indistinctness of its greatly expanded boundaries.

In some ways, this chapter is an homage to these great thinkers of the past who have so changed my mind, heart and life. I introduce and honor some of the many giants upon whose shoulders, as Newton brilliantly analogized and many have appreciated, we stand and see even further vistas. Although Socrates plied his trade before the printing press, Albert Einstein at the rise of mass media, and Michel Foucault his before the age of social media, each of these intellectual maestros was able to create much influence in each of these media. Their thoughts increasingly influence discourse and burble through the timestream of public thought. We are thus talking not only of intellectual enterprise here, but of media.

Netnography is anthropology that conducts research on digital interactions and experiences by focusing upon digital content. Netnography is participant observation. And that netnographic participation is conceptualized as a type of public identity work: personal academic branding. Wandering widely and jetting back and forth between empirical instance and abstract principle, this chapter conceptualizes branding in academia as a multifaceted act that is forever changed by the rise of the Internet and the advent of social media. It is into this social space, I seek to situate netnography, the act of engaging with online others in the technologically mediated and self-reflective research of human individual and collective being.

The essence of netnography is ethnography using sophisticated mass broadcasting social media tools. As such, netnography builds upon the organic and orgasmic origins of academic communicative practice in group oration and writing as a communicative ecstasy and act of strategic public relations. Netnography is ethnography in public, online, where we add publishing, amplifying and ungrading our prior scholarly abilities to gain local, niche, topical, and mass attention through all media, using laptop, mobile phone, one's cell in the network, or any other technologies, of machine or of the self. Netnographers are technologically proficient prosumers. They consume consumer research at the same time as they create it.

Netnography, like ethnography, can and often is a deeply personal affair. Being a reflexive participant in your inquiry means you must include your own self, your experiences, your habitus, your life, as you construct research. Please keep this participative aspect of my writing and scholarship in mind in this chapter as I take you pastiche fashion through a series of discussions of particular academics and their brands, notions of personal branding. As I elaborate ideas of netnography, induced/abducted principles and proposal pertaining to personal academic branding, I am also reflecting very personally upon these experiences, these people, these scholars and ideas and what they mean to me.

THE SHERRY BRAND

In 2012 and 2015, marketing and medical anthropologist and Notre Dame marketing Professor John Sherry presented a couple of poems which brilliantly examine, with the authority of a professionally appointed scribe and chronicler, the story of being an academic. His stories portray living as a scholar today. But they also relate specifically to the terms and practices of the consumer culture theory tribe, to the consumer research tribe, and in many ways the entire groups of tribes of people calling themselves scholars and academics, professionals, those who are legitimated and valorized by international states, with crusty barnacled and ever renewed relevance as professors.

In one of his poems, a situation familiar not only to academics but also to students of almost every age and orientation transpires. A speaker drones on and on, as we assume the position of the listener, who is "sitting in the back of a crowded lecture hall" (the 2015 poem's title). Here are the first few lines (with my capitalizations for emphasis):

The Lecturer drones on,
Hewing to the script,
Siphoning Hope out of the room
Relentless in the piling on of details
Punishing the audience with
Earnest plodding effort,
The Lecturer drones on (Sherry 2015).

Sherry reminds us with humor how our rhetorical reach sometimes exceeds our grasp of audience dynamics. Punishing the audience with earnest plodding effort is poignant and resonant. In the earlier 2012 poem, which is entitled "Fuck You Very Much, Reviewer C," contains barely veiled homicidal fury at one of the three traditional reviewers operating in our tripartite peer review system.

Of each revision you are leery,
'Though I've answered every query.
All attempts to ground my theory
Meet with mockery and scorn.. . .
Too much theory wrung from data,
Need more field notes, less verbata,
Cites become a hot potato
Tossed among anointed few.. . .
At last the fatal flaw's revealed.
Judgment cannot be appealed.
The author is once more annealed
And takes the story on the road. (Sherry 2012)

None of us likes being judged in the peer review process. None of us enjoy the anonymity of the reviewers. None of us likes the somewhat arbitrary comments that repetitively seem to come our way. But Sherry captures these emotions in such a concise manner that we cannot help but feel how "we" (as he points out later in the poem) are both the victim and the aggressor, a part of the system itself as we are doubly-oppressed by it, a poem dedicated, in the poem's last line, "to all of us Reviewer C's".

The poem was presented at the annual *Consumer Culture Theory* conference, an organization that elected John its first President, where it created a bit of a stir among the audience. The poem is published, along with the others in the poetry sessions John organizes every single conference, in a boxer-inspired volume entitled *Clarence Clobbers Tenderly* (2012). The boxer metaphor is suitable, as it evokes Sherry's history and identity as a college boxer. Avuncular and mentorly to me always, he said to me on one occasion that academia was like boxing, that you needed to learn to take a few shots to the head, to get up repeatedly after being knocked down, and to throw a few punches back.

Academia is a strange business, as captured in books such as *Straight Man* by Richard Russo (1998), a hilarious and yet very human novel about an anarchist professor. Another is Donna Tartt's (1992) *The Secret History* and its charismatic and dark professor, and the cultish following he inspires among young and impressionable female students. In terms of insider knowledge and ability to pithily capture academic life and its challenges in poetic verse, John is one of the best.

John has been building his brand as a marketing anthropologist and creative method demonstrator and promoter through five different decades. He knows the game, and he plays it with integrity and honor in the fields of anthropology and consumer research. He is not a big self-promoter. He doesn't blog. He doesn't tweet. He does not have a page that stretches beyond his university page at Notre Dame. John simply persists in doing erudite and intriguing scholarly work, and finds ways to get it published with a range of excellent co-authors like Annamma Joy, Melanie Wallendorf, Stephen Brown, Russell Belk, and Tonya Bradford.

I would also include myself as an eager accomplice. I was inspired by John's poetic openness to create and present *Desert Pilgrim* (Kozinets 1999, 2001), which was the hybrid and reflective videography-poem that started me on the path of alternative forms of research representation. Desert Pilgrim was the first research videography and the first research poem I had created. A production inspired by and about my first ethnographic journey to Burning Man 1999, it represented a personal breakthrough. I presented it first at the 1999 *Heretical Consumer Research* conference, and then twice at the 1999 *Association for Consumer Research* conference that followed it, the second by request from the audience, instead of talking to my slides. I printed some photos from my fieldwork and combined them with the poem, publishing them in the then-new consumer culture journal *Consumption, Markets, and Culture* (Kozinets 2001). Later, I recreated the poem and video experience and shared it on YouTube.

The Heretical Consumer Research group from which sprang the Consumer Culture theory group, according to Sidney Levy, our nonagenarian polymath field-founder, was far more edgy than its successor. Levy was chair of the Northwestern and then later Kellogg School of Management marketing department. He came from a different time than most of us. He was appointed full professor with tenure after being invited there, from a simple PhD to full, with no need to go to assistant or associate levels. He is considered by many to be the founder of the field of consumer culture theory, a field which formally turned ten years old in June of 2015, after its founding in the year after the foundational "CCT article", an academically-encapsulated charter and constitution (Arnould & Thompson 2005) was first presented by Eric Arnould and Craig Thompson to a massive crowd at the *Association for Consumer Research*'s 2014 conference. I was there, watching with admiration and curiosity.

Interestingly, the marketing professors who named this subfield wrote into the chapters and explained what they were doing as academically branding a sub-field. Branding quite consciously, deliberately, and calculatively. Branding for students, for jobs, for the future. To be counted, you need to label. To be included, you need a name. For a long time, in our psychology and economics dominated field of marketing and consumer research, our work was known from its data as "qualitative" or it mysterious paradigm as "postmodern" or "alternative" research. It needed a brand. I know that conversation happened among these scholars, because I was there. Craig Thompson, himself a remarkable brand who uses his research to write about the birth of his child, his natural health interests, his Southern values, his interests in community supported agriculture and hipster coffee shops, and much more, joined with talented anthropologist Eric Arnould to engage in this act of explicit differentiation. They make this branding effort very clear when they write in the introduction to the article:

Over the years, many nebulous epithets characterizing [the research tradition addressing the socio-cultural, experiential, symbolic, and ideological aspects of consumption] have come into play (i.e., relativist, postpositivist, interpretivist, humanistic, naturalistic, postmodern), all more obfuscating than clarifying. Each fails to signify the theoretical commonalities and linkages within this research tradition. They either place too much emphasis on methodological distinctions or they invoke overly coarse and increasingly irrelevant contrasts to a presumed dominant consumer research paradigm. A more appropriate and compelling academic brand would focus on the core theoretical interests and questions that define this research tradition. Accordingly, we offer the term "consumer culture theory" (CCT). (Arnould & Thompson 2005, p. 868).

In the published words of these two marketing academics, founding a now-vibrant sub-field, we see at work the power of branding. Shifting the terms of the labeling to a focus on theoretical abstraction and contribution, and away from an emphasis on method, paradigm or type of data, these two clear-thinking researchers successfully offered the field "a more appropriate and compelling academic brand".

In 2012, I wrote an article for the online journal *Methodological Innovations Online* which looked at the way that the social scientific academic milieu and my actions as a professor, thinker, and writer had interacted and resulted in a number of intentional and unintentional connections (Kozinets 2012). I built first on Thomas Kuhn's ideas that new scientific paradigms are orchestrated productions of contagious social enthusiasm and then extended it to discuss Peter and Olson's (1983) idea that science is a special case of marketing, the marketing of ideas. In that article, I broadened Kuhn, Peter and Olson's ideas to include the methodological practices and terminologies around social scientific work, as well as the theories, axiologies, and ideas of academics. Netnography, I asserted in the article, is a brand that has been, can be, and indeed even maybe *should* be marketed. As all useful and impactful academic brands have been—in their own way.

BRANDING AND THE ACADEMIC ENTERPRISE

Netnography materializes through practices: through reflective participation in the writing of fieldnotes, through active observational techniques that can include data mining and massive search, through the creation and maintenance of a research web page or other social media interface or interfaces, and through analytic coding which includes hermeneutic reading and understanding of human beings' communications as the voices of real people. Within the participative element of netnography, disclosure is required for ethical practice. In ethnography, research purposes disclosed to people result in strange things that can happen to alienate one person from another. As Judith Stacey (1988) wrote, in a classic of feminist methodological inquiry, the power games of ethnography make its fits with humanistic ideals challenging. Her classic article's abstract beautifully summarizes the problem:

Many feminist scholars have identified ethnographic methods as ideally suited to feminist research because its contextual, experiential approach to knowledge eschews the false dualisms of positivism and, drawing upon such traditionally female strengths as empathy and human concern, allows for an egalitarian, reciprocal relationship between knower and known. This paper discusses the irony that ethnographic methods also subject research subjects to greater risk of exploitation, betrayal, and abandonment by the researcher than does much positivist research. Fieldwork and its textual products represent an intervention into a system of relationships that the researcher is far freer than the researched to leave. The paper calls for greater dialogue between feminism and the new ethnography which addresses similar methodological concerns and suggests certain constraints on that dialogue.

As Stacey relates, backing away from such power games is one option. However, it is not the only option for a netnographer who understands the principles of personal branding. We must embrace our own perspective, our own personality and point of view. In ethnography, we must chronicle our own research story in the context in which we find ourselves, whether it is in a eco-village, at a themed hotel in Disney World in Orlando, in the classroom, at a shopping mall, or camping with the Rainbow Family in an annual countercultural gathering. It is an individual effort with collective payouts: myth made manifest in everyday life. That is almost always the core ethnographic story: the universal human story instantiated in a particular time and place, among a particular people and set of persons. Perhaps, those human stories, those stories of humanity can become brands. Ethnographic stories which include the ethnographer, the doing of the research, the tale of fieldwork, become acts of brandings. Branding about the researcher, brands about academic life, about academic story telling and its challenges. Perhaps ethnography on the internet, when it become netnography, becomes something quite different as it changes and adapts to the need of researchers to also be Personal Academic Brands.

A brand's ideology is felt bodily, biologically, through the body. This happens, as Foucault most explicitly theorized in his theory of sexuality, but of course in this always building on Marx, through discursive systems of ideology and consciousness (alienated, false or otherwise). Discursive systems also branch into the world of branding: the social world where emotions like fear, jealousy, greed, and shame often coincide with people's emotions and identities in the moment. Ideology is felt in this way, socially: through systems of meaning and languages, many languages now including machine, application, technology image, and other languages.

Machine Language and Branding

Any good computer scientist will tell you that they can discern something, maybe a lot, about the coder from an examination of the code. A person's basic personality and certainly skill level might shine through. Decoding academics may be even easier. What do we know, from writers like danah boyd, Donna Haraway, Katherine Hayles, Laurel Richardson, Carole Rambo, Linda Scott, Katrien Jacobs, Beth Hirschman, Morris Holbrook, and Shannon Bell? We know that revealing the self makes your work more, not less precise. Coming from a particular social and cultural context allows more, not less, specification of the reasons behind interpretations and the forces underlying extant and often invisible social practices. Personal revelation makes science more, not less, impactful and engaging.

Are you exposing yourself at all in your research? If not, why not? If so, are you doing it strategically? What does your self-revelation say about yourself, your brand, and your method? Think about the connection between impactful researchers, their lives, their images, and their research. Walter Benjamin had his photographs, his self-portraiture, to stand alongside his writing about photography and self-portraiture. Foucault had his pastimes and indulgences, which everyone liked to watch, his notions of self-discipline and self-regulation, sexuality alongside the panopticon in his theory. McLuhan had the catchphrase "the global village" and the slogan "the medium is the message" which became his message as he overtook the media. The meme of Richard Dawkins became his own meme, his tactical labelling of a new concept becoming an entry point into popular discourse.

The personal brand is the new corporate brand: we easily have the personal bandwidth now that entire corporations, at one time even entire countries or economic systems had at their disposal. Do you have something to say, some good use for this bandwidth and processing powers? Are there many who would dare to assess and critique the present? Can we still imagine the academy conducted beyond the walls of the city or the market, beyond surveillance? As academics, do we have something to say about social reality, reflexively and resonantly? Will we reflect on our social and cultural situation, with structure, stress and rigour?

What I believe distinguishes netnographers and other serious students of social media from those who take these phenomena lightly, or at face value, is that they have an axiological axe to grind. And they can tell you quite precisely which axiological axe they intend to grind. Mine is technological. For some, like Donna Haraway, better the cyborg's fate than being even a Goddess. For others, becoming a cyborg represents a major step down into the dark and deadly depths, depths increasingly interlinked with global intelligence (consumer goods producers, government and space providers, military and protection systems, intelligence and legal systems) as well as communications and informational and transactional economies of attention (media), deficit (micro-economic) and disorder (medical) systems. Something, perhaps, for us all to consider, as we decide upon what we can and should say which is relevant to other actors in the world today.

Personal Branding in the Academy

Labreque et al (2011) explore personal branding online, an article that informs us about how people will try to do the right thing and be authentic, as well as try to manipulate things for their own benefit, using social media. What might we learn about how to actually do personal academic branding from this? Simply that academic branding must take place against a context where many more effective celebrity and instant Internet celebrity personalities are able to grab the spotlight of collective attention at the highest mass media levels. The academic Kardashians, as Hall (2014) puts it. What this paper underscores for us, then, is the importance for academics to understand that their public conversations take place in an entertainment, industry, government, and related educated and literate professional setting. This setting is amplified and extended. We enter economies of attention with content: words, ideas, books, seminars, lectures, classes, speeches. Intellectual capital, social capital, cultural capital: words, time, and money. Within all of this is a balance between the frivolous and the trivial, the meaningful and the serious. There are no clear boundaries, and no guidelines in sight. You better be rigorous, but you also should be accessible.

The netnographer must in some sense play a public online academically-enriched round of attention-seeking. Perhaps we are not macro-celebrities, not on the scale of the Kardashians of the world. However, perhaps we can aspire influence a bit wider than the "micro-celebrities" described by Senfft (2008). To do this, you must have a strong sense of what makes you distinctive. This is the essence of personal branding. Hall (2014) cautions personally branded academics to have a proper "Kardashian index". Named for the woman who is more famous for being famous than for any particular accomplishment, academics on social media can become widely shared and quoted in the public sphere without very much accomplishment in the scientific world of peer-reviewed books and journals. The perspective of Hall (2014) is that this is a bad thing: scholars must be legitimated by other scholars and institutions first, and then share their fame outside the world. Like Freud, Eco, or Oppenheimer. The big ones. The real ones. Not like Neil deGrasse Tyson.

We can now as a field write about online interacting and transaction human behaviors as "Consumption in the Net". A Netnographer will use Personal Web Pages as part of the overall repertoire of Building Personal Academic Brand Recognition and Equity. Personal Academics should, in this environment, be able to compete with micro-celebrities, but not quite with celebrities. They exist in a region of recognition and resources between bloggers and real Hollywood, Euro fashion, or world sports celebrity spaces. To do this, however, will require genuine effort, devotion, and a passionate collective effort to contribute to each other and build a network based on shared brand values. People's values should equate to their brand's values. Finding authentic voices and linking them together one to one is how communications should go.

WHAT EXACTLY IS NETNOGRAPHY?

It should at this point be of no surprise that netnography had to recently change for new times. In Kozinets (2015), netnography becomes redefined as something more than the dated study of online culture and community. According to Kozinets (2015, 87), netnography "is the name given to a specific set of related data collection, co-creation, creation, and analysis, ethical, and representational practices, where a significant amount of the data collected and participant-observational research conducted originates in and manifests through" communications and interactions that the researcher her or himself directly conduct in public social media and through private interviews and communication, including mobile, desktop, laptop, tablet, watch, other wearable, implant, whatever technology they have upon or within themselves.

Netnography is about something specific: interactions and experiences through and of the Net and Web and all their related informational and communication experiences, so much of them contributed, contained and controlled by Google and Facebook, the traditional media companies, and ever-changing others, rapidly changing and reconfiguring technocapitalism. Netnography is about human connection. Netnography often means interviews conducted via email, Skype, in person, or using other methods, increasingly I would hope those captured in real time on programs such as Camtasia and then edited and shared on YouTube and Vimeo. Netnography embraces the lore of hermeneutic story telling and the data analysis of natural language word recognition processing, coding, and visualization.

To personally brand in netnography means to focus on communicative acts and interactions flowing through digital connections. These could be textual or visual. Graphic, photographic, audiovisual, or musical. As I say in Kozinets (2015, p. 73), "Netnography begins and ends with an explicitly human window into the rich communicative and symbolic world of the Internet, the Web, and social media."

To truly understand personal academic branding, we must first match up with the core notion of personal branding. What do we do with it? Tom Peters could see it coming at an early stage in the Internet's history. He saw it as a classic marketing principle, applied. Find your point of differentiation first. "Start by identifying the qualities or characteristics that make you distinctive from your competitors — or your colleagues. What have you done lately — this week — to make yourself stand out? What would your colleagues or your customers say is your greatest and clearest strength? Your most noteworthy (as in, worthy of note) personal trait"? (Peters 1997).

Tomas Chamorro-Premuzic wrote in Harvard Business Review in 2013 that *The Future of You* consisted of "showcasing our own individuality is a way that is distinctive, consistent, and meaningful." Hubert Rampersad (2008, p. 34), in an interesting and idealistic early article tells us that our personal brand "should be authentic." Rampersad is guilty of moralizing to us, but somehow it is a guilty pleasure to enjoy being moralized to. "Authentic Personal Branding is a journey towards a happier and more successful life. Your Personal Brand should therefore emerge from your search for your identity and meaning in life, and it is about getting very clear on what you want, fixing it in your mind, giving it all your positive energy, doing what you love and develop yourself continuously. Your Personal Brand should always reflect your true character, and should be built on your values, strengths, uniqueness, and genius." (Hubert Rampersad (2008, p 34).

Rampersad gives us a system for creating and managing our personal brands. We examine our own personal vision, mission, and key roles, and we challenge, plan, deploy, and act in turn on these elements. We define and formulate our personal ambition. They we formulate our personal brand. Then, we work (in outdated managerial terminology) with the personal balanced scorecard system (and early form of data analytics on the topic of personal brand scorekeeping) based on measuring the results of your personal brand, and then you implement, cultivate and I believe integrate your tactical and strategic efforts to achieve your personal ambition through using your personal brand and monitoring and managing it using your analytics system.

Hampersad (2008, p. 38), our moral and spiritual guide on the path to personal brand excellence, tells us that if you should be "branded in this organic, authentic and holistic way your Personal Brand will be strong, clear, complete, and valuable to others. You will also create a life that is fulfilling and you will automatically attract the people and opportunities that are a perfect fit for you. If you are not branded in this unique way, if you don't deliver according to your brand promise, and if you focus mainly on selling and promoting yourself, you will be perceived egocentric, selfish and a unique jerk, and branding will be cosmetic and a dirty business." Watch out. Don't be a "a unique jerk". If you are focused on self-promotion, everyone will be able to tell it. They will stay away. Which brings us to the man who spoke to dolphins: John Lilly.

Academic Brand Lilly

Academic brands are special entities, and this may be difficult for other professions or the public to understand. Academia is a direct popularity contest, but it is about the popularity of your ideas, not of you. People in academic circles seek citation. This is their stock and trade. To write something once, and have people read it but not write about it is not what academia is about.

One of my favorite academic brands, a man who, like his contemporary and far more popular doppelgänger Timothy Leary, the psychiatrist John Lilly wrote an early book analyzing human hardware and especially software using self-experimentation with LSD-25 in water suspension isolation tanks. The result was a surprisingly little-known book entitled "*Programming and Metaprogramming in the Human Biocomputer*", which presented subjective experience and interpretation alongside synthetic theorizing that joined medical education to philosophy, psychology, computer programming and spiritual traditions.

The human brain, a super-biocomputer, as it were, is a parallel processor. . . . In the child, automatic metaprogram implantation (or externally forced meta-programming), persisting as metaprograms below the level of awareness in the adult, can be controlling for the later adult programs, adult thinking, and adult behavior. Energy can be taken from some of these automatic metaprograms and transferred to the self-metaprogram with special techniques and special central states, chemically evoked. (Lilly, 1967, metaprogramming, p. 126-127)

We can even think for a moment of our personal brand as a program in the metaprogram of human society and culture. Our public persona, our social media presence and traces are analogous to "automatic" program implantations of our childhood. We have not really organized or managed them. Much of our life, we simply run the programs that were inserted into us, or which we inserted at one time or another. And similarly, our brand unwinds, bits of traces of ourselves shared and collected on the Net and Web. Yet when we take hold of our own personal academic brand, through netnography or another discipline or means, we become the metaprogrammer. We insert, with effort, with "special techniques" and in special "states" our own deliberate brand programming, communicational metaprograms. We control the energy of attention, the flowing, glowing liquid elixirs of audience engagement and social impact.

As his work became known, Lilly published a number of other books exploring human software and its reprogramming, laying the groundwork for Ken Keyes, Alfred Bandler and the rest of the early neuro-linguistic programmers to put detail and procedure around his theory's contentions of the reprogrammability of people's goals and even sense of reality. *The Center of the Cycle*, Lilly's follow-up "autobiography of inner space" is even more personal and more confessional. Another book Lilly wrote was a startling little pocket book called *Simulations of God*. In it, he presented chapters discussing how each different belief system simulated itself inside of us, as something we desire and worship, for a time, until we move from one to another.

In John Lilly's *The Scientist: A Metaphysical Autobiography*, Lilly offers up some bizarre, drug-induced prophesy about the takeover of AI from humanity in the near future. "As the machines became increasingly competent to do the programming, they took over from Man. Man gave them access to the processes of creating themselves, of extending themselves." (Lilly, 1988, p. 148) Then, he gets specific about our endangerments within the next 80 years and extinction within the next 300 years. "By the year 2100, Man existed only in domed, protected cities in which his own special atmosphere was maintained by the solid-state entity." It is straight out of the Matrix movie, but was written in 1988. By 2300, our AI creations decide they have had enough of Man, who are proving to be troublemakers worth far less than the trouble they are causing in their domed cities. It is unclear whether our machine master leave a few humans alive for curiosity and testing, or if they simply wipe all of us out as a human would thoughtlessly eliminate an entire colony of bees from our backyard or ants from our home. But by 2300, there are no more domed cities. And no more humans. Cast alongside his scientific observations and real-life recounting with dolphin intelligence, Lilly's recountings are chilling.

Who wrote the introduction to Lilly book? Who called him a seer? "No one has gone as far into the future as John C. Lilly and managed to return (reluctantly, we known) with such clarity and good humor" (Leary1988, p. 8). Yes, none other than iconoclast academic Tim Leary, who writes beautiful passages recounting the Four Great Ages of Human Exploration and linking them to psychedelic academic and public voyages of self-discovery.

Standing alongside John Lilly, Timothy Leary reminds us of the successful stigmatization of psychedelic experimentation, revealing our own metaprogrammed biases and the way they effect our opinions, as well as the fulfillment of our life projects: "Going to Vietnam in the late 1960s was about ten thousand times more hazardous to your physical health than dropping acid. The suicide rate, which plummeted during the dope-fiend '60s predictably rocketed in the sober '80s" (Leary 1988, p. 7). Calling the publication of Lilly's *Programming and Metaprogramming in the Human Biocomputer* "a work of inestimable importance", Leary (1988) proceeds to say that "We all saw John C. Lilly as some sort of wizard, a science-fiction starman, a unique back-to-the future alchemist. A new Paracelsus. A veritable Isaac Newton of the Mind."

Among a large group of California New Age intellectuals and pioneers, John Lilly was an academic hero. To others, he was, like Leary himself, dismissed as unstable, a disgraced academic with a serous drug problem. Timothy Leary and Richard Alpert, who also experimented with LDS-25 were able to expand their brands far more effectively than Lilly, becoming household names and famous, as well as infamous, social influencers on a wide scale.

THUS FAR, IN SUM

From John Lilly to John Sherry, Donna Haraway to dana boyd, Plato to Einstein, Michel Foucault to Marshall McLuhan we have examined, pointed to, and discussed a range of popular academic brands and their representations. Throughout this chapter, we might begin to see a number of intriguing patterns arising. In this concluding section to my chapter, I will begin to briefly elaborate and extend some of these patterns as I now discern them. These thoughts hold limitations which I will be exploring in future work. These notions and this chapter mark one of the key beginnings of this explanation and expansion of notions of personal branding. Consider now the following:

1. The role of multiple audiences. Academic brands have internal (scholarly) and external (public, media) audiences. Internal audiences can include a global network of research peers, organizations, publications, and conferences. Another internal audience is students. Another is administrators such as Deans. Externally, we seek to build legitimacy through gaining the attention of resourceful publics such as government agencies, corporate and nonprofit agencies, and the mass media and press. The external and the internal reinforce one another. Media coverage equals legitimacy and publicity, which attracts the attention of deans, as well as students.
2. Social media has become an effective way to address the external audiences, in particular providing content for interested publics and the press. As academics, we can watch and chronicle what is happening, or we can use our talents to help change things, to, for example, publicize the impacts of climate change, conceptualize alternatives, and decelerate our technologically consumerist society.

3. Social media allows a segmenting of audience based on topic. These audiences are pre-sorted and sitting in their seats 24-7 waiting for you to say something. This is an opportunity never before seen, never before conceptualized by academics. Clearly if you are magnetic enough, you could magically draw millions of people to the content of your research and teaching and spread illumination over far greater spaces than ever before. You could do it in a single tweet.
4. There is some concern in personal academic branding that public attention must be matched by legitimate and rigorously researched publications in top journals and by reputable academic publishers. It makes sense many times to work with an existing network in which you have recognition, power, and influence. However, it may also make sense to break free of convention and start something truly new. If you truly believe in yourself, in your idea, then go for it one hundred percent. Why not a poem? Why not a YouTube video? Establish yourself first, so you don't overbalance your Kardashian index. Then experiment. Fuck convention. Grab attention.
5. Internally, personal academic brands are already co-branded by their field, their method, their theory approach, their graduating university, their position, their colleagues, their supervisor, and so on and on. Academics co-brand their work with co-authors. Every citation is a link or hyperlink to another academic brand. Every time an academic's work is cited, this is brand development and extension by another scholar. Remember this if you forget everything else: managing your brand means making and maintaining networked connections for your brand the same way you make those connections for your self. You have a built-in audience. Don't neglect them.
6. Simplicity and uniqueness have been important elements behind the success of personal academic brand managers in the past. The marketing rule of the point of difference used for positioning appears to be relevant to the work of personal academic brands. Be your point of difference. Do not dissuade yourself from blasting ever upwards, like a blazing rocket leaving gravity behind, never ever looking back, as long as you do it based on the intense and paradoxically unified inner polarities of your solo point of difference.
7. Titles are important. I think taking pride in naming things is important. Naming theories, as with Einstein's relativity, Foucault's self-disciplinary gaze, Dawkin's meme, Bourdieu's Practice or LaTour's actor network theories appears to be important to the spread and use of academic theories and their related brands. CCT. Names matter in the branding game.
8. Concision, staging, plot, and design elements may come to play important roles in the efforts of those who promote personal academic brands. Using videos on YouTube, apps on mobile phones, logos on books that spread to LinkedIn pages and Twitter feed, slogans on web-pages, and other graphical, visual, textual and even auditory brand cues may help academic brands distinguish themselves and establish their value in a crowded cultural marketplace of thoughts, images, and ideas.

We can learn many lessons from the past and the present. We begin with Greek philosophers plying their trades in the marketplaces around ancient Athens. We end in the present and near future (your present) with netnographers crafting their message for a global audience on social media. We speculate at how design thinking might be applied to the future of scholarship, in massively online open courses, in open access articles and book published on blogs, in free webinars and massive conferences. Might academia be only a couple of decades behind the music industry, where music labels are now largely nonprofit enterprises enabling musicians to sell their live performances? Watch the business model of academia shift and transform. One future trend that I would bet upon is the value of strong personal academic brands.

This chapter was structured as a series of inductive case studies used to create the foundation for a theory of personal academic branding. This theory relates core academic and theoretical identities to communicative practices and purposes. We have examined a number of academic brands, seeing each as a unique constellation of ideas, terms, locations, names, affiliations, and labels: brands that use other brands to brand and spread influence. In this analysis, branding in academia is a multifaceted act that is forever changed by the rise of the Internet and the advent of social media. Linking personal academic branding to the participative presence of the researcher in online social interactions, netnography is the first social scientific approach to deliberately require and structure the conceptualization of social research as brand communication, and the act of researching, writing, and sharing theories, ideas, and results as personal academic branding enacted through a range of channels including social media. What you do with this in your own work, in your own study, in your own public communications, will become the way you forge your own personal brand.

REFERENCES

Arnould, E. J., & Craig, J. (2005). Consumer Culture Theory (CCT): Twenty Years of Research. *The Journal of Consumer Research*, *31*(4), 868–882. doi:10.1086/426626

Chamorro-Premuzic, T. (2013, July). The Future of You. *Harvard Business Review*.

Hall, N. (2014). The Kardashian Index: A measure of discrepant social media profile for scientists. *Genome Biology*, *15*(7), 424. doi:10.1186/s13059-014-0424-0 PMID:25315513

Kozinets, R. (2005). *Desert Pilgrim*. Available online http://www.youtube.com/watch?v=BM3jrNzt3YI&feature=plcp

Kozinets, R. V. (1999). Desert Pilgrim. In Multimedia Ethnography, Heretical Consumer Research (HCR) 1999 Conference, Columbus, OH.

Kozinets, R. V. (2002). Desert Pilgrim. *Consumption Markets & Culture*, *5*(September), 171–186. doi:10.1080/10263860290015602

Kozinets, R. V. (2012). Marketing Netnography: Prom/ot(ulgat)ing a New Research Method. *Methodological Innovations Online, 1*(7), 37-45. Retrieved from http://www.pbs.plym.ac.uk/mi/index.html

Kozinets, R. V. (2015). *Netnography: Redefined*. London: Sage. doi:10.1002/9781118767771.wbiedcs067

Labrecque, L. I., Markos, E. C., & Milne, G. R. (2011). Online Personal Branding: Processes, Challenges, and Implications. *Journal of Interactive Marketing*, *25*(1), 37–50. doi:10.1016/j.intmar.2010.09.002

Leary, T. (1988). Foreword to *The Scientist*. In *The Scientist: A Metaphysical Autobiography* (pp. 4–8). Berkeley, CA: Ronin Press.

Lilly, J. C. (1967). *Programming and Metaprogramming in the Human Biocomputer*. New York: Julian Press.

Lilly, J. C. (1988). *The Scientist: A Metaphysical Autobiography*. Berkeley, CA: Ronin Press.

Peter, J. P., & Olson, J. C. (1983). Is Science Marketing? *Journal of Marketing*, *47*(Autumn), 111–125. doi:10.2307/1251404

Peters, T. (1997, August/September). *The Brand Called You. Fast Company.*

Rampersad, H. K. (2008). A new blueprint for powerful and authentic personal branding. *Performance Improvement*, *47*(July), 34–37. doi:10.1002/pfi.20007

Russo, R. (1998). *Straight Man: A Novel*. New York: Random House.

Senft, T. (2008). *Camgirls: Celebrity and Community in the Age of Social Networks*. New York: Peter Lang.

Sherry, J. F., Jr. (2012). Fuck you very much, Reviewer C. In Clarence Clobbers Tenderly. St. Bathans, New Zealand: University of St. Bathans Press.

Sherry, J. F., Jr. (2015). Sitting in the back of a crowded lecture hall. In Chickasaw Craft Threnody. South Bend, IN: Aire Lyre Media.

Stacey, J. (1988). Can there be a feminist ethnography? *Women's Studies International Forum*, *11*(1), 21–27. doi:10.1016/0277-5395(88)90004-0

Tartt, D. (1992). *The Secret History*. New York: Alfred A. Knopf.

Chapter 5
Social Media as a Tool, and Tools Offered by Social Media, to Teachers and Researchers: Personal or Professional Use?

Jose-Luis Poza-Lujan
Universitat Politècnica de Valencia, Spain

Ángeles Calduch-Losa
Universitat Politècnica de València, Spain

ABSTRACT

The present chapter provides a clear vision for the social networks environment from the self-promotion point of view. Chapter focuses on organizing tools, audience, and type of publications. Tools are organized to contextualize their use and to give a proper understanding of the relevant contents that can be published. Audience is presented according to the relations and interests with the teacher and researcher. Simultaneously, chapter gives a vision of the privacy scope or the publications, and provides an evaluation mechanism to distinguish the most convenient area of publication depending of the message content. Following submission of these analyses, chapter focuses on the teacher and research activity and how to promote these activities through social networks. The chapter ends with a set of suggestions to make a strategic use of new media with the goal of promoting efficiently personal brand as a teacher and researcher.

INTRODUCTION

Make a simple experiment: look up your name on the Internet; you can use Google, Bing, or any other search engine, probably the results will be similar. Now, ask yourself some simple questions: Do you like the result? How much of that information, that you has obtained, is really about you? Are results in your favour? And the most important question: how much Internet data (about you) are under your control? This is not the first time that this experiment has been done. Many of your colleagues, your

DOI: 10.4018/978-1-5225-0917-2.ch005

students or people interested in your work, have made this experiment previously. And the result is the same result that you have obtained. In the 21st century, there is a law: if you don't control your information on Internet, someone else (probably Google) will do.

The result obtained in the previous experiment is your personal brand. For some time now, to learn more about the work of a teacher or researcher it was necessary to read his written work, his journal and conference papers, books or perhaps through some appearance in radio or TV. In the case of a student, he didn't know anything about his teacher until he went into the classroom. However, nowadays social networks are the means to obtain all information about someone. In the academic environment, where the students are digital natives, teacher's personal image is known before the class, given that students would have searched teachers' data on Google or in Facebook. By means Internet our research papers can be founded easily and our colleges will know our information in ResearchGate before coinciding with us at a conference. Besides students and colleges, there are audiences that gradually use social media to learn more about our work: the whole society.

It is not correct consider social media as a threat. Teachers have in social networks a great opportunity to disseminate their work and to generate an appropriate image appropriate to reality and teachers' interests. Teachers can use social networks from an alternative method to communicate with students to an innovative method to receive financing for their research projects.

So that, it is necessary to know all aspects related with social networks, to use them as a publishing vehicle of our personal branding and to self-promote our work. Besides, this is convenient to know the tools offered by social networks to use them efficiently. To show these aspects the chapter carries out a multidimensional analysis. The first dimension focuses on the context of social networks and tools, from both points of view: social networks as tools and tools for social networks. Second studied dimension corresponds with personal branding and consequently studies the audience typology. Finally, third dimension deals with the teacher activities and the relations with the two previous dimensions. Along the chapter, all concepts are presented with example of the good or bad use of social networks. Examples are based on the author experience, the practise of the authors' colleges and the references.

As far as the first dimension analyzed, the chapter offers a revision and a global overview about social networks. The dimension has been organized in four types, based on the environment of the content published in the social network: social (as Twitter, Facebook or Google+), professional (as LinkedIn), academic (as Edmodo) and research (as ResearchGate or Academia). Initially, chapter review the social network focused on the current deployment in his particular scope of application and the tools provided by the social network. In addition, chapter discuss about the social network contents providers, among others, Instagram, SlideShare or YouTube, these tools (that, in some occasions, are also considered social networks) provide content for publications on social networks so it is necessary to know and use them jointly the social networks to do an efficient communication.

Next, as second dimension, chapter reviews the different audience types that a teacher can found in the social networks. The audience has been organized in five types: students, workmates, unknown colleges (teachers or researchers that don't work usually with you), known personal contacts (usually called real life friends) and the rest of the social network audience. This section matches the user profiles with the different types of social networks. So that, the reader can take their own decision about "social network personal branding", for example what's the frequency that they need to publish his own research results or what's the appropriate social network to use to complement his teach activities. The section considers the external teacher viewpoint when teacher must have into consideration that the interest of the public who access to social networks.

Figure 1. Dimensions of social media to be used by teachers and researchers

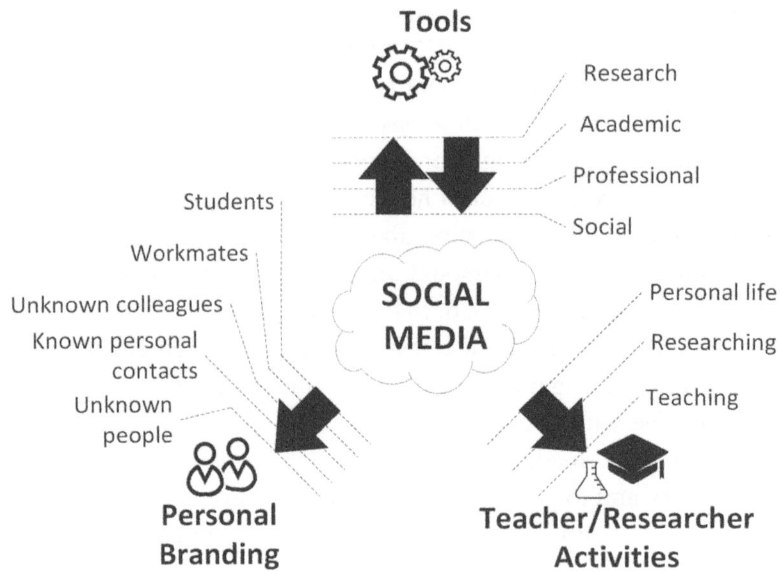

By contrast with the last section, when the initiative lies with the "audience", in this section is considered the opposite case, when the initiative lies with the teacher or the researcher. Consequently, the last dimension considered is the teacher activity: teaching, researching and his personal life. These activities are related with the two previous dimensions (social networks and personal branding) and with this last dimension teacher can decide actions that must take to reach a target audience in a concrete social network. This section corresponds with the self-promotion by means social networks.

Chapter ends with the reflections, based on the author and his colleges' experience, about the current use and trends of social networks in academic and research environment.

The chapter is geared to teachers, especially at University, and researchers that usually don't use the social media (or only use them to personal issues), and they want to use social media to improve his teaching activity or to promote his research results more efficiently. The chapter provides the answers, among others, of the following questions: should I use social networks to interact with my students and my colleges? Should I use the social networks to promote my research results and to increase the interaction and collaborate more with people interested in my research fields? In which media and social networks you should have presence in accordance with your activity? Should I read all messages published in the social networks? What kind of messages I should publish? The chapter covers also the controversial questions, related to the social networks, as: should teachers and students be friends? Should researchers be friend of other researchers instead of they never have worked together? How should interact, using the social networks, teachers and researchers with the general public?

BACKGROUND

Since its emergence, social networks have been used in all life areas, from personal to professional (Ellison, 2007). Even though initially it was not their intention, social media have been used as a means of promoting products, in the case of companies, and provide personal information that can be considered a process of personal self-promotion (Labrecque et al., 2011). In the academic and scientific environment, prestigious publications are used to measure the status and the importance of researchers. However, the emergence of social media has changed the concept of scientific prestige. Some indexes as the h-index (Hirsch, 2005) that are used to compare the impact of researches based on the scientific publications. These kinds of indexes are based only on the number of paper citations. Consequently, these indexes can be easily altered, by means self-citations or cross-citations, and there are some ways to increase the indexes in your favor (Delgado López Cózar et al., 2014). In a social network, the followers are who choose to read a Facebook publication, and to click "I like". Therefore, Social Media introduces a very interesting anonymous factor, because the scientific can't decide who interacts with him. Recent trends show a significant importance in the dissemination of the scientific results by means social networks. One of the most interest indexes is the Kardashian index (Hall, 2014). Kardashian index (K-Index) compares the number of followers of scientist has on Twitter with the number of peer-reviewed citations. A high K-index indicates an over-valuated scientific, and a low K-index suggests an undervalued scientist. Obviously, if you don't have a Tweeter account, your K-index index is zero.

In the case of the university staff, has not been done yet the use of social media to his teaching practice and his research activity. Students make use of social networks more than teachers (Poza et al., 2015). Some examples of the use of social media can be found in (Teruel et al., 2015).

However, social networks are a very important opportunity for teaching in the same context in which students develop their activity (Duart, 2009). Social networking is one of the environments where companies, and future society, are interested, and the university must to participate (Piscitelli, 2001).

The university teacher and researcher is forced to change into a techno-digital profile that integrate digital skills. But this change may be conditioned by the attitude that actors (teachers, students, society) have to social networks and, in general, to the world of Internet. It is necessary to see social media as one opportunity rather than a threat. (Carr N., 2011).

However, recent studios (Poza et al., 2015) demonstrates that a relevant percentage of teachers and researches ignore the existence and usefulness of some of the most important social networks in your area. Most part of social networks users has no problem to communicate all aspects of his life. Provide information on the net is not a feature that has appeared with social networks (Suler, 2004). Why certain groups do not use or are not interested? This question is difficult to be answered.

To ensure that teachers and researcher use social networks with some effectiveness, it is not enough to know the existence of social networks. They should use the right strategy, and also having the right attitude.

How students have assimilated social networks is widely studied. In (Vidal, 2011) attitudes of students for the use of Internet are analyzed. In (Ruiz et al., 2013), some interesting results of student surveys on the use of social networks in academic environment, demonstrates that students are open for use social media.

As far as teachers and researchers concerns, in (Bermudez, 2012) an interesting study about the use of social networks for teachers is presented. In (Sanabria, Hernandez, 2011) is presented the necessary adaptation that teachers (and perhaps also staff and students) should do to integrate into a university environment where social networks are other mass media.

CONTEXT OF SOCIAL NETWORKS AND TOOLS

This section review social networks that are more commonly used in the field of higher education and research. Social networks are presented as technological support to promote personal branding. The use of social networks in the field of business marketing is widely used (Hanna et al., 2011). However, marketing and personal brand is not extended. However, staff at the university level marketing is not as widespread. There are a few cases of professors and researchers who use social networks as part of promoting their work.

The first dimension focuses on the context of social networks and tools, from both points of view: social networks as tools and tools for social networks. The dimension has been organized in four types, based on the environment of the content published in the social network: social, professional, academic, and research. Initially, chapter review the social network focused on the current deployment in his particular scope of application and the tools provided by the social network. In addition, chapter discuss about the social network contents providers, among others, Instagram, SlideShare or YouTube, these tools (that, in some occasions, are also considered social networks) provide content for publications on social networks so it is necessary to know and use them jointly the social networks to do an efficient communication.

Although only a few social media are well known, there is a great amount that can be used to self-promote or to contribute with our own work. Amongst the many social media reviews the Global Social Media Map, by Overdrive (www.ovrdrv.com/social-media-map) organizes all social media.

Figure 2 shows a simple organization of social networks based on the content. On the upper part can be seen the user based social media. These types of social networks are based on the users. Users can share different types of media like texts, images, video, and so on. In this type of social networks, the concept of user profile is important. Consequently, they have more relevance to the formation of our brand.

Social media that share simple text messages are well suited for transmitting single events. These social networks usually are used to broadcast specific events, for example conferences, or short opinions, for example a simple reflexion about a scientific fact. But, event-oriented social media are especially suitable to transmit emotions. At the present, have a good brand to self-promote implies use emotions to define your interests. Is not the same write in a C.V. a classic sentence as "I am an enthusiastic teacher" than write a tweet like "Happy to start a new course" or publish in Facebook a message as "Surprised by the acceptance of my teaching innovation. Good students. Good work". C.V.-based sentences are static and only transmit definitions, short messages in social networks are dynamic, and transmit emotions.

Personal and professional social media are more appropriate to publish long messages accompanied by other media as pictures, Internet links or including videos. Therefore, these kind of social networks are more suitable to transmit details of your life (personal-oriented) or your work (professional-oriented). In such cases, the social network is a powerful tool to promote your work.

Furthermore, current content-based social media are considered as soft social networks. In these cases, the concept of user profile leads to the concept of user contents. You are no longer your own, you are your work materials published. Thus, the multimedia-oriented media are the tools that support the contents to be published by means the user-oriented social networks. Simple content, like audios or pictures can be used to go with a simple publication. For example, the previous commented Twitter publication "Happy to start a new course" may be associated with a picture of the classroom. If the content is more complex, like slides or videos, content can be used for more complex publications. For example, the previous commented Facebook message "Surprised by the acceptance of my teaching innovation. Good students. Good work" is greatly enhanced with a video of the teaching innovation session.

Figure 2. A simple content-based organization of a selection of the most famous social media

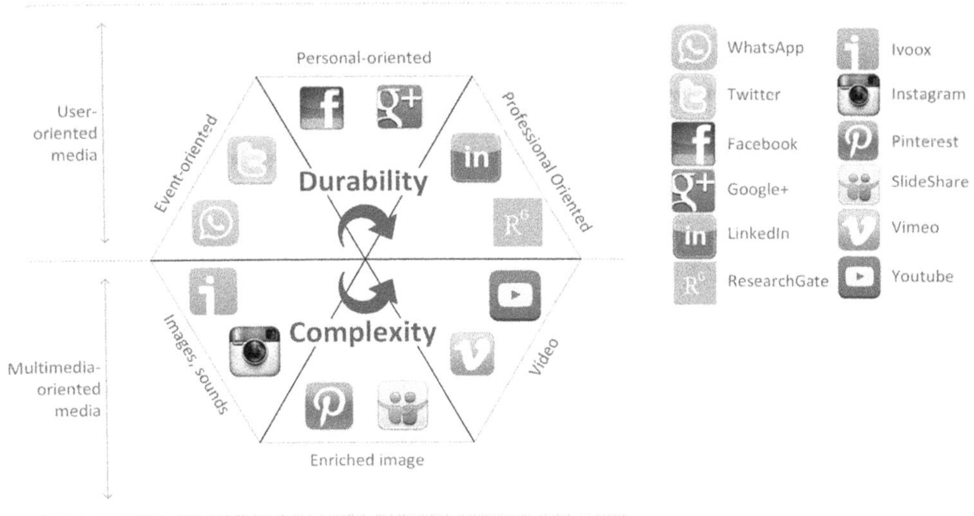

In summary, about user-oriented media, short messages are used to describe yourself and long messages are used to describe what you are doing. Additionally, short-messages use simple multimedia content, and long messages use advanced multimedia content.

AUDIENCE OF TEACHERS AND RESEARCHERS: PERSONAL BRANDING

Second studied dimension corresponds with personal branding and consequently studies the audience typology. Next, as second dimension, chapter reviews the different audience types that a teacher can found in the social networks. The audience has been organized in five types: students, workmates, unknown colleagues (teachers or researchers that don't work usually with you), known personal contacts (usually called real life friends) and the rest of the social network audience. This section matches the user profiles with the different types of social networks. So that, the reader can take their own decision about "social network personal branding", for example what's the frequency that they need to publish his own research results or what's the appropriate social network to use to complement his teach activities. This section considers the external teacher viewpoint when teacher must have into consideration that the interest of the public who access to social networks.

Prior to evaluating the audience, it is necessary consider that social networks are like the real world, but extending the scope of your publications. Consequently, just as in real life, in a social network there are different degrees of confidence. Figure 3, shows a simple and effective approach to organize these degrees.

The degrees shown in Figure 3 are based on a Likert scale (Likert, 1932) associating the five levels to confidence degrees. The most closed level to personal life is the level five, and the totally public life level is the level one. Use these levels like classical alarm levels, from a not alarm state (one) to the maximum level alarm (five). Although it may seem too much complicated to manage five levels, there are two of these levels in which are not necessary publishing: extreme levels. Level five is associated

Figure 3. Recommended confidence levels in social networks

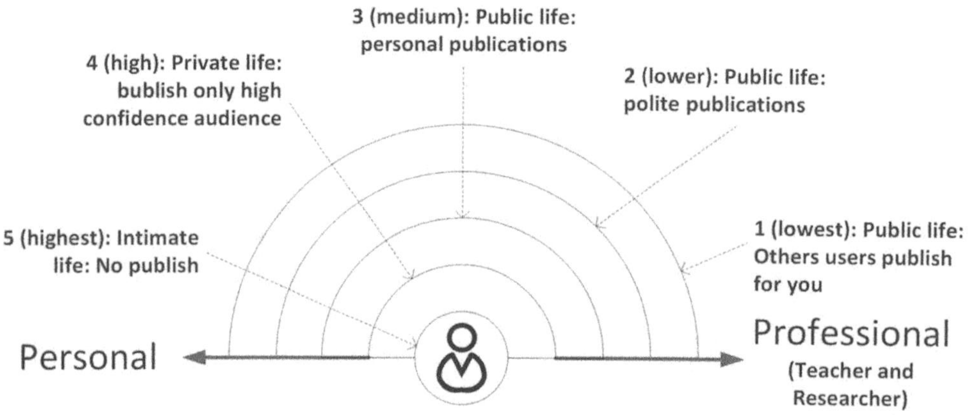

with the intimacy life, whit this level the rule is simple; do not publish in social networks aspects related with your privacy. You decide what privacy is, and where cross the privacy threshold. The other level without maintenance is the first level, public life. Internet has information about you, and you cannot control it. For example, in social networks there are pictures in which you appear because a friend of you posted the picture. One of your co-authors can upload, to their profile, a paper and comment your contribution. Even, some students can discuss details about your teaching abilities in a university forum or a personal blog. Therefore, do not worry about these kinds of publications, you cannot control them. The only thing to be considered is that your attitude in social media should offset the publications that are not under your control.

The three intermediate levels are those in which you must do your publications. Therefore, you should only choose between these three levels. Thus, it is easy to decide where to organize your contacts and where to locate a publication in a social network.

Level 2 is the most important level due this level focuses on your public brand; this is the level to publish polite publications. This is the level that watches people who do not know you. Consequently this is the level to publish your prizes, your position and interesting news about your work. Think about the future, because your future students and professional collaborators will read the information displayed in this level. Level 3 this is the level of people that has previously contacted with you, is the level for your personal brand to people that know you. This level is perfect to ask questions, to publish not compromised personal information. For example, publish a landscape picture or your sport activity.

Level 4 is the real personal level that is used to communicate with the real world close people, your family, and your personal friends. This level is used to committed publications, especially controversial issues like personal opinions about political questions that could face with your workmates or like pictures with your family. One advice, publish only at this level data about your address, trips that you are doing or similar information.

Some social networks simply do not allow users to organize their contacts and all messages are public. This is the case of Twitter or SlideShare. Other social networks, as LinkedIn offers a public profile and a profile for your contacts. In the case of LinkedIn, you can show if other user has viewed your profile. This last characteristic is interesting in a professional social network, but in a generic social network, as Facebook or Google+, it is simply not appropriate and it is precisely these social networks those which offers the possibility to organize your contacts. To classify contacts Facebook offers lists and Google+ offers circles. Both tools, lists and circles, have the same purpose and are similar. In addition, generic social networks also offer the two classic scope levels: user contacts and all users (including those who are not their contacts). All social network users has the same scope that the level 2 of Figure 3. Your contacts, or friends using Facebook terminology, has the same scope that the level 3 of Figure 3. Consequently, at least, create a list or circle to level 4. Despite the latter, social networks are continually changing their characteristics, consequently, you should check with some frequency if every social network has a specific way to organize your contacts.

Social network audience, and unknown colleagues, are placed at level 2 of polite publications. This audience requires neutral publications to offer a good image about your work; because it is difficult to know him in person. Therefore, it is not necessary to transmit personal information, only your professional achievements.

Students and workmates are placed at level 3. They know you in person, so that it is not a bad idea that they can know more about your hobbies or if you practice some sport. As discussed previously, the information placed at this level must allow your contact to maintain a Conversation Café with you. Finally, your personal contacts will be placed at level 4.

RECOMMENDATIONS SOCIAL NETWORKS ACTIVITIES TO TEACHERS AND RESEARCHERS

Finally, third dimension deals with the teacher activities and the relations with the two previous dimensions. By contrast with the last section, when the initiative lies with the "audience", in this section is considered the opposite case, when the initiative lies with the teacher or the researcher. Consequently, the last dimension considered is the teacher activity: teaching, researching and his personal life. These activities are related with the two previous dimensions (social networks and personal branding) and with this last dimension teacher can decide actions that must take to reach a target audience in a concrete social network. This section corresponds with the self-promotion by means social networks.

Before describe the recommendations, it is important the semantic of the publication. Text of the social networks publications can be enriched with meta-information. Meta-information is the key of a good strategy in personal branding. The sentences of a social network publication should relate actors (people or institutions) and keywords. For example, if a colleague receives an important award, the text of the publication must contain (as actor) the name of our colleague and (as a keyword) the word "award". Usually, in the social networks publications, the most extended way to refer an actor is by using the character '@', and the most extended way to refer a keyword is by means of the character '#'.

An important nuance, about publications, is that the message should express emotions. It is preferable to use actions and emotional words as keywords, for example, using the sentence "Happy with the award received by …" instead of a neutral sentence like "I received an award for the paper titled…" Don't forget that you are communicating brief sensations; let the journalism to the professionals.

Keywords are the most prominent topics of the publication. For example, "Award" in the case of a prize, or "congress" or "journal" in the case of a publication accepted. In this case, the point is to decide what number and which specific words are convenient to be used. The criteria are simple: use at least one words and maximum three, consequently the recommendable number is two. Only if the case is justified, it is convenient use more than three words. More difficult is to select the correct word.

Some social media, as twitter, recommend you the keywords. Instead of, there is a rule to select correct keywords to a message and is to use a direct complement of the sentence or, in an opposite viewpoint, don't use the subject or a verb, generally use the same words used in the keywords section of a paper. For example, if you receive an award for a paper, use "award" as a keyword, and don't use your name (I, the subject of the sentence and "receive", the verb), other complementary words, as "paper" depends of the publication audience. If the publication target focuses on your colleagues, perhaps, isn't necessary because their also know that you write paper. But, if you can't find other keyword, use "paper". For social media, two keywords are better than one. However, currently social networks are advancing and in many cases it is not necessary to include a symbol for the actors and for keywords.

Additionally, not only actors and keywords are the meta-information that you must include in your publications. You can use multimedia meta-information as pictures, videos, and links to other Internet pages. Internet links is justified if the page linked continues the topic that you communicate in your publication.

The other common, and important, issue is the publication frequency. Don't forget that every social network has his specific times. Twitter or Instagram are similar that nervous system. These kinds of networks transmit a great amount of small messages (short text or one image, respectively) so they should be used frequently, usually every day. Facebook, or Google+, are similar that short-term memory and this is why these social networks are considered to be used to leisure. In this case, the publication frequency has a wide range, from twice a day publications to once a week. In the case of university teachers or researches, it is recommendable write publications almost three times a week and, usually, in working days. This is because students use social media in weekends almost exclusively to relaxation. Finally, LinkedIn or ResearchGate are similar to long-term memory. These kind of social networks are like information repositories. Consequently the use of these social media is restricted to important information. The publication frequency depends on the content, but usually no more than one publication at week.

First of all, you are a professional. Although you is considered as a teacher or researcher, your lessons or your research work are usually set within a professional field. This is because your students will be employed by a company or your scientific findings will be applied in a product of a company. Therefore, you should consider that your personal brand must also have a professional orientation.

Teach Activity

Teaching activity is so interesting than research activity for a general audience. So that social networks are the most adequate media to share all teaching activities and innovations. Obviously, all teaching activities can't be published.

One of the characteristics of teaching activity, related with social media, is that it is difficult to determine which content is the most suitable to be shared in social network. However, the teaching activity has a variety of publications to be disseminated: from the beginning of the academic course to the finish of a session that has gone particularly well. The most part of teaching publications have a great impact. General public remembers their school days and allows audience return to their younger years. In addition, the teaching publications can show how the methods used in the classroom are renewed. Therefore, publish messages in social networks about innovative teaching methods is a good way to elf-promote.

However, these innovations occur, perhaps, a couple of times every academic year. This publication frequency is very low. It is convenient to maintain higher activity in relation which teaching activity, and the content providers for publications of social networks can help us. Can be useful broadcast a slideshow that has been deposited on SlideShare or to link a TED conference related with teaching activity.

Research Activity

In the case of research activity, the target audience defines the type of the publication. Consequently, it is necessary establish a strategy. Initially, you can think that the target audience is your workmates and perhaps unknown colleagues. However, they have other information sources. Thus, your target audience is your personal contacts, or a general social network audience. First, you should evaluate your main role as a researcher. Are you an information generator or an information bounder?

If you publication is a forward of a scientific information, be sure that the research field is close to your research field. That is, if you forward a general science publication, you are sending information that, certainly, will be reached to your contacts by other channels. Unless if you want to be a scientific commentator, don't forward publications about general scientific news. Except if the information is directly related with one of your research fields. Consequently, you must generate most part of the contents of your own publications. Don't forget that every publication contributes to your personal brand. Therefore, try to write about your topics.

Next step lies on establish the filter about what publish. Don't doubt you to publish press information in which you are implicated. For example, forward a Call for Papers (CFP) of a congress only if you are part of some congress committee, because the CFP sure that will arrive at your colleagues by means e-mail. You are really disseminating that you are part of a committee, not the congress CFP.

Personal Life

A student can found one of his teachers in the street, in a restaurant or in a store. In the same way, the student can found teachers in social networks. Contrary to popular belief, teachers and researches must also show part of their personal life. This is a delicate matter, because it is difficult to decide where privacy life begins.

Absolutely, social networks are used mainly to publish private opinions, leisure aspects and trivial questions. This should not be an obstacle to do similar publications, but not do the same as everyone else. The point is to have personal publications to provide in social networks enough information that allows audience to talk to you about a non-professional subject. The strategy is simple. Your co-workers know your professional aspects, but not your personal taste. If you publish any personal information on social networks, you will provide a talking point for elevator speeches.

At this point, it is important to note that your personal publications provide too much information. If you sign a political petition and share it on social media, you are giving signs of your political ideology. It's your choice to determine the visibility of these issues, especially when the publication can be read by unknown people. If you want to make public information, choose neutral aspects as your hobbies.

Make a Strategy, Not a Plan

Although it can be considered that the activity of a teacher or researcher in Social Media is similar to a Community Manager, it is necessary avoid adding a new set of tasks. If you believe you should publish every day, or every hour, a new message in a social network, probably your presence in social media is reduced to a few days, because it is very difficult to maintain a high rate of publications. Community managers planning its publications well, because is their job, but a teacher or researcher do not work as community manager, consequently i should not be thought about contents to publish. Is recommended to wait for the events come to you. Figure 3 shows a simple strategy to publish in Social Media. The main concept to be considered is "publication bubble". A publication bubble is a simple and isolate question, event or good new that is appropriate to be published. Like real soap bubbles, publication bubbles have similar characteristics:

- **Different Sizes:** Based on the content, there are very short publications, for example if you communicate a "good morning" message. In other occasions, it is necessary to express more information, for example if you upload the slides of a presentation and put a link in Facebook. In this case, it is necessary to give some explanations about the slides content. Usually hard-content publication, as political opinions needs more words. In LinkedIn, is not enough write about one conference presentation, is good to communicate the abstract, include links to the slides, consequently the size of the bubble in these kind of social networks are usually big.
- **Uniqueness in Time:** Every publication is unique and should not relate to the others. This is because a publication is not connected with other publications. It's important because in social networks is very difficult follow a line of publications related between them. Some examples of this question: is not a good idea write in a publication sentences like "as I say in my last publication..." or "As I published a few days before…" Nobody is going out to look for a specific publication in your profile.
- **Brief Life**: You must consider that a publication has a little presence in every social network. In the case of short-term social media, as Twitter, publications only are considered in minutes and in some cases a few hours. In the case of medium-term social networks, as Facebook or Google+, publications can be reared by our contacts ion the day of its publication or, in some exceptional occasions, the next day. Long-term social networks, as LinkedIn, allow users to show their publications more than one day.
- **Density**: The density of a publication refers to how the publication can compromise your personal brand. Light publications are those that everybody smile when read it because every person agree the content. In resume, the density is the neutrality (in opinion terms) of the publication. For example, a landscape picture shared by means Instagram, a message on twitter saying that classes begin and wishing good luck to everyone. A hard density publication is one that communicates a personal data about you. For example, a political opinion can cause opinions either for or against. Professional publications tend to be dense due to this kind of publication has a strong impact on our personal brand.

Figure 4 shows how the publications are placed at the level that corresponds to their target audience, depending on the d characteristics described above. In Figure 4, usually small size bubbles have low density, and are located in event-oriented social networks when the publication frequency is high. High density bubbles have less publication frequency and that is good because this kind of publications need to measure the words.

Based in the bubbles of the Figure 4, a small white bubble represents a short text with a polite content. For example, a simple and small sized message like "Hi everyone, this is a good day for have good ideas" is perfect to Twitter. This message has not a concrete audience, and the message's effect is independent

Figure 4. A simple strategy to publish in social media without perish

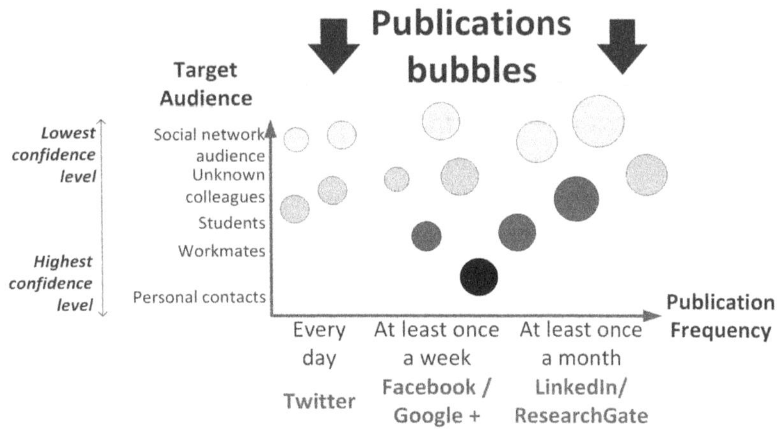

from the reader. A simple, but big, bubble represents a long text publication like a description about your last paper publisher or a report about a recent research. This kind of publications, in contrast to the previous example, has a specific audience, and the massage effect is foreseeable, mainly due to the fact that the audience think like you. In contrast to previous examples, a black bubble represents a committed message, like a political opinion about a recent fact. This kind of messages, have an effect difficult to predict, because you do not know the opinion of all potential readers. Consequently, it is convenient to publish committed messages in a confidence scope when readers can know your political opinion (in this case) or you can predict the reader reaction.

As a conclusion about publications, some advices about how to make it. Although there are many sources to document about the publications, it is necessary to know some basic issues.

- Organize your contacts into groups not by the target audience criteria, but because by content because the contents create the self-image that you want offer them.
- Write in the language more used by the target audience. But when the target audience is unknown people, and the content of the publication is related with your professional profile, use English.
- Transmit emotions instead information. Social networks are perfect for generating feelings. Through these feelings you will be known better.
- The more time you need to write a publication, more committed is the publication. A polite publication does not require many reflections. Simple publications do not generate responsibilities and, even, can improve your personal brand.
- Do not lie. Social networks multiply the good things, but also multiply your mistakes.
- Use social networks to ask questions to your contacts. An interesting experiment is to ask your contacts about an advice, for example for an accommodation in a city that you will visit in the future. You will be surprised that the experience of your contacts becomes your own experience.

As a final conclusion, just remember the first question: personal or professional use? The answer is both, but do not forget that personal brand in a social network is a question of time, and it is necessary to dedicate publications to all kind of audience. The key is to use the same strategy for a long time. The results come after a long time.

CONCLUSION

In the same way that at the end of the 20th century hardly anyone was predicted the impact of Google in our lives, currently it is very difficult to predict the impact of Social Media in our lives.

Rather than worrying about our privacy in Internet, teachers and researchers should consider social media as an appropriate powered speaker. By means of a Social Network, a teacher can increase the look projected. Their students may create a positive preconception and their colleagues will appreciate his work. If the teacher's publication has a public domain, society will value positively your activity. As a researcher, a social network will provide an excellent mean to share your projects, to disseminate your publications, and to establish contacts and potential working relationships.

As teachers we are required to know the context in which students work. Students have come out of the classrooms and continue their work on social networks. It is wrong to think that a social network is used only for entertainment; students also use the social networks to collaborate between them, to coordinate their activities. Teachers are accustomed to coincide in space and time with students by means of classroom lessons, laboratory activities, or mentoring meetings in the office. However, students have their own common spaces to work, and social networks give them support without coincide in the space (as the chat) and without coincide in time (by means the publications in the social networks). To some extent, it is the same case with the traditional face-to-face education in contrast to the current trend based on the online teaching. Have the skills to teach in classrooms does not guarantee have social networks skills.

Does not seem that the trend is that the students stop use social networks, everything suggests that probably will appear new tools and new uses. Teachers must know how students work and communicate with them in the same technological language. Therefore, use social networks, will allow teachers to obtain the necessary experience to apply them in their work.

In the same way, using social networks, as researchers, will allow us to acquire the skills to improve the dissemination of our research and learn more characteristic about the researchers can refer our articles.

Introduction section invited you to look yourself on the Internet. You can now create and modify your social network profiles. You can update the research articles and not leave all the work to Google. You can also do the same experiment suggested in the introduction section, but with other colleagues, or with the main author to reference in your next paper. It is possible that using the social networks, you can discover some common aspects with your students, colleagues or authors referred in your papers. If some day you find one student, or one colleague, you can speak with them about more things and more relaxed conversation, sure that social networks will make us more human.

REFERENCES

Aguilar, M. G., Campos, S. R., & Batlle, P. F. (2012). El uso académico de las redes sociales en universitarios. *Comunicar*, *19*(38), 131–138.

Bermúdez, A. M. R. (2012). *Profesores 2.0 en la universidad del siglo XXI.* Criterios para la integración educativa de la web social en la universidad.

Carr, N. (2011). *Superficiales ¿Qué está haciendo internet con nuestras mentes?* Madrid: Ed. Taurus.

CINAIC 2015

Delgado López-Cózar, E., Robinson-García, N., & Torres-Salinas, D. (2014). The Google Scholar Experiment: How to index false papers and manipulate bibliometric indicators. *Journal of the Association for Information Science and Technology*, *65*(3), 446–454. doi:10.1002/asi.23056

Duart, J. M. (2009). Internet, redes sociales y educación.RUSC. Universities and Knowledge Society Journal, 6(1).

Ellison, N. B. (2007). Social network sites: Definition, history, and scholarship. *Journal of Computer-Mediated Communication*, *13*(1), 210–230. doi:10.1111/j.1083-6101.2007.00393.x

Hall, N. (2014). The Kardashian index: A measure of discrepant social media profile for scientists. *Genome Biology*, *15*(7), 424. doi:10.1186/s13059-014-0424-0 PMID:25315513

Hanna, R., Rohm, A., & Crittenden, V. L. (2011). Were all connected: The power of the social media ecosystem. *Business Horizons*, *54*(3), 265–273. doi:10.1016/j.bushor.2011.01.007

Hirsch, J. E. (2005). An index to quantify an individuals scientific research output. *Proceedings of the National Academy of Sciences of the United States of America*, *102*(46), 16569–16572. doi:10.1073/pnas.0507655102 PMID:16275915

Labrecque, L. I., Markos, E., & Milne, G. R. (2011). Online personal branding: Processes, challenges, and implications. *Journal of Interactive Marketing*, *25*(1), 37–50. doi:10.1016/j.intmar.2010.09.002

Likert, R. (1932). A technique for the measurement of attitudes. *Archives de Psychologie*.

Mack, D., Behler, A., Roberts, B., & Rimland, E. (2007). Reaching students with Facebook: data and best practices. *Electronic Journal of Academic and Special Librarianship, 8*(2), 4.

Piscitelli, A., Adaime, I., & Binder, I. (2010). *El Proyecto Facebook y la Posuniversidad. Sistemas operativos sociales y entornos abiertos de aprendizaje*. Colección Fundación Telefónica.

Poza-Luján, J. L., Calduch-Losa, Á., Albors, A., Cabrera, M., Teruel, D., Rebollo, M., & Somavilla, R. D. (2015). Propuesta de parámetros y caracterización de los grupos de las redes sociales orientados a la docencia universitaria: experiencia y resultados. *Revista de Educación a Distancia*, (44).

Roblyer, M., McDaniel, M., Webb, M., Herman, J., & Witty, J. V. (2010). Findings on Facebook in higher education: A comparison of college faculty and student uses and perceptions of social networking sites. *The Internet and Higher Education*, *13*(3), 134–140. doi:10.1016/j.iheduc.2010.03.002

Ruiz, R. M., López, F. C., & Millán, J. (2013). Experimentando con las redes sociales en la enseñanza universitaria en ciencias. *Revista Eureka sobre enseñanza y divulgación de las ciencias, 10*(3), 394-405.

Sanabria Mesa, A. L., & Hernández Jorge, C. M. (2011). Percepción de los estudiantes y profesores sobre el uso de las TIC en los procesos de cambio e innovación de la educación superior. *Aloma: Revista de Psicologia, Ciències de l'Educació i de l'Esport*, (29).

Sarkar, A., Agarwal, S., Ghosh, A., & Nath, A. (2015). Impacts of Social Networks: A Comprehensive Study on Positive and Negative Effects on Different Age Groups in a Society. *International Journal (Toronto, Ont.)*, *3*(5).

Suler, J. (2004). The online disinhibition effect. *Cyberpsychology & Behavior*, *7*(3), 321–326. doi:10.1089/1094931041291295 PMID:15257832

Teruel, L., Poza-Lujan, J.-L., Cabrera, M., Calduch-Losa, A. Rebollo, M., Diez-Somavilla, R., & Lloret-Romero, N. (2015). *Is it possible to use social media in university teaching?* CINAIC 2015. III Congreso Internacional sobe Aprendizaje, Innovación y Competitividad

Thelwall, M., & Kousha, K. (2015). ResearchGate: Disseminating, communicating, and measuring Scholarship? *Journal of the Association for Information Science and Technology*, *66*(5), 876–889. doi:10.1002/asi.23236

Vidal, C. E., Martínez, J. G., Fortuño, M. L., & Cervera, M. G. (2011). Actitudes y expectativas del uso educativo de las redes sociales en los alumnos universitarios. RUSC. *Universities and Knowledge Society Journal*, *8*(1), 171–185.

Chapter 6
Academic Community Manager:
Manager of the Academic Community

Ariana Daniela Del Pino Espinoza
Escuela Superior Politécnica del Litoral, Ecuador

ABSTRACT

The Academic environment is changing, especially with the nominated web 2.0 and science 2.0. The increase in new academic social network used for researchers to publicize their knowledge, research and contributions in science, can not take lightly. For this and everything that comes with being a researcher, the academic community manager comes to manage the social networks and academic social network, care workflow, generate networking and disseminate results, among others. Increasingly universities seeking to spread the results generated by their students, teachers, masters, doctors or research groups. This research is intended to show the important role that will have the academic community manager inside the group of researchers.

INTRODUCTION

It is imperative to be immersed in the evolving technological environment in order to use company branding for positioning, communicate with users, and to increase sales. Companies may ask themselves: Who can coherently and relevantly promote my brand online? The answer is the community manager (CM). This chapter explores the role of the CM within an organization, as well as compares and contrasts the roles of the CM and the academic CM (ACM).

BACKGROUND

What is a CM? According to José Antonio Gallego, president of the Spanish Association of Managers of Online Communities (AERCO), the CM "is the one who is in charge of caring for and maintaining the community of followers that the brand or company have and be the link between the requirements

DOI: 10.4018/978-1-5225-0917-2.ch006

of the clients and the possibilities of the company could give them. To do this, one must be an expert in the use of social media tools". (p. 4)

According to David Coghlan, the CM "is the art of the efficient management of the communication of other online in the different tools suitable for the type of conversation that we see fit with our potential clients (either a blog, a community measure, a Twitter account, a Fan Page on Facebook …). This is the face of the Brand." (p.5). Connie Besson states: "The CM must be the voice of the company outside, and the voice of the customer inside.". (p.5).

As expressed above, the CM is the caregiver and maintains the followers of the brand. The CM manages marketing strategies, builds social network relationships around the brand, is informed about topics of interest to the community, generates creative content, and manages strategies for potential crises. Finally, the CM successfully manages administration tools and communicates effectively with other departments within the company. The position of a CM is one of great responsibility.

Currently, Web 2.0 is necessary for researchers, teachers, and academic managers. The CM handles the online brand image and understand the need of the customer. The academic CM, or manager of the academic community, does not manage social networks within university. Rather, the academic CM manages the flow of scientific research of a university, magazines, congresses, presentations, a research center, or a group of academics and researchers.

According to Lluís Codina (2010), "there is a third component, which refers to the idea that Web 2.0 can make not only new forms for the popularization of science, but also new methods of evaluation of the articles and scientific reports. This third component is very controversial because it includes proposals such as the elimination of anonymity in the review process of the scientific publications known as peer review."

Codina references the work of sociologist Diana Crane, stating that "documented the patterns of communication in science of a small group of scientists working online."

The sociologist determined that scientists researching themes in particular and are generally known among each other, in addition to keep abreast of the respective work. Crane called this informal community of scientists the "invisible college." This group works on the same subject and exchanges information, acting as a means of socialization and education of new scientists who seek its beginnings of work in a particular line of research.

Also Codina in 2009, talk about scientific ditigal social network. He explains how web 2.0 is linked to social networks, and as they were born before the advent of the internet. In his article on Science 2.0, speech on the first 8 tools that appear to the scientific field that enable collaboration between professionals and its use to share results. He makes a reference how work the ResearchGate tools, at that time it was one of the best, and stili so on.

DIFFERENCES BETWEEN ORGANIZATIONS AND THE ACADEMIC ENVIRONMENT

Organizations

If enterprises, institutions, organizations, and universities do not maintain profiles in the network, they will be considered "digitally dead." Formerly, people asked trusted colleagues for references on products such as education in a university. Today, people are certainly looking for those colleagues in the network, through online profiles and websites. In many cases, one uses technology to provide more realistic experiences. It is here that the CM determines how the company's digital role is most useful to their users.

Academic Environment

At one time, researchers and academics only worried about writing articles and the topics to prepare for their classrooms. Today, that issue remains. However, participating in forums and conferences brings a lot to the academic scientific environment of your area. One must consider if the profiles that you show in the network are necessary to meet the group's needs. The ACM should help to generate and maintain these profiles. In turn, the group can generate their own network of contacts and research groups.

Currently, the science is shared and applies new social technologies to the scientific process. Hence, it is assigned the term "Science 2.0." Web 2.0 brings more tools to researchers, which can be used as part of the production, methodology, or dissemination of results. This allows a flow of communication of their productions, an implementation of activities, or organization of an investigation. Says Merlo (2010) in his article of Science 2.0: "innovation applied to the research is also serves as a participatory and open resource technologies."(p. 5).

All this, involves the idea of the use of open technologies, to build collaborative information disinterested. A point in favor, is the joint participation in a physical location is or from remote points. Enabling shared data, management of bibliographic information, create collaborative links between researchers. Merlo notes three strengths:

- **Share Research:** Researchers share their work in various phases, where they can receive criticisms and comments.
- **Share Resources:** Share data, references, achievements, results, links, information, documentation, etc.
- **Share Results:** Dissemination of results of the open form or repositories through academic social networks, blogs, web pages, etc.

Differences

The main differences are shown in Table 1.

Table 1. Differences between organizations and the academic environment

Feature	Organization	Academic
Type	Companies use their profiles for commercial purposes	Universities or academic groups use their profiles for scientific purposes and for non-profit purposes
Responsible	Use a CM to know user needs and future trends of the market	Using an ACM, focus on the dissemination of information, create links and manage social networks to contribute to documentation
Purpose	Monetary contribution	Acquisition and dissemination of knowledge

New Environments 2.0. for Scholars

Digital environments are increasingly frequent for academics in order to compose dynamic presentations or materials for their students and research groups. They look for tools to host classes or communicate with students. It is imperative to choose well because these environments can be very advantageous for sharing knowledge.

2.0 environments are most frequently used to share information in education and research. These environments have varying features from basic presentations to research galleries to video evolution content.

Without a doubt, these environments will continue to evolve and revolutionize the scientific experience. It is probable that with the time, then just make investigations in shared time in line, starting from a web server and connected by servers, unlike that now that we still have the offline versions, let us not forget that the cloud has more users.

For the ACM, this is part of their daily life as a manager of bibliographic information, content, and blogs.

Table 2. New environments 2.0. for scholars

Name	Web	Feature	Type
Slidestory	www.slidestory.com/?page=home_	Stories with images	Academic
Zoho Show	www.zoho.com/docs/show.html	Online presentations, linked with Flickr	Academic
PhotoShow	www.photoshow.com/home/start_	Online presentations that are narrated, support with blogger, movable type, and typepad	Academic, Research
Gickr	www.gickr.com/_	Online presentations	Academic
Movie Masher	www.moviemasher.com/_	Make videos	Academic
Travelistic	www.travelistic.com/_	Make videos linked to GoogleMaps	Academic
Mojiti	www.mojiti.com/	Allows user to add text to video from YouTube or GoogleVideo, provides codes to publish in blogs or online	Academic, Research
Fliptrack	www.fliptrack.com/_	Program with videos from photos, provides codes to publish in blogs or online	Academic, Research
Bookr	www.pimpampum.net/bookr/_	Digital books linked with Flickr, generate books to share knowledge	Academic, Research
Animoto	www.animoto.com/	Videos with music, implements payment, For publish photos and more used for business.	Academic
Ning	www.ning.com/_	Generate social networks, social integration, online communities	Academic, Research
Scribd	www.scribd.com/	Share all types of documents	Academic, Research
Schoolrack	www.schoolrack.com/_	Generate web pages to provide information, blogs, and online classes	Academic, Research
Slideshare	www.slideshare.net/_	Share presentation documents, link to LinkedIn	Academic, Research
CiteULike	www.citeulike.org/_	Save academic credentials, search for articles and research, communicate with your community	Academic, Research

One of the most recommended is Scribd, which is a page that allows you to share information and research documents. An additional feature charges for the purchase of such information so that your time can have a great remuneration.

Ning helps to generate an individual social network on a topic of your interest. The user can maintain many people at the global level who comment on the posted research. The user can generate a large amount of interesting data.

CiteULike, like Ning, can host a community. Another advantage of CiuteULike is the ability to search and save bibliographic references, and carry out academic work with students.

Searching for an appropriate network environment will help the user save time. In addition, the user will gain information for future research and feedback from statistical data. Remember that you are not only researcher, but also a professor and these networks or sites on the web can transform you in the academic 2.0 that is required, connected with their students and that facilitates its own resources at the time of giving classes.

CHARACTERISTICS OF THE ACM

How do you know who will be the perfect ACM of the group? This would sound better in group. A great advantage of Web 2.0 is the ability to combine information remotely from different locations. This means that valuable information is easily available to scholars, therefore saving time. We need to cover some edge, in this case, must not only be a designer, but must have an instinct or a degree of natural-born investigator, who knows how to investigate and promote events, presentations, conferences, and lectures. The ACM should be aware of new social networks that spread information, share data, and collaborate online from different points of the planet.

It is recommended that the ACM has a lot of empathy with the workgroup and facilities of words for contact with characters representing the middle or areas of research.

The ACM must be:

- **Honest:** In research, the veracity of the data is important for the credibility of the institution. It is very likely that you discover that what you wrote or shared was generated by another group of researchers. Therefore, do not forget to perform new investigations or continue with a pre-made investigation by the group.
- **Humble:** When sharing information, one must know that things may not go as planned and one is open to criticism by the experts. One must learn from others at all times and from one's own mistakes.
- **Respectful:** In this role, one collaborates with international experts. It is important to be courteous and polite, use appropriate language, and be culturally sensitive. These qualities allow for richer learning.
- **Generous:** If other groups of researchers share information with your team, remember to share data with them. Promote researchers through your networks, giving more dissemination in little time. However, do not expect anything in return.
- **Grateful:** Be reciprocal with others and with your computer. Commend the work of others and your connections.

- **Collaborative:** Enhance research through peer review, consistently participate with different people, that aided by the technology will provide the co-creation of works cooperatives.
- **Open Data:** Open information creates better learning environments because you do not always keep all the information generated by your group. Once information is published in journals or indexed, it begins to share by digital environments, social networks own, management tools, or portals of research.

Important points for a better job include:

- **Curious and Creative:** Curiosity is something within an individual. This includes research. The most creative people have had the most experiences, and can quickly forecast and resolve problems.
- **Knowledge:** One of the expertise characteristics is to know the sectors in which you want to play and the group of researchers that you want to engage. It is important to strengthen the material's credibility, and have plans for marketing and advertising.
- **Empathy with the Working Group:** It is good to be able to place yourself in another's shoes. One must keep in mind that the participants are not only researchers, but also teachers and individuals with personal lives. One must have good character, knowing how to defend the ideas of themselves and the group. One must understand and respect the views of their colleagues and the community related to the research topic.
- **Communicative, Ease of Word:** A good communicator and good conversationalist is important. One must know how to listen and respond at any moment, in spite of criticism. It is important to have an open mind, appreciating the diversity of comments. One must be clear and able to correct other researchers in order to learn more about a topic. Remember that one should always be polite when answering.
- **Intuitive:** It is sometimes good to listen to the inner voice that lets you detect opportunity. Intuition leads to partnerships between researchers, communicators, institutions, universities, and members of the community.
- **Soul of Researcher and Self-Taught:** motivate itself, to find more and better result, encourages colleagues to continue, this attentive to possible investigations in group or for your department. Is accessible with their peers.
- **Good Organization:** Every good researcher must be clear about the time that you want to use in your project. It is very important to correct answers quickly and be ready to help peers in the project. You have some important tools for the organization of file like Google Drive or DropBox, it's let you to have different folder with the information you need and share it with all your workgroup.
- **Project Planner:** A project planner can recognize when a team member will be able to head the research, lead group participation, and look for pairs to collate the research. One strives to create a suitable atmosphere, where creativity and scientific development flows. A project planner seeks to make sure that everyone on the team has the same amount of workload. In this case, the use of Basecamp 3 or GSites will be a great option, because allows you to mark deadlines.
- **Aware of New Social Networks, Systems, or Technologies for Academics:** One is always aware of new technological trends, looking for innovations in the field. Using the internet and Web 2.0,

one must explore new applications and systems before they leave the market. It is important to remain connected 24/7.

- **Concise Spelling and Writing:** Grammar is important and must be strong. One must be able to write, compose, and express cohesive thought. One must be able to translate scientific materials in a language that is enjoyable for all users.
- **Promotion in Presentations and Lectures on the Work Carried Out by the University or Group of Researchers:** In promoting the best results of the research group, one must encourage colleagues to share positive comments, reviews, and references. It is important to create a dynamic profile space.
- **Online Communication:** Contacts with good references help to improve reviews of the results. One must get to know the best channels for sharing information.
- **Personal Brand Management of Researchers:** An individual express one's own beliefs, and enthusiastically promotes the brand, the research group, and the university. It is important to positively champion colleagues.

MANAGEMENT TOOLS FOR AMCs

At first, the internet was used to exchange messages between people and news through forums. Users also sent e-mail. This changed when we reached the informatics revolution, which created a flow of accurate and relevant information.

The current goal of the ACM is to manage, coordinate, and lead teams, academics and suppliers. One will be able to decide in which educational events (i.e., conferences or presentations) to could participate with the data obtained. The ACM is responsible for the flow of internal and external communication. The ACM must be aware of new and improved collaboration tools, it must be tested, which can be for free or charging.

Codina said that ''Scientific social networks are a new meeting place for researchers and a major instrument for the promotion of knowledge. Their use is increasing in the scientific community, as they are practical tools to disseminate research results and to share knowledge.'' (p. 641)

The following information includes the most effective tools.

When looking to share information online, the ACM must choose the best platform for the developing project. He or she will be the person who be aware of the evolution of the information. The best thing is to not only share the progress made, but that controls the time of realization in person.

Google has large amounts of applications to stay connected. It can be used on smartphones, tablets, and computers. It is easily installed and maintains a social network. It also allows you to create your own web page as a project organizer.

Basecamp can also be a great tool, you can work all together, organize ideas and discuss everything about the project that is developing. It let you, to invited people with you stay in touch all the time with messages, email, and other stuff.

It would be best to use a large amount in order to eventually choose one that is the most useful to your team. Another option is to combine several in order to perform simultaneous investigations.

Table 3. Tools for AMCs

Tool	Advantages	Type	Purpose
Basecamp 3 www.basecamp.com/	Perfect for distance teamwork, but also useful for a group in the same physical location. Allows a follow-up to an article or research project. Team can be assigned roles or tasks. Great alloy to record comments, attach files, and organize the calendar and deadlines. Discerns who can see certain tasks and activities. Recommended by 98% of its users. Has many strengths that make it easier to keep current with all the work you have to do. Understands multiple languages. Is available in Spanish. Has a mobile version.	Try it for free. Cost varies between \$50 and \$150 dollars per month. Unlimited number of users.	Basecamp 3 allows you to keep everything in one work space, is oriented to the investigation, remote work and control deadlines.
GSites www.apps.google.com/intx/en/products/sites/	Google always surprises us with more services and this is no exception. Creates pages. Registers the work of a team or an ongoing investigation (depending on the template that you use). Does not requiring programming knowledge in order to create pages. Use it to show progress and completed work. Assign tasks, dates of delivery, store files, receive e-mail alerts, and choose preferred language. Has fewer options than Basecamp but allows for simple projects and is economical.	Free and charging.	GSites allows you to keep everything in one work space, is oriented to the investigation, remote work and control deadlines.
Google Drive www.drive.google.com	Works with a Gmail account. Includes a calendar to remember dates and deadlines. Work on documents online with multiple participants. Share achievements and discoveries through social network G+. Generate surveys, link files, and talks. Space is a disadvantage. Limited group space of approximately 15 GB. However, combine the effort of the whole group, to achieve optimal space. If this is not sufficient, the value to cancel by larger amount of space is quite reasonable, something which may be annual and if that were not enough, allows you to have a corporate mail account when it warrants, mostly give you 30 GB of space per user, depending on your plan. Desktop application allows to save information to individual computers.	Free and charging.	Google Driver allows you to have multiple folder with information for the project, is oriented to the Web, remote work and share informaction.
Skype www.skype.com/en/	Ideal for lectures at a distance. Free Video-calls across the globe, when part of your team lives on the other side, it is pertinent to hold meetings to maintain a relationship pleasant. On certain occasions cannot come to an agreement with the entire group or if an investigation deviates from the topic, not always the mails are understood correctly, the comments or times are relevant. If you want to economize recalls that with WIFI you can achieve long chats.	Free and charging	Skype 3 allows you to talk with different person at the same time, is oriented to the commnucation, remote work meeting and share files.
Dropbox www.dropbox.com/	Initially, a nuisance due to space. However, if entire team uses it and shares files, each user is given 2.5 GB. More space will be needed as partners are added to the project. Information stored in the cloud. Synchronizes online. Includes a desktop application. Ability to purchase additional space.	Free and charging.	The purpose of DropBox is file management.
Wetransfer www.wetransfer.com/	Helpful for sending information in a timely manner. Use for free or Wetransfer Plus for multiple users. Includes 2 GB of space. Premium account includes more space.	Free and charging.	The purpose of Wetransfer is share files.

SOCIAL MEDIA FOR ACADEMICS OR VECTOR COMMUNITIES FOR THE ACM

Social networks that form part of Web 2.0 play a role in many lives. At present, companies and universities manage their own profiles. Social networks have become communities for shared ideas, solutions, and knowledge.

The dissemination of knowledge is a valuable communication opportunity. Real-time communication improves the overall process and increases research on similar subjects. As stated by Codina (2009) "It is a social structure that can take many forms and various characteristics," (n.7) by which we should not be surprised that the networks are used as a means of investigation, experimentation, or academic dissemination.

According to Echavarria (2010): "now scientists in different areas will meet with these resources [academic social networks to share their progress] sharing advances and discoveries, through its publications and opinions, are made visible and extend the goal of the invisible colleges to gather scientists in a common theme, to create networks of experts, but opening the doors to the free access to information."

Elements of Social Networks

Social network interaction use is advantageous for academics, teachers, researchers, and students. It increases the possibilities of collaboration, resulting in better, faster results. Codina shares the most important components in his article that must be mentioned, plasma the ideas of two important aspects from the general to the specific part of the was 2.0 digital.

General Components of Web 2.0

In the general aspect of the academics social networks are derived on seven aspects that Codina has determined by the following and that we will explain below:

- **Home Page:** The home page is where a user logs into the system. The home page can also include notifcations, news, messages, events, member information, and new additions.
- **Profile:** A profile should include a user's time and post information. A user should not be humble when listing skills; the user's skills may be very useful to others.
- **Messages:** Messages allow users to share information, and communicate between one or several members.
- **Searches:** All users create individual profiles. Carefully search for members to join your community. Create your own network of contacts. In some cases, you can define parameters based on interests, themes, institutions, cities, etc.
- **Groups:** A user can create groups to socialize with professionals based on certain themes. Codina (2009) indicates that, "in academic social networks, this component acquires a special value because it ends becoming stakeholder's groups that allow the development necessary for a academis dynamics." (n.7).
- **Communication and Collaboration:** The aforementioned tools facilitate communication and collaboration of all your networks, contacts, and research groups.
- **Preferred:** As in most, there are the favorites, trafficking of guards the largest amount of sites, pages, resources at the beginning, as you find your topics of interest, see deleting those which

don't provide no nutritional value for the research of your computer. Share with your computer so that they can use them as a means of consultation.

Specific Components of Science 2.0

In the specific aspect of the academics social networks are derived on three aspects that Codina has determined by the following and that we will explain below:

- **Documentary Management:** Codina has defined three important variables in science 2.0 to:
  - **Personal Manager:** It is best to organize by folders or lists, entered manually or automatically imported from databases/ digital repositories. Codina recommends EndNote bibliographic and reference manager. Zotero or Mendeley are also good options. It also supports the generic formats BibTex i RIS, but only encoded in XML.
  - **Consultation to Databases:** This searches thematic papers that are centralized in external sources. You can include specific topics of interest to your personal library. When open data is limited, one must find repositories that allows for useful data at no charge. The digital repositories that Codina recommends include: arXiv, CiteSearch, IEEE Xplore, NTRS, PubMed, and RePEc.
  - **Semantic Search:** It is always a good idea to review the suggestions that we made in your own network, not only of individuals but of necessary documentation from the analysis of other documents or our profile or our publications, etc.
- **Academic Profile:** Do not forget that between more information proportions, it is more feasible than others in the area can contact you to conduct investigations. Includes publications both books as articles, groups of collaboration and data on research projects.
- **Groups:** Groups include those of your academic world and those who share comments, knowledge, etc.

Academic Social Networks and Benefits

The ASN are your best allies when exposing the information that you have generated with your colleagues. But these on the web do not exist, currently all persons and entities relate through simultaneous digital media, are becoming more and more users. Some of them, allow you to share information, documents, among others. They are a good option to work in a team, where they can share comments in real time, subjects of interest, experiences and assumptions, among others.

An advantage is that we can now find specific platforms for researchers and operate almost equal to the daily use of social networks like Facebook or Twitter. The difference is that we include only research projects and involving people from these areas.

According to Merlo, in his article on "The science 2.0": "Social networks are excellent virtual labs since they offer all the services that a research group demand: communication systems, means to share resources, warehouse of documents, and discussion forums.". (Rebiun, 2010, p.7)

Next, the chapter will describe academic social networks oriented to the scientific community. It is important to note that not all social networks are RSA and may not be useful for your team.

Table 4. Academic social networks

Social Media	Advantages	Type
2Collab www.2collab.com	2collab was created by Elsevier, integrated in the database of Scopus. It helps to share references and facilitates the work of collecting bibliographical references of Science Direct and Scopus.	Free. High circulation and multidisciplinary.
Academia.edu www.academia.edu	Academia.edu was formed by academics and PhD students in the United States and England. Members include scientists of the stature of recent Nobel Laureate Paul Krugman, as well as of Stephen Hawking and Richard Dawkins. It is experiencing a remarkable boom. It is considered the "Facebook of Researchers."	Free. High circulation and multidisciplinary.
Academici www.academici.com	Academici is the initiative of a private English company. Users can upload pictures and videos, as well as create thematic networks or social networks.	Free and collection. Low coverage and focused in the areas of humanities, social and legal sciences.
ACS Network www.communities.acs.org/welcome	Formed by the American Chemical Society, it is reliable for the veracity of its publications and has a large number of affiliates.	Free and collection. High circulation and focused in the areas of chemistry and the sciences.
Bloomberg www.bloomberg.com/europe_	Formed by NYC, website is very versatile, you should not take into account to read the material provided, but you have to publish.	Free and collection. Media dissemination and focused in the areas of economy, technology, market, policy, etc.
CARET's www.caret.cam.ac.uk_	• It is a center for applied research in educational technologies (Center for Applied Research in Educational Technologies), generated by the University of Cambridge. It has its own collection of data and a policy of protection of intellectual property rights.	Free. High circulation and focused in the area of educational technology.
Epernicus Network www.epernicus.com	Created by a company specializing in software and web solutions for research, it is linked with MIT and Harvard. It allows a user to create networks or groups to communicate with colleagues. It has a more complete service, Epernicus Solutions, that must be purchased.	Free and recovery. Low dissemination and focused in the area of biomedical and biohealth.
EXIT www.directorioexit.info/_	It is linked to a magazine by the information professional (a publication indexed in ISI). It is of Spanish origin. It is a set of professional communication, information, and documentation.	Free. Media dissemination and focused on the area of directories of experts of the management of information.
Fuse www.qsensei.com	Fuse helps to build applications that aid in the search, analysis, and monitoring of information or data. It belongs to the German-American Q-Sensei.	Free and recovery. High circulation in Germany. Multidisciplinary.
Kisswin.de www.kisswin.de_	Located in Germany, it provides information and consultations for academic careers. It maintains a community of young researchers.	Free. High circulation in Germany. Multidisciplinary.
Laboratree www.laboratree.org/pages/home_	Form part of the University School of Medicine in Indiana, it focuses in the computational biology and bioinformatics through processes and development of new medications, clinical researchers, and healthcare.	Free. Media dissemination and focused in the areas of computational biology and bioinformatics.
Labroots www.labroots.com_	A scientific social network, it provides virtual education, attempts to amplify global communities, and focuses on collaboration and science learning. It is part of the largest scientific learning community in the world.	Free. High circulation and focused in the area of science and virtual learning.
Labspaces www.labspaces.net	Addressed to the scientific community, it shares scientific news and allows for socializing with colleagues.	Free. High circulation. Multidisciplinary.
LinkedIN www.linkedin.com	It is one of the largest social networks, where a user creates a profile to exchange messages and create networks of contacts.	Free. High circulation. Multidisciplinary.

continued on following page

Table 4. Continued

Social Media	Advantages	Type
Mandeley www.mendeley.com	It allows a user to have an application on a computer. It has many positive features, including organization of research in a single place, bibliographic citation generation, work display, contact collaboration. It is directed to students and researchers.	Free. High circulation. Multidisciplinary.
MyExperiment www.myexperiment.org	This social network shares information of groups and finds others who contribute on the same topic. A user can publish a workflow and allow others to contribute to the work.	Free. High circulation. Multidisciplinary.
Nature www.nature.com/index.html_	Nature is a community that connects scientists from all over the world. It is recognized because its members have won awards and recognitions. It have another network that is the Scitable (Http://www.nature.com/scitable) responsible for teaching/learning.	Free. High circulation. Multidisciplinary.
Ozmosis www.ozmosis.com/home	Physicians can share healthcare experiences and knowledge of medicine. It transforms the wisdom in reliable information to all members.	Free. High circulation and focused in the area of medicine.
PLoS ONE www.journals.plos.org/plosone	This is a publishing company and social network for academics. Its worries about publishing with substantiated criteria, in magazines of great quality reviewed by peers.	Free. High circulation. Multidisciplinary.
ResearcherID www.researcherid.com	ResearcherID network is linked to Thomson Reuters, producers in the ISI database (also known as the web of knowledge).	With record. High circulation. Multidisciplinary.
ResearchGATE www.researchgate.net	Researchers and scientists use this to share research, post responses, and seek contributors. This is one of the largest academic social networks.	Free. High circulation. Multidisciplinary.
Scholar Universe www.scholaruniverse.com_	This site allows uers to find school experts and researchers in any topic. It has a large database of multiple sources.	Free. High circulation. Multidisciplinary.
SciBlogs www.sciblogs.co.nz	It is a set of bloggers who create scientific analysis, allow for topical discussions, and facilitate the interaction of readers.	Free. High circulation and focused in the areas of science, biology, medicine, environment, and technology.
Scientific Solutions www.scientistsolutions.com_	It promotes advances in science and biotechnology. It allows for the exchange of information and ideas between researchers.	Free. High circulation and focused in the areas of science and biotechnology.
Scitizens www.scitizen.com_	It allows a user to create magazines and become an editor.	Free. High circulation and focused in the areas of science.
Alive www.vivo.cornell.edu/_	Of the University of Cornell, design its own res social researchers as a discovery tool to collaborate in all disciplines.	Free (but must be part of the university). High circulation. Multidisciplinary.
Ways Http://ways.org/_	It is an academic social network of young scientists. It is a nonprofit with a large base of knowledge, research in sciences, interactivity of messages, and a community.	Free. High circulation and focused in the areas of science.

This chapter has listed several academic social networks. Social networks continue to grow and acquire new research. It is here where the ACM makes decisions to participate in networks that are linked in certain áreas of research. The majority of those listed focus on the sciences and others multidisciplinary areas.

Among the best are Academia.edu and ResearchGATE. Not only are they the easiest to use, but they include large numbers of researchers and groups.

Contribution of the Normal Social Networks to Academics

All the researchers or academics must have their own personal brand, so the ACM will continue to be linked to the common social networks, helping the researcher to keep your profile and disseminate their achievements.

Currently, according to Gonzalez (2016) the 92% of people have or use a social network. Facebook (ABC, 2016) has 900 million users, followed by MySpace, and, in Spain, Tuenti. "Professionals" use LinkedIn and Xing (Uruena, Blano, & Valdecasa, 2011). Snapchat won big among 18 to 35 year olds. Its information lasts 24 hours and has been a challenge for companies to generate content. For researchers, it can be of great help to disseminate research and to generate greater expectation.

Thus, as there are academic social networks we are also subject city and business. Therefore, search for the ASN that can benefit your research in the town that was going to be made. Another point that you can plant, is the use of Livestreaming, that can be used as a daily or logs for investigations, you can use Tumblr (tumblr.com) or Posterous (posterous.com).

Private environments by research groups are common, as is the case of Yammer and Social Cast. These have more functionality than Twitter. Twitter only allows a user to send comments with a maximum length of 140 characters. Do not forget YouTube to publish videos of the progress or a collection of research.

BLOGS

For the CM, blogs full of information or inspiration are very important. The ACM needs feeds to be informed of research, up-to-date topics, and trends in academic social networks.

The feeds can save search time if the user customizes reading materials. Next, compared information required for future investigations and above all we can have most reliable sources.

Table 5. Tools for feeds

Tool	Advantages	Web
Feedly	It is a good system to be current with news. It is fast and reliable. It combines Evernote and OneNote.	Free.bwww.feedly.com/i/welcome
AOL Reader	It combines mobile messaging with the user's chosen feeds and trending videos.	Free. www.reader.aol.com/
The Old Reader	It was intended as an alternative to Google Reader.	Free. www.theoldreader.com/
Netvibes	It monitors information and allows a user to see what has happened 24/7. It can measure metrics and impacts of a campaign.	Recovery. www.netvibes.com/
LinkedIn Press	• When our investigation time is limited, this tool can be of great use. It shows all the novelties of subscribed sites, organizing information by pages or columns. It shares and archives.	Free. www.play.google.com/store/apps/details?id=com.linkedin.pulse =search_Result

[1]Gonzalez, G. (2016, August 4). 85% of adults in the world uses Facebook services. Retrieved April 13, 2016, fromhttp://hipertextual.com/2016/04/uso-de-facebook-en-el-mundo

[2]Facebook Messenger reaches 900 million users. (2016, July 4). Retrieved April 13, 2016, fromhttp://www.abc.es/tecnologia/redes/abci-facebook-messenger-alcanza-900-millones-usuarios-201604071756_noticia.html

[3]Urueña, A., Annie, F., Blanco, D., & Valdecasa, E. (2011). Social network in internet. Retrieved fromhttp://www.ontsi.red.es/ontsi/sites/default/files/redes_sociales-documento_0.pdf

It is important that the ACM generate blogs or wisely choose a feedback system. When starting, it is essential to know communities that feed daily or weekly. There are other applications or pages that allow a user to store reading pages. However, some are difficult to use and others charge for the service.

Wordpress allows a user to generate a unique blog.

SOLUTIONS AND RECOMMENDATIONS

Program Content

Maintaining a research project calendar can improve programmed timing of the spread of information. As research phases progress or culminate, one must present the results in carefully chosen academic social networks.

CONCLUSION

The ACM can prove to be the best alloy of the research group. Even more, because it can be of great support for the leader of the same, taking the time, organizing meetings, reading recently published research subjects and seeking the means of dissemination relevant as congresses, indexed journals, prestigious publishers, among others.

In addition, to knowing the most renowned researchers of the east, which can collaborate or be part of the group of investigators, which is responsible for all kinds of internal and external relationship that needs the group to be more efficient.

REFERENCES

AERCO & Creative Territory. (n.d.). *The role of the Community Manager*. Retrieved April 7, 2016 from http://www.maestrosdelweb.com/images/2010/04/community-manager.pdf

Carbellido, C. (2013, November 3). *The profile of the good community manager. Not any serves*. Retrieved from http://www.uncommunitymanager.es/perfil-community-manager/

Codina, L. (2010, October 11). *Science 2.0: Social networks and online applications for scholars*. Retrieved April 7, 2016 from http://www.upf.edu/hipertextnet/numero-7/ciencia-2-0.html

Echavarria Ramirez, A. F. (n.d.). *Academic Social Networks ... the boom of the academic Web 2.0 | so-called gatic* [Blog of the support group in the use of ICTS for the Universidad ICESI]. Retrieved April 7, 2016 from http://www.icesi.edu.co/blogs/egatic/2010/11/18/redes-sociales-academicas/

Facebook Messenger reaches 900 million users. (2016, July 4). Retrieved April 13, 2016, from http://www.abc.es/tecnologia/redes/abci-facebook-messenger-alcanza-900-millones-usuarios-201604071756_noticia.html

Gonzalez, G. (2016, August 4). *85% of adults in the world uses Facebook services*. Retrieved April 13, 2016, from http://hipertextual.com/2016/04/uso-de-facebook-en-el-mundo

González-Díaz, C., Iglesias-García, M., & Codina, L. (2015, September - October). Presence of Spanish universities on scientific digital social networks: case of communication studies. *The Information Professional, 24*, 640 - 647.

Lambrechts, D. (2011b, September 13). *Analytical tools for social networks*. Retrieved from http://www.maestrosdelweb.com/guia-community-manager-herramientas-redes-sociales/

Lambrechts, D. (2011a, August 30). *Guia community manager*. Retrieved March 1, 2016, from Http://cursosnet.es/wp-content/uploads/2014/05/community-manager.pdf

Mejia Armijo, R. (2011, July 18). *Tools to work together without having!!* Retrieved from http://www.maestrosdelweb.com/herramientas-para-trabajar-titanpad-googledocs-basecamp/

Rebiun, R. B. U. E. (2010). *Science 2.0: Implementation of the social web to research* (Report). Retrieved from http://eprints.rclis.org/3867/

Ruiz King, F. J. (n.d.). *Web 2.0. A New Learning Environment in the Network*. Retrieved April 7, 2016 from http://dim.pangea.org/revistaDIM13/Articulos/pacoruiz.pdf

Stanford. (n.d.). *Science 2.0: The Design Science of collaboration*. Retrieved from https://www.youtube.com/watch?v=nxBmKkLJCEc#t=50

Urueña, A., Annie, F., Blanco, D., & Valdecasa, E. (2011). *Social network in internet*. Retrieved from http://www.ontsi.red.es/ontsi/sites/default/files/redes_sociales-documento_0.pdf

KEY TERMS AND DEFINITIONS

Academic Community Manager (ACM): The person who is responsible for handling all academic social networks of a group of relevant research.

Academic Social Networks: This ASN are special for the academics, who needs to have their own networking around the world and in specific areas. Know as science social network as well.

Online Communities: Groups of people that communicate over the Internet about a common interest. Groups can form in forums, social platforms, mail groups, blogs, etc.

Chapter 7
Brand Journalism and Personal Branding in Academia

Hada M. Sánchez Gonzales
Universidad de Sevilla, Spain

ABSTRACT

The book chapter seeks to highlight the importance of brand journalism and personal branding within the university context. First, the focus is on determining how universities are adapting to the changing needs of their audience by utilising new narrative techniques. Second, we must determine whether brand journalism in turn allows for the construction of personal branding. If so, we would like to know whether teachers coming from the Schools of Communication / Information of universities in Spain have created their personal brands, in addition to what criteria and tools had to be taken into consideration for its creation. In conclusion, a lack of progress in regards to the development of a personal brand as well as personal branding in academia is observed.

INTRODUCTION

The chapter of the book discusses the importance of brand journalism in universities and the use of personal brands by Spanish university professors. Firstly the term brand journalism is emphasised, taking into consideration the evolution of journalism in relation to the needs of the audience which has converted from an ignored mass into a fifth power. The public decides what, how, when and where they will consume news due to a suppressed need to consume quality information.

Given this scenario, multinational companies have seen a clear advantage when publishing their own news without the need for intermediaries (the traditional media). Companies known around the world such as Coca Cola, Red Bull and Microsoft have become editors of their own content. Second, the chapter will delve into how universities are seeking to transmit the importance of knowledge to users as well as their contributions to improve the progress of society using brand journalism. To do this, specific cases regarding use of university brand journalism will be analysed at both the international and national level, although commitment is timid due to academic institutions' lack of knowledge regarding the matter at

DOI: 10.4018/978-1-5225-0917-2.ch007

hand. In addition, the chapter will present academics' opinions concerning the term in question and its applications in educational settings.

Secondly we will insist on the necessity for Spanish university professors to utilise personal branding. Clearly, the presence of the Internet and the emergence of the Social Web have helped different professionals to more closely reach their audience. A journalist does the same in order to attain greater visibility, given that "the visibility that the net gives us has enormous potential when it comes to sharing content and reaching other audiences which weave a community around the professional, in turn building credibility" (Yuste & Cabrera, 2014: 22).

Undoubtedly, the need for academics to create personal branding on the Internet is increasingly important. The secret is in adding extra value to professional actions. This leads us to understand that each task or project performed by the teacher should be considered a rewarding challenge capable of transmitting an emotional attitude that adds value to their brand and likewise, benefits the University and in turn its brand. To determine the use that professors make of personal branding, we have supported our study with quantitative and qualitative techniques while keeping in mind the theoretical review, document analysis and surveys given to academics.

The main objective of the chapter focuses on promoting the importance of brand journalism in the educational context and the use of personal branding by professors who teach at Schools of Communication in Spain. We begin on the following premises which we will try to respond to throughout the chapter: on one hand, we will reveal questions about the importance of brand journalism in the educational context. Do universities use brand journalism to become known and obtain recognition? Do they try to directly report to a segmented and specialised public? Is it considered a new way of communicating? Does it promote the employment of unemployed journalists? Does it build a reputation or is it trying to sell? Should brand journalism be implemented as class content within the curriculum?

Likewise, the chapter attempts to answer the following questions about teachers' use of personal branding: what criteria should be taken into consideration when creating the personal brand? Do you think that what the Internet says about you and your work is important? What tools do you consider important for building a digital identity? Should a professor's personal brand convey an emotional stimulus to anyone interested in their work?

BACKGROUND

Brand journalism is assuming greater prominence in business and media, while the attitude of the audience has changed and the advertising practice has lost efficiency. However, in no way is this a new form of communicating. Approximately one hundred and twenty years ago, according to Abiad (2014), The Furrow emerges in 1895 to attractively publicize agricultural content with an added value. "John Deere - the world's leading agricultural machinery manufacturer - had the good idea of setting up a free publication for his potential clients", which helped him to convert Deere tractors into the most highly demanded tractors on the market. The magazine is still being published today. It is obvious that another technology was used at that time to publicize such content. Today, with the help of the Internet and web 2.0 and 3.0, the technological change is indisputable owing to the fact that the magazine contents can be more widely spread at a cheaper price. In any case, the way of telling the story to reach the audience prevails.

Years later, in 2009, Adobe CMO commits to the change in one of its portals. According to Barciela (2013a), "The editors, who claim that CMO is the world's leading marketing web site, explain that their material is obtained from about 100 forefront sources in the industry. The web offers professionals the opportunity to publish their articles". Undoubtedly it is a commitment to brand journalism using marketing and interacting with the audience through forums, interviews, etc.

Later on between 2011 and 2012, some multinational companies decided to embrace the change due to technological advances and economic crisis. Web sites like The Network (owned by Cisco), financial sites such as The Financialist (Credit Suisse) and business sites like Businesswithoutborders (HSBC), OpenForum and American Express are some examples we can mention. Coca-Cola Journey would be another example, which, according to The New York Times itself, "is a real news site" (Barciela 2013a).

Regarding university brands, from a corporate point of view we can say that they were and are still seen as an asset that can help an institution stay competitive as well as reap benefits that both the organization and its members can enjoy. Authors like Takaki, Bravo and Martinez (2015) have identified significant shortcomings in university brand management owing to various factors, including lack of a brand that makes them stand out from the rest and little social outreach from an academic and research point of view.

It is noticeable that many companies, media and universities have embraced brand journalism because it is a "descendant of a traditional marketing strategy - the company's magazines" Martin, Sanchez and Hernandez (2014: 1614). Brand journalism gains strength due to the information overload, the weakening of traditional media and the crisis of the economic model. However, it is positioned as a journalistic practice because it uses the techniques of traditional journalism. In other words, it consults resources, contrasts information and spreads it through stories. This form of journalism has the added benefit of making a story directly known using different formats and channels, while encouraging interaction with the audience unlike in traditional media.

On the other hand, in relation to personal branding we could say that its history originates in 1996 at the hands of Derek Lee Armstrong and Kim Wai Yu in the book entitled: "The Persona Principle: How to Succeed in Business with Image". The authors conceived the personal brand concept by comparing it to the corporate brand rather than as an isolated idea. Later, in 1997, Tom Peters published the article entitled "The Brand Called You" in Fast Company magazine. The article seeks to demonstrate that each of us belong to our own company called "ME, S.A." in the work environment in which we live.

Since 2007 in Spain, Andres Perez[1] has presented the concept of personal branding and states that "a personal brand takes time and persistence to be developed. It is based on the perceptions that people have of people" (Marcos, 2008). This means that it cannot occur in isolation. Likewise, Tom Peters (1997) asserts that we are responsible for ourselves and therefore must make the most of ourselves and communicate it for our and others' benefit. Similarly, we can highlight Neus Arques Salvador (2007) with the book "What Brand Are You?": 12 keys to managing your personal reputation. We can also point out Roberto Alvarez del Blanco (2008), author of the book "You, Personal Brand": how to skillfully manage your life and turn it into an extraordinary experience, among others.

Regarding the digital identity of professors, there are no studies on the subject despite its relation to the brand of the university. However, "professors must be conscientious and empowered when assessing and considering the possibility of independently offering their work" (Casellas 2014).

MAIN FOCUS OF THE CHAPTER

Brand Journalism in the University Setting

It is evident that brand journalism is gaining greater importance in our society despite being a controversial term among journalism and public relations professionals, as is well described and analysed by Barciela (2013b) in his article "El último desafío, el 'brand journalism'" (The Latest Challenge, Brand Journalism). The author states that "brands create their own internet news portals to avoid the middleman - traditional media - baptising them as "brand journalism"".

Controversy aside, we believe that brand journalism is becoming established as a new way of communicating as well as something that companies need in order to optimise their reputation. It draws on the changes produced on the Internet and on the flood of social media in order to communicate and convey news to an audience. News is communicated using storytelling through its own channels, and without intermediaries.

According to Iván Pino (2014:3), brand journalism stems from the combination of two factors: first, "audiences breaking away" from mass media, and second, prosumers' interest in being active players in "demanding that companies be more transparent, responsible and trustworthy".

Flagship newspapers such as "The Guardian", "The New York Times" and "El País" are supporting brand journalism by offering new storytelling services. At the same time, universities around the world have taken on the challenge. A clear example is the web site Futurity.org, a non-profit project that started in 2009 and is now gaining strength as a reference point for university publications. In this case, the goal is focused on showing society how universities improve people's lives by applying knowledge and using news stories to connect with users. It is evident that U.S. universities (about 71) are the main characters of this change that benefits everyone. Users can go to Futurity.org and search by university or by topic (sci-tech, culture, health and environment) depending on what interests them. It also goes without saying that the information is available on social networks.

Figure 1. Futurity news and the academic world
Source: https://twitter.com/FuturityNews

In Spain we can take a look at the Clínica Universitaria de Navarra which has served society for about 63 years. Their goal is to offer patients the best personal and medical treatment possible. The web site stands out due to the fact that story protagonists are given voices in their stories section. The article "I beat cancer: Hugo's battle" is a clear example. A young boy (12 years old) has a long story to tell. When he was six years old in his hometown (Tarifa, Ecuador) he was diagnosed with Acute Leukaemia. However, Hugo has never felt sad because of his disease and says that "it is just the complete opposite. I have been enthusiastic and my friends have always supported me, especially Pau who has always helped me with my homework". Today, thanks to the Haematology and Hemotherapy Service specialists as well as economic collaboration from "Niños contra el cáncer" (Children against cancer), Hugo can tell his story and "encourage more kids so that they can keep fighting", he emphasises.

The contents published by the Clínica Universitaria de Navarra as well as those from Futurity.org are undoubtedly a gamble on brand journalism as Obispo (2015) notes in his article "Una historia detrás de cada paciente. Cómo aplicar el periodismo de marca a la comunicación de salud" (A story behind each patient. How to apply brand journalism to health communication). They strive to connect with audiences by appealing to their emotions and to the story, rather than focusing on the technical information that, in many cases, cause people to feel lost. This change is due to user response on social networks and on their own web site.

Figure 2. Hugo's battle
Source: https://twitter.com/ClinicaNavarra/http://www.cun.es/

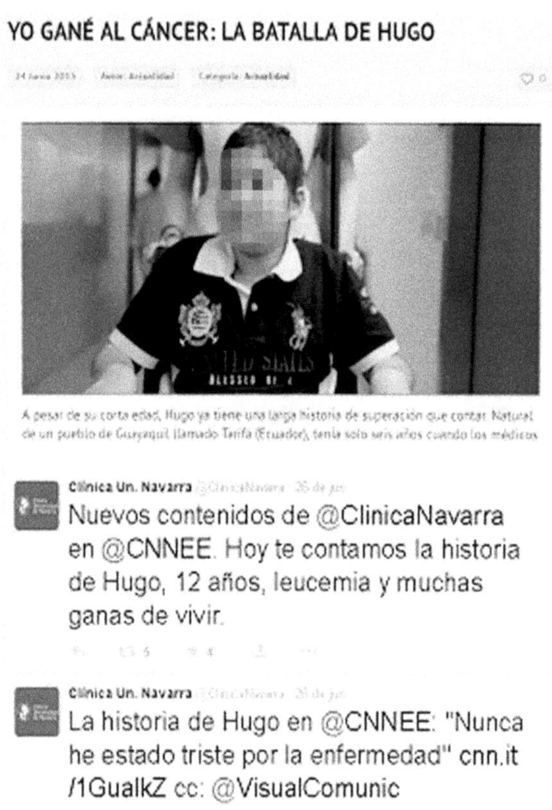

From Brand Journalism to Personal Branding for Professors

It is evident that we cannot overlook personal branding after having learned about the branding strategies that businesses and institutions use to make their name known. Some authors are already saying it - well-positioned personal branding makes the company/institution visible. Nevertheless, how important are brand journalism and personal branding for professors in the university setting?

To answer this question, a survey was given professors who teach at Schools of Communication (Journalism) at Spanish universities. The academics who took the survey (60) come from the following universities (21): Zaragoza (7), La Laguna (1), Católica de Murcia (3), Pompeu Fabra (14), Rey Juan Carlos (1), CEU Cardenal Herrera (4) (Valencia) (1), Complutense de Madrid (1), País Vasco (3), Málaga (2), Navarra (1), Camilo José Cela (1), Sevilla (6), Católica San Antonio de Murcia (2), CEU San Pablo (1), Santiago de Compostela (1), Valladolid (1), Autónoma de Barcelona (1), Antonio de Nebrija (1), Pontificia de Salamanca (1), Universidad Nacional de Educación a Distancia (1) and 6 did not respond. Of the participants, 43.3% are female (26) and 56.7% are male (34). 73.3% of the participants are 40 years of age or older, 23.3% are between 25 and 40 years of age and 3.3% are between 18 and 25 years of age.

The survey was given in the form of a self-administered questionnaire using a Likert scale with 14 questions. It is divided into two parts: the first part involves questions regarding brand journalism (questions 1-7) and the second part involves questions regarding personal branding (questions 8-14) in the university setting. The questionnaire was sent digitally keeping in mind that the interviewer's presence can produce a lack of bias.

The professors indicate that the main objective of brand journalism is company/institution reputation management with 55.2% (32), and with 3.4% (2) indicating that it is used to manipulate audiences (see Figure 3). At the same time, 94.8% (55) indicate that a new debate concerning information ethics could be opened with 60.3% (35) in agreement and 34.5% (20) in total agreement, while 5.2% (3) indicated disagreement regarding the possibility of a new debate. We consider this data while keeping in mind that 83.3% (50) of the professors indicate that they know the term in question, compared to 16.7% (10) that recognised that they had no knowledge concerning the term at hand.

Figure 3. Objective of brand journalism
Source: Elaborated from survey data.

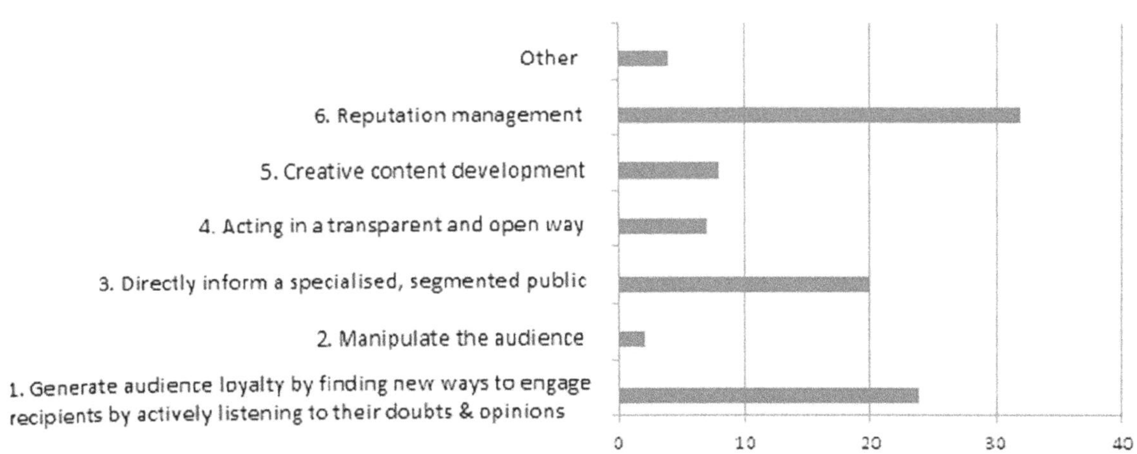

From figure 3 we can see that brand journalism seeks to generate audience loyalty by finding new ways to engage recipients 41.4% (24) as well as directly inform a specialised, segmented public 34.5% (20). These could be some indicators that favour development alongside the fact that businesses and institutions manage without the mass media, but not without journalists. That being said, could brand journalism help to foster employment for journalists? Establishing a scale from one to ten, 45.7% (27) of the professors rated this idea with a 'B' (7-8 points), while 30.5% (18) rated it with a 'C' (5-6 points). 11.9% (7) professors rated it with a 'D' (1-4 points) or an 'A' (9-10 points). General opinion (88.1% (52)) indicates that brand journalism is a means of eliminating unemployment among journalists. So, could we be talking about a new professional profile? This is a question that poses a series of doubts, especially when "the scope of new profiles within the journalism profession as well as their social value is growing. This can be proven with two rationales: a surge in labour demand to cover these types of positions with professionals who have associate's or bachelor's degrees in Journalism, Marketing and Publicity, as well as the fact that the journalism profession recognises this growth through the use of social media on a daily basis" (Sánchez and Méndez, 2015, 56-57).

Another aspect that has been assessed is companies' and institutions' use of brand journalism to make their name known and to gain recognition. For that reason, those surveyed were asked the question: how well do you think universities are making use of said communication strategy? 34.5% (20) of those surveyed answered with a 'B' (7-8 points), 22.4% (13) of those surveyed answered with a 'C' (5-6 points) and 12.0% (7) of those surveyed answered with an 'A' (9-10 points). At the same time, 31% (18) of those surveyed answered with a failing mark (1-4 points). It is evident that there is a disparity in the data. Nevertheless, the professors' favourable opinions prevail with 68.9% (40).

In accordance with the data obtained, we believe it is best to discuss the future of brand journalism due to the controversy that the term in question has raised. The information that we have gathered is shown in Table 1. Vertically, five opinions are listed. Horizontally is a scale from one to three, with 1 being of least importance and 3 being of greatest importance. Professors from the 21 universities indicate that brand journalism will coexist with traditional journalism with 61.4% (35), on a scale of two. Similarly, on a scale of two, 60.7% (34) indicate that brand journalism will promote employment and new forms of communication, thus earning it a promising future. To a lesser extent, on a scale of one, 56.4% (31) indicate that journalism's days are numbered. These percentages support the data obtained in previous paragraphs regarding brand journalism as a means of employment for unemployed journalists.

Table 1. The future of brand journalism

	1		2		3	
a) Brand journalism's days are numbered.	56.4%	31	40%	22	3.6%	2
b) Brand journalism has always and will always exist, only now it's the companies themselves who are spreading their contents.	9.1%	5	49.1%	27	41.8%	23
c) Brand journalism is the future of the profession.	48.1%	26	48.1%	26	3.7%	2
d) Brand journalism will coexist with traditional journalism.	7%	4	61.4%	35	31.6%	18
e) Brand journalism promotes employment and new forms of communication, thus earning it a promising future.	12.5%	7	60.7%	34	26.8%	15

Source: Elaborated from survey data.

After having verified the importance of brand journalism in companies/institutions with the professors' opinions, we believe it desirable to find out whether the professors think that brand journalism should be implemented in the Schools' of Communication plans of studies. 56.9% (33) of academics consider that its implementation is either important (41.4% (24)) or very important (15.5% (9)). To a lesser extent, 32.8% (19) estimate that its introduction into course curriculum as a subject is of little importance, compared to 10.3% (6) who think that it is not important at all.

Regarding the second part of the chapter - personal branding in the academic setting - 62.1% (36) agree that brand journalism allows for the simultaneous construction of personal branding; 25.9% (15) are in total agreement with this opinion, thus equalling a sum of 88.0% (51) between the two (agreement and total agreement). To a lesser extent, 12.0% (7) are either in disagreement (10.3% (6)) or total disagreement (1.7% (1)) with this opinion. For this reason, the professors consider what the Internet says about them and their work to be important, especially when public image is increasingly reflected on the Internet and Social Media. Keeping in mind a rating from one (not at all important) to five (very important), 75.9% (44) of professors surveyed think that it is very important with a rating of 4 and 5 due to the fact that someone will always be searching for you on the Internet. Meanwhile, 20.7% (12) gave a rating of 3 and medium importance, in comparison to 3.4% (2) who gave a rating of a 1 or 2 with no importance. In spite of the results obtained, it is interesting that 56.9% (33) of professors have not created a personal brand, while 43.1% (25) have (see Figure 4). This data proves to be meaningful and invites us to reflect on the underlying causes of this effect.

It is evident that the Internet has become an undeniable tool for professional development, as well as a showcase of opportunities to make your name known, create synergies and look for employment. The professors who have created personal brands (43.1%) indicate a number of criteria that should be kept in mind in order to establish a successful personal brand. Among the most relevant results are:

Figure 4. Creation of a personal brand for professors
Source: Elaborated from survey data.

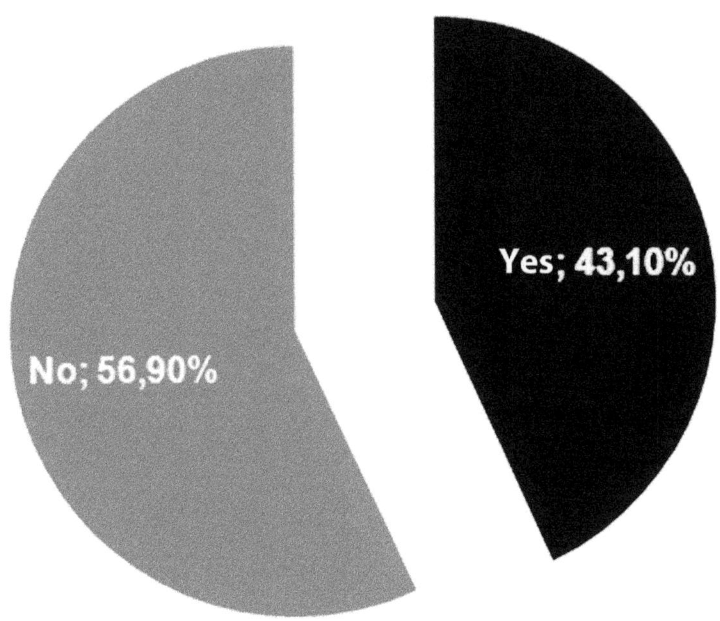

1. A consistency in subject matter,
2. Coherence with image, messages and the journalism profession,
3. Originality,
4. A respectful attitude and a willingness to engage in dialogue,
5. Correction,
6. A clear style,
7. A presence and visibility control,
8. A capacity to respond quickly,
9. Deterrence from narcissistic attitudes,
10. Thoroughness, honesty and transparency,
11. Being proactive in terms of content generation,
12. Determination and a communication strategy, and finally,
13. Content specialisation.

What is more is that 77.6% (45) think that a professor's personal brand should transmit an emotional stimulus to anyone interested in their work, compared to 20.7% (12) and 1.7% (1) that did not respond.

Apart from the professors' criteria, it is important to keep in mind the use of tools that allow for the creation of a personal brand in order to optimise development of a digital fingerprint. In this respect, it is necessary to select which tools are useful in creating the greatest visibility on the Internet as well as leverage your activity in order to design your own digital strategy and attain a strong digital reputation. We can mention four social media tools that are substantially more important compared to others. Of those surveyed, 56.9% (33) consider Twitter to be the platform which helps to construct a personal brand, followed by LinkedIn with 44.8% (26), Academia.edu with 43.1% (25) and Facebook with 36.2% (21) (see Figure 5).

In light of the answers from professors regarding which tools are most conducive to reaching users, it is necessary to assess and auto-diagnose the presence and definition of personal branding concerning professors from the Spanish Universities Schools of Communication/Information. In addition, it is also

Figure 5. Social media and construction of a personal brand
Source: Elaborated from survey data.

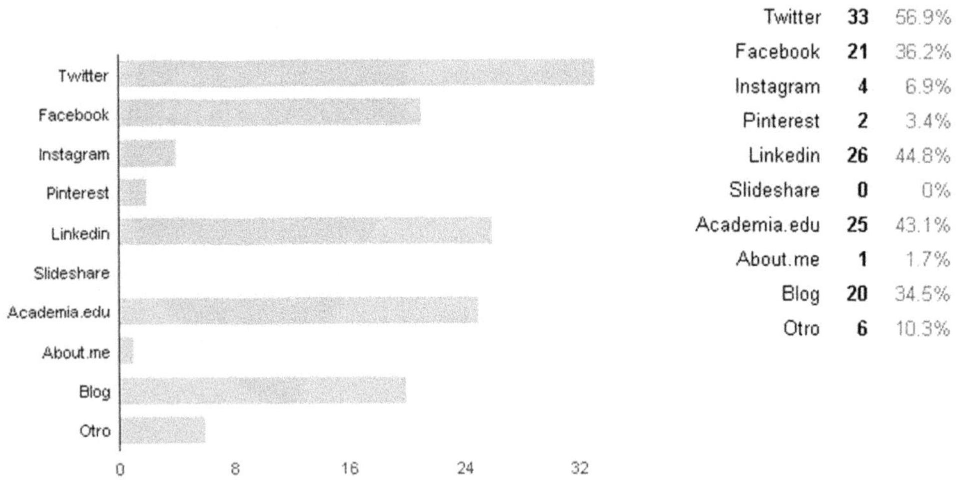

Figure 6. Construction of a personal brand: Schools of communication professors
Source: Elaborated from survey data.

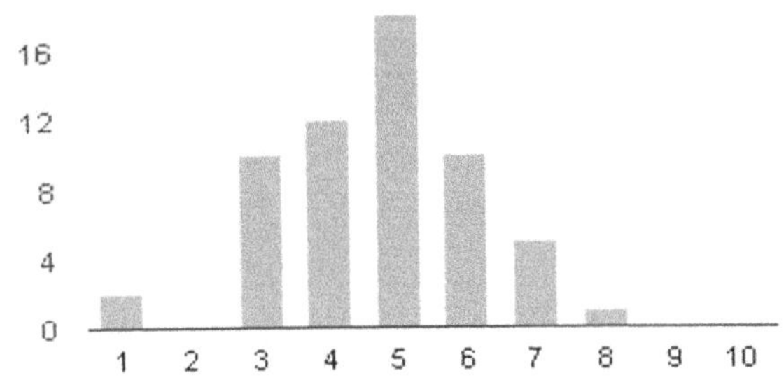

1	**2**	3.4%
2	**0**	0%
3	**10**	17.2%
4	**12**	20.7%
5	**18**	31%
6	**10**	17.2%
7	**5**	8.6%
8	**1**	1.7%
9	**0**	0%
10	**0**	0%

important to keep in mind the necessity of contributing something of added value in order to attain a digital reputation and a committed audience. Of those surveyed, only 41.3% (24) would give a failing mark (1-4 points) due to professors' lack of digital fingerprints, 48.2% (28) would give a 'C' (5-6 points) and 10.3% (6) would give a 'B' (7-8 points). These percentages are significant results that university professors should keep in mind (see Figure 6).

Before discussing specific cases of professors who stand out for having left their digital fingerprint on the Internet as a result of content contribution and management of personal brand/digital reputation, we would like to name some academics whose names were mentioned more than twice by those surveyed and who could be examples of personal branding: Carlos Alberto Scolari, José Luis Orihuela, Ramón Salaverría, Lluís Codina and Juan Luis Manfredi Sánchez.

SOLUTIONS AND RECOMMENDATIONS

It is possible to see that brand journalism in the university setting has timidly made its way into Spanish Universities in contrast to other countries such as the United States. The web site Futurity.org brings together different universities to spread knowledge using a different, informative approach in order to benefit society and attain recognition. In Spain there are some isolated cases of brand journalism in the university setting such as the University of Navarra with the Clínica Universitaria de Navarra, but this same type of work is needed in other universities that have the same or similar social strategies, either individually or in groups as is the case of the Futurity.org web page. However, 68.9% (40) of academics rate using brand journalism in Spanish universities with an 'A'. This percentage proves to be contradictory given the fact that the academics surveyed failed to mention examples of brand journalism in Spanish universities, but did mention its use in foreign universities such as the University of Cambridge.

Notwithstanding, it is important to highlight that professors of the Spanish Schools of Communication/Information indicate that brand journalism simultaneously allows for the creation of personal branding with 88.0% (51). However, low presence of brand journalism in the university setting may confirm the results we obtained regarding a lack of creation of digital fingerprints. Compared to 43.1% (25) of survey participants who have created a digital fingerprint, more than half of those surveyed (56.9%) (33)

have not yet created one due to a lack of information, knowledge, training and/or time. Similarly, when evaluating and/or self-evaluating the creation of professors' personal brands, 41.3% of those surveyed gave it a failing mark, while 58.5% assessed it as positive - another contradictory finding due to the fact that 56.9% have yet to construct a personal brand.

These findings lead us to propose possible solutions such as: increasing awareness concerning the importance of a personal brand in every university, or workshops that allow professors to be up-to-date with new advancements regarding creation of personal branding. Finally, these types of studies help to envisage the shortfalls which, due to a lack of knowledge or funding, may be resolved.

FUTURE RESEARCH DIRECTIONS

The chapter at hand forecasts future topics for investigation - actions regarding the question at hand. Firstly, Universities in Spain or even the Conference of Vice-Chancellors of Spanish Universities, (CRUE), would have to be introduced to the possibility of contributing information to society using a different form of communication. For instance, utilising brand journalism. The objective is none other than finding a way to benefit society and said institutions in terms of their social and informative strategies. The proposal would be focused on creating a platform that brings together different Spanish universities with the goal of connecting to society through knowledge, thus improving people's lives.

Similarly, in the second section of the chapter, personal branding, it would be advisable to carry out a study on each school, first locally, then regionally and finally nationally, with the aim of determining which schools at which universities really foster the use of personal branding. This aims at performing a general analysis in order to offer effective solutions to those cases in which professors are slow to incorporate personal branding.

CONCLUSION

Brand journalism has evolved and has a promising future as indicated by 60.7% of the academics surveyed. However, it leaves much to be desired in the educational setting due to its weak presence in Spanish universities in contrast to other countries. An example is the United States where a specific platform was launched (Futurity.org) in order to spread knowledge through specialised information. This is the aim of brand journalism, according to 34.5% of the professors, or, in other words, more than half of those surveyed.

It is apparent that the need to create a 2.0 platform exists - one that strives, in the first place, to bring together Spanish universities, and in second place, to offer society a service using the techniques and strategies of brand journalism. Finally, the platform should strive to communicate and interact with its audience. Universities would see that it would be worth the effort thanks to better visibility and reputation management - the purpose of brand journalism according to 55.2% of those surveyed.

Another underlying fact is that 58.3% of the professors who responded to the survey think that brand journalism should be incorporated into course curriculum. This is especially important when talking about a field which fosters employment and new forms of communication according to academics (60.7%). In any case, brand journalism coexists with traditional journalism, causing institutions and businesses themselves to be the ones who disseminate information as a service to society.

At the same time, 88.0% of the academics surveyed indicate that it is important to recognise the relationship that exists between brand journalism and the creation of a personal brand. It is probable that only 43.1% of those surveyed have created a personal brand owing to a lack of training, knowledge and/or time. In any case, it is necessary to establish a series of criteria that should be kept in mind when it comes time to creating a personal brand, such as keeping a proactive emotional attitude, transmission of specialised, information of interest in a rigorous, transparent and clear way, as well as being original, respectful and open to dialogue. Likewise, certain 2.0 tools need to be kept in mind such as: Twitter (56.9%), LinkedIn (44.8%) and Academica.edu (43.1%), among the first to be noted.

For all of the abovementioned reasons, professors from the Schools of Communication in Spain should become acquainted with techniques concerning brand journalism. In addition, they should maintain a more proactive attitude in terms of creating a personal brand, whether it is because of a lack of knowledge and/or training that they have not created one yet.

REFERENCES

Abiad, P. (2014). *El abuelo del periodismo de marca cumple 120 años.* Retrieved from http://www.pabloabiad.com/2014/10/el-abuelo-del-periodismo-de-marca.html

Álvarez de Blanco, R. (2008). *Tu, marca personal: Como conseguir gestionar tu vida con talento y convertirla en una experiencia extraordinadira.* Madrid: Prentice-Hall.

Arqués, N. (2007). *Y tú qué marca eres?: 12 claves para gestionar tu reputación personal.* Barcelona: Alienta.

Barciela, F. (2013a). *El 'boom' del periodismo de marca.* Retrieved from http://economia.elpais.com/economia/2013/02/22/actualidad/1361540029_041048.html

Barciela, F. (2013b). *El último desafío, el 'brand journalism'.* Retrieved from http://www.apmadrid.es/images/stories/125-136.pdf

Casellas, V. (2014). *Marca personal para profesores y marketing educativo.* Retrieved from http://www.artecasellas.es/marca-personal-para-profesores-y-marketing-educativo-por-vicky-casellas/

Marcos, T. (2008). Marca personal, máximo valor de las personas. In Marca Personal: Cómo convertirse en la opción preferente. ESIC.

Martín, R., Sánchez de la Nieta, L., & Hernández, M. A. (2014). *Es periodismo el "brand journalism"?: análisis de los portales informativos creados por multinacionales de referencia en los últimos dos años. Asociación Española de Investigación de la Comunicación (AE-IC).* Universidad del País Vasco /Euskal Herriko Unibertsitatea.

Obispo, M. (2015). *Una historia detrás de cada paciente. Cómo aplicar el periodismo de marca a la comunicación de salud.* Retrieved from http://periodismodemarca.llorenteycuenca.com/una-historia-detras-de-cada-paciente-como-aplicar-el-periodismo-de-marca-a-la-comunicacion-de-salud/

Obispo, M., Sánchez, H., & Méndez, S. (2015). Medios sociales en el periodismo español: Demanda, expectativas y oferta universitaria. *Cuadernos.info,* (36), 155-171. doi: 10.7764/cdi.36.636

Pérez Ortega, A. (2008). Marca Personal: Cómo convertirse en la opción preferente. Ed. ESIC.

Pino, I. (2014). *Hacia un hexálogo del Periodismo de Marca. In Periodismo de marca y reputación corporativa*. Madrid: LLORENTE & CUENCA.

Takaki, M., Bravo, R., & Martínez, E. (2015). La gestión de la identidad corporativa en la Universidad: Análisis y consecuencias desde la perspectiva del profesorado. *Revista Europea de Dirección y Economía de la Empresa*, *24*(1), 25–34. doi:10.1016/j.redee.2014.05.001

Yuste, B., & Cabrera, M. (2014). *Emprender en Periodismo. Nuevas oportunidades para el profesional de la información*. Barcelona: UOC.

ENDNOTES

1 Creator of the web site 'personal brand' in Spanish (www.marcapropia.net), as well as several publications.

Chapter 8
Utilising Content Marketing Metrics and Social Networks for Academic Visibility:
Content Marketing for Academic Impact

Mark Anthony Camilleri
University of Malta, Malta

ABSTRACT

There are numerous assumptions on research evaluation in terms of quality and relevance of academic contributions. Researchers are becoming increasingly acquainted with bibliometric indicators, including; citation analysis, impact factor, h-index, webometrics and academic social networking sites. In this light, this chapter presents a review of these concepts as it considers relevant theoretical underpinnings that are related to the content marketing of scholars. Therefore, this contribution critically evaluates previous papers that revolve on the subject of academic reputation as it deliberates on the individual researchers' personal branding. It also explains how metrics are currently being used to rank the academic standing of journals as well as higher educational institutions. In a nutshell, this chapter implies that the scholarly impact depends on a number of factors including accessibility of publications, peer review of academic work as well as social networking among scholars.

INTRODUCTION

Academic contributions start from concepts and ideas. When their content is of a high quality and is relevant to other scholars, they could be published in renowned, peer-reviewed journals. Many researchers are resorting to online full text databases, institutional repositories or online open access journals in order to disseminate their findings. The internet has brought an increased engagement among peers, over email or video communications. In addition, the web and online academic networks have surely helped to enhance the fruitful and collaborative relationships among researchers. Notwithstanding, researchers are increasingly sharing their knowledge with colleagues as they

DOI: 10.4018/978-1-5225-0917-2.ch008

present their papers in seminars and conferences. After publication, their contributions may then be cited by other scholars, including students.

The researchers' visibility does not solely rely on the number of publications they produce. Both academic researchers as well as their institutions are continuously being under scrutiny as they are rated and classified by independent reviewers. Very often citations appear in publications in highly reputable journals or well-linked homepages that promote scholarly content (Thelwall, 2008, 2009; Bonzi & Snyder, 1991). Publications are usually ranked through bibliometrics that assess the individual researchers as well as their organisational performance (Thelwall, 2008; Delgado López-Cózar, Robinson-García & Torres-Salinas, 2014). Such metrics measure the effectiveness of academic publications. Citation analysis, impact factor (IF), h-index, webometric analysis, altimetrics and social academic networks are some of the most commonly used measures that assess the quality and relevance of scholarly work. High bibliometric scores and numerous citations are widely believed to constitute the academic reward system. The most cited authors are usually endorsed by peers for their significant contribution to knowledge. As a matter of fact, citations are at the core of scientometric methods. They have been (and are being) used to measure the academic visibility and impact of scholarly work (Kousha & Thelwall, 2007; Moed, 2005, 2006; Borgman, 2000).

This chapter's objective is to critically review and analyse extant literature on the subject of academic branding. It clearly explains how the content marketing of individual researchers' and of their educational institutions' publications may lead to improvements in their academic standing and increased reputation. The author re-examines the communication structures and processes of scholarly communication (Thelwall & Kousha, 2015a, 2015b, 2015c; Kousha & Thelwall, 2007; Borgman & Furner 2002) among academic institutions. At the same time, this contribution presents relevant literature on the most popular scholarly metrics. It also reports how the evolution of WEB2.0 and the ubiquity of online academic networks have helped individual researchers to engage with their peers. In conclusion, this chapter implies that there is potential for scholars to use content marketing metrics for their academic standing (Thelwall & Kousha, 2015a, 2015b; Orduna-Malea & Ontalba-Ruipérez, 2013; Adler & Harzing, 2009; Thelwall & Harries, 2004; Wilson, 1999; Tague-Sutcliffe, 1992).

BACKGROUND

The independent evaluation of any scholarly work is a notoriously difficult and arduous task. Ideally, academic contributions ought to be scrutinised by qualified experts in their field. These scholars are recognised and accredited for both quality and quantity according to previously established rules. Essentially, such evaluation is called peer-review and is usually carried out by research committees. Yet, very often certain journal reviewers could lack specialist knowledge to assess research, particularly empirical data. Hence, committees use secondary criteria including crude publication counts, number of downloaded documents, journal prestige, the reputation of individual authors and their institutions; as well as other metrics in order to estimate the importance and relevance of particular research fields. Hence, it is the scope of this chapter to shed light on academic marketing which relies on the use and application of different metrics to build the reputation and standing of scholars, institutions and research centres. Authoritative, multidisciplinary content covers over 10,000 of the highest impact journals worldwide, including open access journals and over 110,000 conference proceedings

(Courtault, Hayek, Rimbaux & Zhu, 2010; Thomson Reuters ISI Web of Science, 2010). References are sorted in such a way in order to reveal how often any given publication has been cited within a given time period, and by whom. Afterwards, these results are published through the Thomson Reuters Science Citation Index (SCI). Hence, SCI gives a breakdown of the annual citation rate of papers, by author or by research group. SciVerse Scopus is another commercial database which was launched in November 2004. Its providers claim that their system provides the largest abstract and citation database containing peer-reviewed research literature: scientific journals, books and conference proceedings (SciVerse Scopus, 2015). The journal impact factors which are published by SCI or in Scopus are widely regarded as quality rankings for journals. Such metrics are used extensively by leading journals in their content marketing endeavours. Other excellent databases include the Social Science Research Network (SSRN) database, which contains a very large number of working papers and publications in the realms of social sciences (including economics, finance, accounting, and business); research papers in economics database for economics; the Scopus subscription-based database, and free Internet databases, such as Google Scholar (Orduna-Malea, Ayllón, Martín-Martín & López-Cózar, 2015; Labbe 2010), Researchgate (Kousha & Thelwall, 2014), Academia.edu, Mendeley and the like (Haustein, Peters, Bar-Ilan, Priem, Shema & Terliesner, 2014). Each of these databases have their own strengths and weaknesses.

Citation Analysis of Academic Papers

The majority of academic papers, notes, reviews, corrections and correspondence published in scientific journals contain citations to other academic contributions (Eysenbach, 2011; Kousha & Thelwall, 2007; Brooks, 1986; Garfield, 1979). These citations comprise the author(s)' name(s), date, title, journal name, volume (issue), pages and url (or doi). Thousands if not millions of studies have used such data from citation indices (Thelwall, 2008; Bonzi & Snyder, 1991). These bibliographic applications extract, aggregate and analyse quantitative aspects of bibliographic information. Therefore, citations constitute formal and explicit links between papers that usually share common features (Shuai, Pepe & Bollen, 2012; Brooks, 1986). A citation index is built around these links as the latest academic contributions cite and identify previous publications. Thanks to citations a literature search can find from one to dozens of additional papers on a subject just by knowing one that has been cited. Every paper provides a list of new citations from which to continue the search (Athar, 2014; Garfield, 1979). Relevant data for citation analysis is derived from in-text references and bibliographies of publications. Indicators of scholarly 'impact', 'influence' or 'quality' are applied both in the study of academic communication and when assessing the researchers' performance (Moed, 2005). Hence, citations represent the documents' content (Athar, 2014; Brooks, 1986). They are an extensive dimension of retrieving publications that have been cited in previous works (Knothe, 2006).

The citation analysis offers enormous possibilities for the tracing of trends and developments across different research areas (Bonzi & Snyder, 1991). This metric has become the de-facto standard in research evaluation. In fact, publications can be simply evaluated on the number of citations they receive; due to their relative availability and accessibility (Knoth & Herrmannova, 2014). Yet, one should bear in mind that citations represent only one of the attributes of publications. By themselves, they do not provide adequate and sufficient evidence of impact and quality. This may be due to a wide range of characteristics they exhibit:

1. Including the semantics of the citation itself (Knoth & Herrmannova, 2014),
2. The motives for citing (Nicolaisen, 2007),
3. The variations in sentiment (Athar, 2014),
4. The context of the citation (He, Pei, Kifer, Mitra & Giles, 2010),
5. The popularity of topics, the size of research communities (Moed, 2006),
6. The time delay for citations to show up (Priem & Hemminger, 2010),
7. The skewness of their distribution (Seglen, 1992),
8. The difference in the types of research papers (Seglen, 1997a) and finally,
9. The ability to game / manipulate citations (Arnold & Fowler, 2011).

The Journal Impact Factors (IFs)

The citations and journal impact factors (IFs) are quantitative and objective indicators that assess published science (Seglen, 1997a). Scholarly impact is a measure of frequency in which an "average article" has been cited over a defined time period in a journal (Glänzel & Moed, 2002). Journal citation reports are published every year by Thomson-Reuters' Institute of Scientific Information (ISI). These reports feature data for the ranking of the Immediacy Index of articles. This index is the average number of times an article is cited in the year it is published. It is calculated by dividing the number of citations to articles published in a given year by the number of articles published in that year. The ISI (IF) that is published in journal citation reports (JCR) is a ratio of received citations (in its numerator) to published articles, notes or reviews in a journal (in the denominator). Hence, the IF represents the mean of a skewed citation distribution of frequency counts per cited journal title. It is calculated by dividing the number of citations in the most recent calendar year (e.g., 2015) by the total number of articles published in the previous 2 years (i.e., 2013–2014). If a journal recorded an IF of 1.0 in 2010; the articles published in 2014 or 2013 have been cited, on average, one time in 2015 (Hodge & Lacasse, 2011). The IF is therefore biased towards journals revealing a rapid maturing or decline in citation impact (Moed, 2005). The JCR database continuously records scientific citations as represented by article reference lists that are drawn from a large number of scientific journals (Seglen, 1997b).

The journals' IFs are often determined by technicalities that are unrelated to the scientific quality of articles. Academic authors receive different citation counts for their numerous contributions. For instance, they could have published a single paper in a particular research area that is cited very often by other scholars. Notwithstanding, the IFs also depend on the research field. Their citation rates will determine the journals' IFs, and not vice versa (Seglen, 1997a). Journals with high IFs usually cover large areas of basic research with a rapidly expanding but short lived literature; that use many references per article. The number of references per article and their age distribution varies considerably among subfields. Such differences could distort IFs to the extent that journals from different subfields cannot be compared with one another. It may appear that a journal IF indicator takes into account such differences in reference practices among subfields, by dividing a journal's impact by the citation 'average' in the subfields covered by the journal (Seglen, 1997a). Review journals tend to have higher citation rates than other journals (Brooks, 1986). This can be taken into account by an advanced normalised impact indicator that disaggregates journal citation impact and subfield averages by type of document (Seglen, 1997a).

In highly dynamic research fields, such as biochemistry and molecular biology, the research papers may become obsolete in a relatively short period of time. A large proportion of citations are captured by the short term index that is used to calculate journal IFs. On the other hand, there are certain topics whose content is durable in literature. For instance, it is often the case that there is a smaller fraction of short term citations on mathematics. This would translate to lower journal IFs as most research fields could be considered as self-contained. For instance, clinical medicine draws heavily on basic science, but not vice versa. The result is that basic medicine is cited three to five times more than clinical medicine, and this is clearly reflected in the journal IFs (Seglen, 1989). Notwithstanding, in young and rapidly evolving research areas, the number of publications making citations is large relative to the amount of citable material. This leads to high citation rates for articles and high journal impact factors (Seglen, 1997a).

Hodge and Lacasse (2011) held that publishers of core scientific journals could evaluate their prospective contributions by using IF indicators. These authors indicated that in the Nordic countries, researchers were being evaluated according to the journal IFs. Moreover, their use was recently reinforced in Italian universities in order to remedy for the purported subjectivity and bias in academic appointments in higher educational insitutions. Hodge and Lacasse (2011) reported that in several other countries the allocation of resources toward tertiary education is often based on IFs. Therefore, the increased awareness of journal IFs, and their potential use for evaluation is encouraging researchers to publish papers in journals with maximum impact. This is happening at the expense of specialist journals that might actually be the most appropriate vehicles for the research in question.

On the other hand, the impact factor has often been subject of ongoing criticisms by researchers for their methodological and procedural imperfections. Whilst a higher impact factor may indicate journals that are considered to be more prestigious, it does not necessarily reflect the quality or impact of an individual article or researcher. This may be attributable to a wide array of journals and research contributions (Hodge & Lacasse, 2011). Top influential journals are usually given higher impact factors. Therefore, researchers in a small field may lack access to journals of equally high citation impact. This puts them at a disadvantage when compared with colleagues in larger fields (Seglen, 1997a).

The article citation rates are measured in comparison to citation impact which is specific to a specific field (Garfield, 1972). Such field corrections range from simply dividing a single article's citation rate by the impact factor of its journal (Moed et al., 1987). Seglen (1997a) claimed that this punishes publication in high impact journals to the use of complex, author specific, field indicators based on reference lists. However, field corrections cannot be simply applied to journal impact factors, as many research areas are dominated by one or a few journals. Yet, the case corrections might merely generate relative impact factors of unit value. Even within large fields, the tendency of journals to subspecialise with certain subjects is likely to generate significant differences in journal impact (Seglen, 1997a).

Journal IFs depend on the research field. It is widely assumed that publication in a high IF will enhance the impact of an article. Researchers publish in high impact journals as their academic peers often judge the quality of their papers by their "wrapping" rather than by their content. The journals' academic clout is considered as a valid evaluation criterion. However, journals cannot in any way be taken as representative of the articles they contain. Even if they could, the journal impact factors would still be far from being quality indicators.

Arguably, the citation impact is primarily a measure of scientific utility rather than one of scientific quality. Very often, the authors' selection of references is subjective and is unrelated to quality. Seglen (1997a) hinted that there is no alternative to having an objective evaluation of qualified experts reviewing the scientific quality of publications. Perhaps, the scientific community ought to concentrate its efforts on quality measures rather than on developing ever more sophisticated versions of useless indicators. Ultimately, it is the scientific content of research papers and their significant contribution to knowledge is what really matters to academia.

Seglen (1997a) maintained that 15% of the most cited articles account for 50% of all citations. The most cited 50% of these articles account for 90% of citations. In other words, the most cited half of articles are cited on average 10 times as often as the least cited half (Seglen, 1997a). This is exactly the opposite of what evaluations are meant to achieve. A few highly cited articles will predominantly determine the value of the journals' impact factor. However, the researchers do not always publish their most citable work in journals of the highest impact, nor do their articles necessarily match the impact of the journals they appear in. Some authors may consider other factors including the journals' subject area; their relevance to the author's speciality; the fairness and rapidity of the editorial process; the probability of acceptance, publication lag and publication cost (page charges) among other issues.

In spite of its wide use, the impact factor has often been the subject of numerous criticisms (Brumback, 2009; Favaloro, 2009). The use of the 2-year citation window, the reliance on the mean in computing impact factors, and the limited number of journals indexed by Thomson ISI are some of the main limitations of the impact factor (Hodge & Lacasse 2011).

The H-Index for Individual Scholars and Institutional Impact

Hirsch (2005) proposed a new measure of scientific achievement. His h-index attempts to calculate the citation impact of academic publications of researchers. It measures the scholars' productivity by taking into account their most cited papers and the number of citations that they have received in other publications. The h-index assesses the significance and impact of the scholars' cumulative research contributions. This index can also be applied to measure the impact and productivity of academic journals, as well as research groups including university departments or countries (Orduña-Malea & López-Cózar, 2014; Bornmann & Daniel, 2005). The journals' h-index value can be calculated with data from Thomson ISI, Elsevier's Scopus or Google Scholar (Labbe, 2010). It mostly used with the latter one.

The h-index consists of a single number that reports on the authors' academic contributions that have at least the equivalent number of citations (Bornmann & Daniel, 2007). Hirch's h-index is neither entirely dependent on the number of citations nor on the number of publications. It synthesises both measures as one needs to publish many papers with many citations each in order to achieve a high h-index. The higher the index, the greater the number of significant papers published by an author and the higher the significance of the papers. This index was initially designed to overcome the limitations of other measures of quality and productivity of researchers. It has received a lot of attention from the scientific community in the last few years due to some of their good properties including the easiness of its computation, balance between quantity of publications and their impact and so on (Alonso, Cabrerizo, Herrera-Viedma & Herrera, 2009). As the h-index is an objective indicator it may play an important role when making decisions about promotions, funding allocations and awarding academic prizes (Costas & Bordons, 2007). Vanclay (2007) suggested that the h-index is robust as it is insensitive to a set of lowly cited papers (the difficulty of increasing the h-index grows exponentially as the researcher has to receive new citations to obtain a higher index).

The h-index metric is not limited to a fixed time period. Hence, the citation window can be set at whatever time frame that could be the most appropriate for any given discipline (Hodge & Lacasse, 2010). h-index values have been calculated using various citation windows, including 1 year (Braun et al., 2006), 2 years (Bador & Lafouge, 2010), 5 years (Harzing & van der Wal, 2009), and even longer (Olden, 2007; Vanclay, 2008). Secondly, the h-index is not based on the mean. Therefore, it could attenuate the effect of highly cited articles on computations of journal quality. Indeed, the h-index is unaffected by those few articles that are highly cited. It accurately reflects the unit of analysis (i.e. journals) (Harzing & van der Wal, 2009).

The h-index values that are derived from Thomson ISI and Google Scholar tend to exhibit modest to strong correlations, although variations do exist from discipline to discipline. For instance, the management journals have reported coefficients of .55 to .66 (Harzing & van der Wal, 2009). A correlation of .61 was registered in disciplinary computer science journals. 0.78 was recorded for the interdisciplinary computer science journals (Franceschet, 2010). Google Scholar produces high h-index values due to its wide coverage of academic source material (Orduna-Malea et al., 2015; Labbe, 2010; Hodge & Lacasse, 2010). This index is increasingly being used in citation-based analyses as it is perceived as accurate by many researchers (Moussa & Touzani, 2010; Lee, Kraus & Couldwell, 2009; Mingers, 2009). Therefore, Google Scholar's h-index may be a better measure of journal quality than Thomson's ISI impact factor (for the contributors of social sciences (Orduña-Malea & López-Cózar, 2014). The h-index's strengths lie in its flexible time frame, its computational method that emphasise the quality and quantity of papers, as well as its source coverage. Courtault and Hayek (2008) indicated that when authors increase the number of research papers to journals, they increase their h-index. They went on to suggest that the value of social scientists cannot be fully evaluated with single results. Very often, researchers are awarded for several outstanding contributions, including their complete works. Alternatively, they may be appraised by their peers for initiating a new sub-discipline. Therefore, the h-index is also a useful characterisation that compares the different contributions of scholars.

However, the h-index could also present some drawbacks that have been pointed out in relevant literature. Hirsch (2005) himself noted that the h-index should not be used to compare scientists from different disciplines. Moreover, Hirsch noted that there exist some technical limitations, such as the difficulty to obtain the complete output of scientists with very common names (this problem decreases as citation databases improve their records and search engines). Moreover, the h-index depends on the duration of each scientist's career because their pool of publications and citations tends to increase over time (Kelly & Jennions, 2006). So it might not be suitable to compare scientists at different stages of their career. Costas and Bordons (2007) held that highly cited papers are important for the determination of the h-index. They went on to suggest that extremely cited papers may have a similar or equal h-index as researchers with moderate or high cited papers. They admitted that the research performance is a complex multifaceted endeavour that usually cannot be assessed adequately by means of a single indicator. The best papers in terms of quality will be the most cited in academia. Therefore, the h-index may be considered as a compound measure of productivity and quality, as measured by the number of citations received by the published papers (Courtault & Hayek, 2008). The h-index considers the papers' citations as proxies that assess their quality. It may appear that there is a good correlation between the prestige of a review and its impact factor. Presumably, authors receive many citations because their papers are published in renowned journals. Of course, these authors may have also published high quality papers in less 'impact' journals for many reasons. Authors are often willing to shorten the publication delay or they may simply experience inconsiderate refereeing (Courtault & Hayek, 2008).

Webometrics for Higher Education Institutions and Research Centres

Although research impact metrics can be used to evaluate individual academics, there are other measures that could be used to rank and compare academic institutions. There are currently several international ranking schemes for universities, some of which use citations to estimate the institutions' impact

(Buela-Casal, Gutiérrez-Martínez, Bermúdez-Sánchez, & Vadillo-Muñoz, 2007). Nevertheless, there have been ongoing debates about whether bibliometric methods should be used for the ranking of academic institutions (e.g., van Raan 2005; Ioannidis et al., 2007). The science of webometrics (which is also known as cybermetrics) is still in an experimental phase (Thelwall, 2004; Kretschmer & Aguillo, 2004; Thelwall, 2002; Thelwall & Price, 2003). Early webometric studies explored how hyperlinks to research papers were used to generate impact indicators. In fact, these studies were structurally similar to citations (Ingwersen, 1998; Smith, 1999).

Relevant theoretical underpinnings have shown that the number of links to British university web sites were also correlated to their research productivities (Thelwall, 2001). The most productive universities were increasingly enclosing the link to their papers online (Thelwall & Harries, 2004). Yet, many commentators argued that these hyperlinks were unreliable indicators of journal impact (Kenekayoro, Buckley & Thelwall, 2014; Smith, 1999; Vaughan & Hysen, 2002). Yet, several universities were experimenting with different methodological innovations such as mapping to popular web sites through links (Thelwall, Vaughan, and Björneborn, 2005; Vaughan and Shaw, 2005, 2007). Apparently, the scope of webometrics was to link the research organisations with firms, intermediary groups including professional associations and government agencies. Hence, the webometrics were intended to compliment traditional bibliometrics as they kept a track record of publications. Notwithstanding, the web helps to promote research funding initiatives and to advertise academic related jobs. The webometrics could also monitor the extent of mutual awareness in particular research areas (Thelwall, Klitkou, Verbeek, Stuart & Vincent, 2010).

Aguillo, Granadino, Ortega and Prieto (2006) reported how webometrics were used to rank the academic standing of universities. In this case, the webometrics measured the higher educational institutions' wider impact rather than just their research impact. Moreover, there were other uses of webometric indicators in policy-relevant contexts within the European Union (Thelwall et al., 2010; Hoekman, Frenken & Tijssen, 2010). The webometrics refer to the quantitative analysis of web activity, including profile views and downloads (Davidson, Newton, Ferguson, Daly, Elliott, Homer, Duffield & Jackson, 2014). Therefore, webometric ranking involves the measurement of volume, visibility and impact of web pages. These metrics seem to emphasise on scientific output including peer-reviewed papers, conference presentations, preprints, monographs, theses and reports. They also analyse other academic material including courseware, seminar documentation, digital libraries, databases, multimedia, personal pages and blogs among others (Thelwall, 2009; Kousha & Thelwall, 2015; Mas-Bleda, Thelwall, Kousha & Aguillo, 2014a; Mas-Bleda, Thelwall, Kousha & Aguillo, 2014b; Orduna-Malea & Ontalba-Ruipérez, 2013). Thelwall and Kousha (2013) have identified and explained the methodology of five well-known institutional ranking schemes:

- "QS World University Rankings aims to rank universities based upon academic reputation (40%, from a global survey), employer reputation (10%, from a global survey), faculty-student ratio (20%), citations per faculty (20%, from Scopus), the proportion of international students (5%), and the proportion of international faculty (5%).
- The World University Rankings: aims to judge world class universities across all of their core missions - teaching, research, knowledge transfer and international outlook by using the Web of Science, an international survey of senior academics and self-reported data. The results are based on field-normalised citations for five years of publications (30%), research reputation from a survey (18%), teaching reputation (15%), various indicators of the quality of the learning

environment (15%), field-normalised publications per faculty (8%), field-normalised income per faculty (8%), income from industry per faculty (2.5%); and indicators for the proportion of international staff (2.5%), students (2.5%), and internationally co-authored publications (2.5%, field-normalised).

- The Academic Ranking of World Universities (ARWU) aims to rank the "world top 500 universities" based upon the number of alumni and staff winning Nobel Prizes and Fields Medals, number of highly cited researchers selected by Thomson Scientific, number of articles published in journals of Nature and Science, number of articles indexed in Science Citation Index – Expanded and Social Sciences Citation Index, and per capita performance with respect to the size of an institution.
- The CWTS Leiden Ranking aims to measure "the scientific performance" of universities using bibliometric indicators based upon Web of Science data through a series of separate size- and field-normalised indicators for different aspects of performance rather than a combined overall ranking. For example, one is "the proportion of the publications of a university that, compared with other publications in the same field and in the same year, belong to the top 10% most frequently cited" and another is "the average number of citations of the publications of a university, normalised for field differences and publication year."
- The Webometrics Ranking of World Universities Webometrics Ranking aims to show "the commitment of the institutions to [open access publishing] through carefully selected web indicators": hyperlinks from the rest of the web (1/2), web site size according to Google (1/6), and the number of files in the website in "rich file formats" according to Google Scholar (1/6), but also the field-normalised number of articles in the most highly cited 10% of Scopus publications (1/6)" (Thelwall & Kousha, 2013).

Evidently, the university ranking systems use a variety of factors in their calculations, including their web presence, the number of publications, the citations to publications and peer judgements (Thelwall and Kousha, 2013; Aguillo, Bar-Ilan, Levene, & Ortega, 2010). These metrics typically reflect a combination of different factors, as shown above. Although they may have different objectives, they tend to give similar rankings. It may appear that the universities that produce good research also tend to have an extensive web presence, perform well on teaching-related indicators, and attract many citations.

On the other hand, the webometrics may not necessarily provide robust indicators of knowledge flows or research impact. In contrast to citation analysis, the quality of webometric indicators is not high unless irrelevant content is filtered out, manually. Moreover, it may prove hard to interpret certain webometric indicators as they could reflect a range of phenomena ranging from spam to post publication material. Webometric analyses can support science policy decisions on individual fields. However, for the time being, it is difficult to tackle the issue of web heterogeneity in lower field levels (Thelwall & Harries, 2004; Wilkinson, Harries, Thelwall & Price, 2003). Moreover, Thelwall et al., (2010) held that webometrics would not have the same relevance for every field of study. It is very likely that fast moving or new research fields could not be adequately covered by webometric indicators due to publication time lags. Thelwall et al. (2010) argued that it could take up to two years to start a research and to have it published. This would therefore increase the relative value of webometrics as research groups can publish general information online about their research.

Online Academic Networks

Webometric measures ought to be combined with other approaches, whenever possible (Thelwall et al., 2010). Prolific authors publish both in high impact as well as in lower impact journals. They also use online academic networks to engage with other scholars who share their same research interests. Web 2.0 applications can be turned into productive social research tools (Thelwall & Kousha, 2015a, 2015b; Henning & Reichelt, 2008). Multi-purpose social software, such as wikis, blogs, and social networks are increasingly being utilised and evaluated by a number of researchers and academics. At the same time, many individuals are developing and publishing their user generated content, online. In a similar, it has never been so easy for academics to engage with their peers. Many scholars are creating their own personal web sites and blogs to enhance the visibility of their publications. This medium has improved the academic marketing in addition to the traditional bibliometrics (Camilleri, 2015). Besides, the influence of social media has also changed the academic publishing scenario. For instance, scholars share ideas, common interests, as well as promoting their scientific findings on social media (Davidson et al., 2014). Evidently, there has been a rapid proliferation in the uptake of Twitter as it is used for networking purposes (Thelwall et al., 2013). Moreover, other academic social networking sites including Academia.edu and ResearchGate as well as reference sharing sites including Mendeley, Bibsonomy, Zotero and CiteULike scholars are also publicising academic contributions.

During these last few years, several academics have created their own profiles on social academic networks. They list their publications and interact with each other. These online networks provide a new way for scholars to disseminate their work as they change the dynamics of informal scholarly communication. Interestingly, Thelwall and Kousha (2014) found that the rankings based on ResearchGate's statistics were moderately correlating with other rankings of academic institutions. These results indicated that the RG score from ResearchGate had correlated well with other university rankings (Thelwall & Kousha, 2015c). Yet, this may not be the case with other social networks. However, the social media sites are free to use, and they may provide a more democratic way for universities to reach out to the various audiences and interest groups.

Forkosh-Baruch and Hershkovitz's (2012) indicated that "the potential of SNS [social networking sites] as means of sharing academic knowledge in higher education institutes had not been actualised at the time of their study. Nevertheless, Vaughan and Romero-Frías (2014) noticed that the highly ranked universities have attracted more attention in Google Trends. These authors indicated that a great amount of searches for the US universities came from outside the US, whilst only a few searches for the Spanish universities came from outside of Spain. These findings may have also reflected the international positions of the two sets of universities. Thelwall and Kousha (2015c) took another approach to study universities' online presences as they investigated whether the usage of ResearchGate and its publications were somehow related to the "academic hierarchies" of different university rankings. Thelwall and Kousha (2015c) held that the RG score "broadly reflected the traditional academic capital".

It may appear that the universities' usage of social media and the attention they receive there is still relatively unclear. Academic visibility in social media is measured by various metrics (Bollen, Van De Sompel, Hagberg & Chute, 2009; Lin & Fenner, 2013). These new social media metrics, the so called altmetrics, could potentially give a more nuanced view of the attention towards research outputs (Priem, Piwowar & Hemminger, 2012). The underlying premise is that, for example, mentions in blogs, number of re-tweets or saves of articles in reference management systems, may be a valid measure of the use of scientific publications (Holmberg, 2015). As a matter of fact, recently there has been an increased recognition for additional measures of scholarly contributions that are published online (Delgado López-Cózar et al., 2014). For instance, Bornmann (2014) suggested that altmetrics could provide indicators for the societal impact of research or provide some insight on the research interests of other audiences outside academia (Haustein, Peters, Bar-Ilan, Priem, Shema & Terliesner, 2014). On the other hand, Thelwall, Haustein, Larivière and Sugimoto (2013) identified some of the weaknesses of altmetrics. They noted that the older articles were not experiencing high altmetric scores as opposed to the latest articles. In the past, there academic journals did not have social media plugins. They argued that very often scholars were searching for the most recent articles in academic literature.

CONCLUSION AND FINAL REMARKS

The content marketing of academic material involves a series of metrics that measure the researchers' or their institutions' 'impact', 'influence' or 'quality' of their contributions. There are various sources of bibliometric data, and each possess their own strengths and limitations. For example, the emerging field of altmetrics is concerned with impact measures rather than scholarly communication itself (Thelwall et al., 2013; Lin & Fenner, 2013; Priem et al., 2012). For the time being, there is no single bibliometric measure that is perfect. Multiple approaches to evaluation are highly recommended. Moreover, bibliometric and webometric approaches should not be the only measures upon which academic and scholarly performance ought to be evaluated. Sometimes, the use of bibliometric indicators could reduce the publications' impact to a quantitative, numerical scores. Many commentators have argued that when viewed in isolation these metrics may not necessarily be representative of a researcher's performance or capacity. Nonetheless, bibliometrics still have their high utility in academia. It is very likely that the metrics that were mentioned in this contribution will still to continue to be in use, in the foreseeable future. They represent a relatively simple and accurate data source. Undoubtedly, bibliometrics are an essential aspect for the measurement of academic clout and organisational performance. This chapter has identified a number of systematic ways of assessment; including citation analysis, impact factor, webometrics and social academic networks among others. The researchers' and their academic institutions' outputs are continuously being evaluated as independent reviewers frequently measure the quality and quantity of their academic publications. It transpires that citations are conspicuous in highly reputable journals or well-linked web sites that contain relevant scholarly content. This contribution has indicated that the most cited authors are recognised by their peers for their significant findings. In a similar vein, the highly reputable journals that have high impact factors are renowned for their contribution to knowledge, as their publications are sought by numerous researchers and scholars for their academic standing.

This chapter suggests that recognition by peers can help to boost the researchers' and their educational institutions' productivity levels. It has mentioned many issues that revolve on the research evaluation of academic content. A thorough literature review on scholarly impact and academic reputation indicated that the most cited contributions, journals and educational institutions are utilising a wide array of metrics to raise their standing. Researchers have become increasingly aware of the potential of personal branding. Nevertheless, their scholarly impact depends on a number of factors including the accessibility of publications, the peer review of their academic work as well as ongoing social networking and fruitful collaborative relationships with other scholars. Notwithstanding, the ongoing changes in academic behaviours and their use of content marketing on internet seem to have challenged the traditional metrics. Evidently, there is more to the measurement of impact than citation metrics. In this digital era, researchers are getting acquainted with WEB2.0. The ubiquity of online academic networks is also helping individual scholars' to establish performance and impact. At the same time, they are continuously sharing resources as they engage with colleagues. Social networks such as ResearchGate, Academia.edu, among others; have become new outlets for the publication of academic research.

In conclusion, this chapter has critically reviewed the extant literature on bibliometrics, webometrics, altimetrics and mentioned a few social networks for academic marketing. It clearly explained how content marketing of individual researchers' and educational institutions' contributions may lead to better academic standing and increased reputation.

FUTURE RESEARCH AVENUES

This chapter has presented relevant theoretical underpinnings on metrics that measure academic clout. It has also shed light on particular social networks that are increasingly disseminating research, online. The author believes that further research ought to re-examine the communication structures and processes of academic communications among researchers and their institutions. For instance, at the moment we are witnessing an evolution of webometrics. Researchers are increasingly using the web to enhance the visibility of their publications. Future empirical studies could reassess the importance of the web as a communication medium. Nowadays, social academic networks are already hosting a myriad of academic resources; ranging from journal articles, chapters, books, teaching resources, presentations and the like. Given the proliferation of accessible sources of research, there are limitless possibilities for scientists to raise their academic profile among their peers. Therefore, additional research could possibly shed light on the latest techniques and issues that are being used to promote scholarly research across different fields and subfields. Notwithstanding, this research area is relatively wide as it includes educational institutions, research entities such as intermediary groups like professional and trade associations as well as government agencies.

REFERENCES

Adler, N. J., & Harzing, A. W. (2009). When knowledge wins: Transcending the sense and nonsense of academic rankings. *Academy of Management Learning & Education*, *8*(1), 72–95. doi:10.5465/AMLE.2009.37012181

Aguillo, I. F., Granadino, B., Ortega, J. L., & Prieto, J. A. (2006). Scientific research activity and communication measured with cybermetrics indicators. *Journal of the American Society for Information Science and Technology*, *57*(10), 1296–1302. doi:10.1002/asi.20433

Alonso, S., Cabrerizo, F. J., Herrera-Viedma, E., & Herrera, F. (2009). h-Index: A review focused in its variants, computation and standardization for different scientific fields. *Journal of Informetrics*, *3*(4), 273–289. doi:10.1016/j.joi.2009.04.001

Arnold, D. N., & Fowler, K. K. (2011). Nefarious numbers. *Notices of the AMS*, *58*(3), 434–437.

Athar, A. (2014). *Sentiment analysis of scientific citations.* University of Cambridge, Computer Laboratory, Technical Report (UCAM-CL-TR-856).

Bador, P., & Lafouge, T. (2010). Analyse comparative du facteur d'impact et du h-index pour les journaux de pharmacologie. *Therapie*, *65*(2), 129–137. doi:10.2515/therapie/2009061 PMID:20478245

Bar-Ilan, J. (2004). A microscopic link analysis of academic institutions within a country-the case of Israel. *Scientometrics*, *59*(3), 391–403. doi:10.1023/B:SCIE.0000018540.33706.c1

Björneborn, L., & Ingwersen, P. (2004). Toward a basic framework for webometrics. *Journal of the American Society for Information Science and Technology*, *55*(14), 1216–1227. doi:10.1002/asi.20077

Bollen, J., Van de Sompel, H., Hagberg, A., & Chute, R. (2009). A principal component analysis of 39 scientific impact measures. *PLoS ONE*, *4*(6), e6022. doi:10.1371/journal.pone.0006022 PMID:19562078

Bonzi, S., & Snyder, H. W. (1991). Motivations for citation: A comparison of self citation and citation to others. *Scientometrics*, *21*(2), 245–254. doi:10.1007/BF02017571

Borgman, C. L. (2000). Digital libraries and the continuum of scholarly communication. *The Journal of Documentation*, *56*(4), 412–430. doi:10.1108/EUM0000000007121

Borgman, C. L., & Furner, J. (2002). *Scholarly communication and bibliometrics*. Academic Press.

Bornmann, L. (2014). Do altmetrics point to the broader impact of research? An overview of benefits and disadvantages of altmetrics. *Journal of Informetrics*, *8*(4), 895–903. doi:10.1016/j.joi.2014.09.005

Bornmann, L., & Daniel, H. D. (2005). Does the h-index for ranking of scientists really work? *Scientometrics*, *65*(3), 391–392. doi:10.1007/s11192-005-0281-4

Bornmann, L., & Daniel, H. D. (2007). What do we know about the h index? *Journal of the American Society for Information Science and Technology*, *58*(9), 1381–1385. doi:10.1002/asi.20609

Braun, T., Glänzel, W., & Schubert, A. (2006). A Hirsch-type index for journals. *Scientometrics*, *69*(1), 169–173. doi:10.1007/s11192-006-0147-4

Brooks, T. A. (1986). Evidence of complex citer motivations. *Journal of the American Society for Information Science*, *37*(1), 34–36. doi:10.1002/asi.4630370106

Brumback, R. A. (2009). Impact factor wars: Episode V-the empire strikes back. *Journal of Child Neurology*, *24*(3), 260–262. doi:10.1177/0883073808331366 PMID:19258283

Camilleri, M. A. (2015a). Using Big Data for Customer-Centric Marketing. In *Handbook of Research on Open Data Innovations in Business and Government*. IGI Global. Retrieved April 15, 2015, from http://www.researchgate.net/publication/275349565_Using_Big_Data_for_Customer-Centric_Marketing

Costas, R., & Bordons, M. (2007). The h-index: Advantages, limitations and its relation with other bibliometric indicators at the micro level. *Journal of Informetrics*, *1*(3), 193–203. doi:10.1016/j.joi.2007.02.001

Courtault, J. M., & Hayek, N. (2008). On the Robustness of the h-index: A mathematical approach. *Economic Bulletin*, *3*(78), 1–9.

Courtault, J. M., Hayek, N., Rimbaux, E., & Zhu, T. (2010). Research in economics and management in France: A bibliometric study using the h-index. *Journal of Socio-Economics*, *39*(2), 329–337. doi:10.1016/j.socec.2009.12.006

Davidson, P. M., Newton, P. J., Ferguson, C., Daly, J., Elliott, D., Homer, C., & Jackson, D. et al. (2014). Rating and Ranking the Role of Bibliometrics and Webometrics in Nursing and Midwifery. *The Scientific World Journal*. PMID:24550691

Delgado López-Cózar, E., Robinson-García, N., & Torres-Salinas, D. (2014). The Google Scholar Experiment: How to index false papers and manipulate bibliometric indicators. *Journal of the Association for Information Science and Technology*, *65*(3), 446–454. doi:10.1002/asi.23056

Eysenbach, G. (2011). Can tweets predict citations? Metrics of social impact based on Twitter and correlation with traditional metrics of scientific impact. *Journal of Medical Internet Research*, *13*(4), e123. doi:10.2196/jmir.2012 PMID:22173204

Favaloro, E. J. (2009). The journal impact factor: Don't expect its demise any time soon. *Clinical Chemistry and Laboratory Medicine*, *47*(11), 1319–1324. doi:10.1515/CCLM.2009.328 PMID:19769542

Franceschet, M. (2010). Ten good reasons to use the Eigenfactor™ metrics. *Information Processing & Management*, *46*(5), 555–558. doi:10.1016/j.ipm.2010.01.001

Garfield, E. (1972). *Citation analysis as a tool in journal evaluation*. American Association for the Advancement of Science. Retrieved April 15, 2015, from http://www.elshami.com/Terms/I/impact%20 factor-Garfield.pdf

Garfield, E. (1979). Is citation analysis a legitimate evaluation tool? *Scientometrics*, *1*(4), 359–375. doi:10.1007/BF02019306

Glänzel, W. (2006). On the h-index-A mathematical approach to a new measure of publication activity and citation impact. *Scientometrics*, *67*(2), 315–321. doi:10.1007/s11192-006-0102-4

Glänzel, W., & Moed, H. F. (2002). Journal impact measures in bibliometric research. *Scientometrics*, *53*(2), 171–193. doi:10.1023/A:1014848323806

Harzing, A. W., & Van Der Wal, R. (2009). A Google Scholar h-index for journals: An alternative metric to measure journal impact in economics and business. *Journal of the American Society for Information Science and Technology*, *60*(1), 41–46. doi:10.1002/asi.20953

Haustein, S., Peters, I., Bar-Ilan, J., Priem, J., Shema, H., & Terliesner, J. (2014). Coverage and adoption of altmetrics sources in the bibliometric community. *Scientometrics*, *101*(2), 1145–1163. doi:10.1007/s11192-013-1221-3

He, Q., Pei, J., Kifer, D., Mitra, P., & Giles, L. (2010). Context-aware citation recommendation. In *Proceedings of the 19th international conference on World wide web* (pp. 421-430). ACM. doi:10.1145/1772690.1772734

Henning, V., & Reichelt, J. (2008, December). Mendeley-a last. fm for research? In *eScience, 2008. eScience'08. IEEE Fourth International Conference on* (pp. 327-328). Retrieved January 22, 2015, from http://blog.mendeley.com/wp-content/uploads/2009/05/henning-reichelt-2008-mendeley-a-lastfm-for-research-ieee-e-science.pdf

Hirsch, J. E. (2005). An index to quantify an individual's scientific research output. *Proceedings of the National Academy of Sciences of the United States of America, 102*(46), 16569–16572. doi:10.1073/pnas.0507655102 PMID:16275915

Hodge, D. R., & Lacasse, J. R. (2010). Evaluating journal quality: Is the H-index a better measure than impact factors? *Research on Social Work Practice*.

Hoekman, J., Frenken, K., & Tijssen, R. J. (2010). Research collaboration at a distance: Changing spatial patterns of scientific collaboration within Europe. *Research Policy, 39*(5), 662–673. doi:10.1016/j.respol.2010.01.012

Holmberg, K. J. (2015). *Altmetrics for Information Professionals: Past, Present and Future*. Chandos Publishing.

Ingwersen, P. (1998). The calculation of web impact factors. *The Journal of Documentation, 54*(2), 236–243. doi:10.1108/EUM0000000007167

Kelly, C. D., & Jennions, M. D. (2006). The h index and career assessment by numbers. *Trends in Ecology & Evolution, 21*(4), 167–170. doi:10.1016/j.tree.2006.01.005 PMID:16701079

Kenekayoro, P., Buckley, K., & Thelwall, M. (2014). Hyperlinks as inter-university collaboration indicators. *Journal of Information Science, 40*(4), 514–522. doi:10.1177/0165551514534141

Knoth, P., & Herrmannova, D. (2014). Towards Semantometrics: A New Semantic Similarity Based Measure for Assessing a Research Publication's Contribution. *D-Lib Magazine, 20*(11), 8.

Knothe, G. (2006). Comparative citation analysis of duplicate or highly related publications. *Journal of the American Society for Information Science and Technology, 57*(13), 1830–1839. doi:10.1002/asi.20409

Kousha, K., & Thelwall, M. (2007). Google Scholar citations and Google Web/URL citations: A multi-discipline exploratory analysis. *Journal of the American Society for Information Science and Technology, 58*(7), 1055–1065. doi:10.1002/asi.20584

Kousha, K., & Thelwall, M. (2015). Web indicators for research evaluation. Part 3: books and non standard outputs. *El profesional de la información, 24*(6), 724-736.

Kretschmer, H., & Aguillo, I. (2004). Visibility of collaboration on the Web. *Scientometrics, 61*(3), 405–426. doi:10.1023/B:SCIE.0000045118.68430.fd

Labbé, C. (2010). Ike Antkare one of the great stars in the scientific firmament. *ISSI Newsletter, 6*(2), 48-52.

Lee, J., Kraus, K. L., & Couldwell, W. T. (2009). Use of the h index in neurosurgery: Clinical article. *Journal of Neurosurgery, 111*(2), 387–392. doi:10.3171/2008.10.JNS08978 PMID:19392590

Lin, J., & Fenner, M. (2013). Altmetrics in evolution: Defining and redefining the ontology of article-level metrics. *Information Standards Quarterly, 25*(2), 20. doi:10.3789/isqv25no2.2013.04

Mas-Bleda, A., Thelwall, M., Kousha, K., & Aguillo, , I. (2014). Successful researchers publicizing research online: An outlink analysis of European highly cited scientists' personal websites. *The Journal of Documentation, 70*(1), 148–172. doi:10.1108/JD-12-2012-0156

Mas-Bleda, A., Thelwall, M., Kousha, K., & Aguillo, I. F. (2014). Do highly cited researchers successfully use the social web? *Scientometrics*, *101*(1), 337–356. doi:10.1007/s11192-014-1345-0

Mingers, J. (2009). Measuring the research contribution of management academics using the Hirsch-index. *The Journal of the Operational Research Society*, *60*(9), 1143–1153. doi:10.1057/jors.2008.94

Moed, H. F. (2005). Statistical relationships between downloads and citations at the level of individual documents within a single journal. *Journal of the American Society for Information Science and Technology*, *56*(10), 1088–1097. doi:10.1002/asi.20200

Moed, H. F. (2006). *Citation analysis in research evaluation* (Vol. 9). Springer Science & Business Media.

Moed, H. F., Burger, W. J. M., Frankfort, J. G., & Van Raan, A. F. (1985). The use of bibliometric data for the measurement of university research performance. *Research Policy*, *14*(3), 131–149. doi:10.1016/0048-7333(85)90012-5

Moussa, S., & Touzani, M. (2010). Ranking marketing journals using the Google Scholar-based hg-index. *Journal of Informetrics*, *4*(1), 107–117. doi:10.1016/j.joi.2009.10.001

Nicolaisen, J. (2007). Citation analysis. *Annual Review of Information Science & Technology*, *41*(1), 609–641. doi:10.1002/aris.2007.1440410120

Olden, J. D. (2007). How do ecological journals stack-up? Ranking of scientific quality according to the h index. *Ecoscience*, *14*(3), 370–376. doi:10.2980/1195-6860(2007)14[370:HDEJSR]2.0.CO;2

Orduna-Malea, E., Ayllón, J. M., Martín-Martín, A., & López-Cózar, E. D. (2015). Methods for estimating the size of Google Scholar. *Scientometrics*, *104*(3), 931–949. doi:10.1007/s11192-015-1614-6

Orduña-Malea, E., & López-Cózar, E. D. (2014). Google Scholar Metrics evolution: An analysis according to languages. *Scientometrics*, *98*(3), 2353–2367. doi:10.1007/s11192-013-1164-8

Orduna-Malea, E., & Ontalba-Ruipérez, J. A. (2013). Selective linking from social platforms to university websites: A case study of the Spanish academic system. *Scientometrics*, *95*(2), 593–614. doi:10.1007/s11192-012-0851-1

Priem, J., & Hemminger, B. H. (2010). Scientometrics 2.0: New metrics of scholarly impact on the social Web. *First Monday*, *15*(7). doi:10.5210/fm.v15i7.2874

Priem, J., Piwowar, H. A., & Hemminger, B. M. (2012). *Altmetrics in the wild: Using social media to explore scholarly impact*. Cornell University Working Paper. Retrieved May 2, 2016 from http://arxiv.org/abs/1203.4745

Rimbaux, E. (2010). *Research in Economics and Management in France*. Retrieved April 10, 2015, from http://core.ac.uk/download/pdf/6508361.pdf

SciVerse Scopus. (2015). *About Scopus*. Retrieved April 15, 2015, from http://www.elsevier.com/solutions/scopus

Seglen, P. O. (1989). From bad to worse: Evaluation by Journal Impact. *Trends in Biochemical Sciences*, *14*(8), 326–327. doi:10.1016/0968-0004(89)90163-1 PMID:2799903

Seglen, P. O. (1992). The skewness of science. *Journal of the American Society for Information Science, 43*(9), 628–638. doi:10.1002/(SICI)1097-4571(199210)43:9<628::AID-ASI5>3.0.CO;2-0

Seglen, P. O. (1997a). Why the impact factor of journals should not be used for evaluating research. *BMJ (Clinical Research Ed.), 314*(7079), 497. doi:10.1136/bmj.314.7079.497 PMID:9056804

Seglen, P. O. (1997b). Citations and journal impact factors: Questionable indicators of research quality. *Allergy, 52*(11), 1050–1056. doi:10.1111/j.1398-9995.1997.tb00175.x PMID:9404555

Shuai, X., Pepe, A., & Bollen, J. (2012). *How the scientific community reacts to newly submitted preprints: Article downloads, twitter mentions, and citations*. Academic Press.

Smith, A. G. (1999). A tale of two web spaces: Comparing sites using web impact factors. *The Journal of Documentation, 55*(5), 577–592.

Tague-Sutcliffe, J. (1992). An introduction to informetrics. *Information Processing & Management, 28*(1), 1–3. doi:10.1016/0306-4573(92)90087-G

Thelwall, M. (2001). Results from a web impact factor crawler. *The Journal of Documentation, 57*(2), 177–191. doi:10.1108/EUM0000000007081

Thelwall, M. (2002). Conceptualizing documentation on the Web: An evaluation of different heuristic-based models for counting links between university Web sites. *Journal of the American Society for Information Science and Technology, 53*(12), 995–1005. doi:10.1002/asi.10135

Thelwall, M. (2004). *Link analysis: An information science approach*. Elsevier, UK: Academic Press.

Thelwall, M. (2008). Bibliometrics to webometrics. *Journal of Information Science*. Retrieved April 24, 2016 from http://jis.sagepub.com/content/early/2008/06/13/0165551507087238.short

Thelwall, M. (2009). *Introduction to webometrics: Quantitative web research for the social sciences*. Morgan & Claypool.

Thelwall, M., & Harries, G. (2004). Can personal web pages that link to universities yield information about the wider dissemination of research? *Journal of Information Science, 30*(3), 240–253. doi:10.1177/0165551504044669

Thelwall, M., Haustein, S., Larivière, V., & Sugimoto, C. R. (2013). *Do altmetrics work? Twitter and ten other social web services*. Academic Press.

Thelwall, M., Klitkou, A., Verbeek, A., Stuart, D., & Vincent, C. (2010). Policy-relevant Webometrics for individual scientific fields. *Journal of the American Society for Information Science and Technology, 61*(7), 1464–1475. doi:10.1002/asi.21345

Thelwall, M., & Kousha, K. (2015a). Web indicators for research evaluation. Part 1: Citations and links to academic articles from the Web. *El profesional de la información, 24*(5), 587-606.

Thelwall, M., & Kousha, K. (2015b). Web indicators for research evaluation. Part 2: Social media metrics. *El profesional de la información, 24*(5), 607-620.

Thelwall, M., & Kousha, K. (2015c). ResearchGate: Disseminating, communicating, and measuring Scholarship? *Journal of the Association for Information Science and Technology*, *66*(5), 876–889. doi:10.1002/asi.23236

Thelwall, M., & Price, L. (2003). Disciplinary differences in academic web presence–a statistical study of the UK. *Libri*, *53*(4), 242–253. doi:10.1515/LIBR.2003.242

Thelwall, M., & Tang, R. (2003). Disciplinary and linguistic considerations for academic Web linking: An exploratory hyperlink mediated study with Mainland China and Taiwan. *Scientometrics*, *58*(1), 155–181. doi:10.1023/A:1025487826207

Thelwall, M., Vaughan, L., & Björneborn, L. (2005). Webometrics. *Annual Review of Information Science & Technology*, *39*(1), 81–135. doi:10.1002/aris.1440390110

Thomson Reuters ISI Web of Science. (2010). *Web of Science*. Retrieved April 17, 2015, from http://thomsonreuters.com/en/products-services/scholarly-scientific-research/scholarly-search-and-discovery/web-of-science.html

Vanclay, J. K. (2008). Ranking forestry journals using the h-index. *Journal of Informetrics*, *2*(4), 326–334. doi:10.1016/j.joi.2008.07.002

Vaughan, L., & Romero-Frías, E. (2014). Web search volume as a predictor of academic fame: An exploration of Google trends. *Journal of the Association for Information Science and Technology*, *65*(4), 707–720. doi:10.1002/asi.23016

Vaughan, L., & Shaw, D. (2005). Web citation data for impact assessment: A comparison of four science disciplines. *Journal of the American Society for Information Science and Technology*, *56*(10), 1075–1087. doi:10.1002/asi.20199

Vaughan, L., & Shaw, D. (2007). A new look at evidence of scholarly citation in citation indexes and from web sources. *Scientometrics*, *74*(2), 317–330. doi:10.1007/s11192-008-0220-2

Wilkinson, D., Harries, G., Thelwall, M., & Price, L. (2003). Motivations for academic Web site interlinking: Evidence for the Web as a novel source of information on informal scholarly communication. *Journal of Information Science*, *29*(1), 49–56. doi:10.1177/016555150302900105

Wilson, C. S. (1999). Informetrics. *Annual Review of Information Science & Technology*, *34*, 107–247.

Chapter 9
From Universities to Private Companies:
A Measurable Route of Linkedin Users

Enrique Orduna-Malea
Universitat Politècnica de València, Spain

Cristina I. Font
Universitat Politècnica de València, Spain

José-Antonio Ontalba-Ruipérez
Universitat Politècnica de València, Spain

ABSTRACT

In publishing their education background together with the professional experience, users make LinkedIn a privileged web source for understanding "University-Industry" connections. Precisely, the main goal of this study is to test LinkedIn as a valid source for analyses oriented to the quantification of the university-industry interactions. To this end, the authors propose two different procedures (method A: direct through the URL mentions between LinkedIn profiles; and Method B: indirect through the information from LinkedIn University Pages), comparing them against the direct procedure based on URL mentions between official websites (Method C). To do this, the authors have selected the whole Spanish academic system. The results show that method A is unusable yet due to the low web connectivity between LinkedIn profiles, while method B provides reliable though too volatile data that complements method C, which reveal in turn relations of different nature.

INTRODUCTION

LinkedIn is a professional social network launched in May 2003 which has over 400 million users (October 2015) from more than 200 countries.[1] More than 80% of its users are aged between 36 and 45 years, although the average is decreasing over time (Archambault & Grudin, 2012).

DOI: 10.4018/978-1-5225-0917-2.ch009

This platform focuses on the professional information, encouraging users to build an abbreviated curriculum and establishing connections with other people or entities. People can apply and recommend to other members as well as to control the amount of information to share either to the general public or their professional connections.

LinkedIn also offers a section with the most influential people, indicating the number of followers they have. Finally, since job or any personal information may change, LinkedIn provides an address book updated automatically (Skeels & Grundin, 2009).

In the business environment, the degree of LinkedIn usage is associated with the size of the company and the industry in which this firm operates. In this sense, the companies belonging to the quaternary sector are noteworthy and, particularly, those related to Technology, Telecommunications and Financials (Bonsón & Bednárová, 2013; Kim, Kim & Nam, 2014).

Showing social and professional skills, LinkedIn users not only increase their professional value in the labour market but can also improve the image of the corresponding company (Van Dijck, 2013), improving thereby the interest and participation of employees in the social network, which reaches in the case of Microsoft up to 80% of the staff (Archambault & Grudin, 2012).

Since the information displayed on LinkedIn profile pages is characterized by a high level of integrity and accuracy (Brouer et al., 2015), where curricula have been demonstrated to be less misleading than traditional one (Guillory & Hancock, 2012), this social net is used as a recruiting tool because the company is capable to reach both active and passive job seekers, as users describe and share their educational background (Bonsón & Bednárová, 2013).

Despite being more popular outside than inside the academic environment, LinkedIn is also used by researchers to publish information related to their professional activities, including details about research publications and related activities (Mas-Bleda & Aguillo, 2015), although its value as a source for Altmetrics is limited to date (Thelwall *et al.*, 2013).

In any case, LinkedIn offers an enormous amount of valuable data about education and careers of its members, which can be explored to find success profiles, to identify specific professional groups, to detect education trends and itemize professional careers, or even to predict a user's degree of professional success (Dai et al., 2015). In short, LinkedIn allows access to information from users about universities and companies not available from another source (Unkelos-Shpigel, Sherman & Hadar, 2015). In publishing their education background together with the professional experience and skills, professional users make LinkedIn a privileged web source for understanding "University- Company" connections.

Otherwise, the vast amount of information (primarily professional) makes LinkedIn profiles (both personal and institutional) new information sources of quality, that can be visited as well as linked to by external websites, aspect that is favoured by the fact that LinkedIn has a great web authority in Google general search engine (Kelly & Delasalle, 2012). Since LinkedIn profiles are accessible online via URLs, this also converts these personal profiles into units suitable for link analysis (Thelwall, 2004; Orduna-Malea & Aguillo, 2015).

Thereupon, LinkedIn appears as a cybermetric feasible web source to analyse university-company interactions under two different approaches:

1. Directly through the URL mentions between the institutional LinkedIn profiles of universities and companies; and
2. Indirectly through the information contained in the personal LinkedIn profiles of particular users.

These two methods constitute an alternative to the one classic utilised in link analysis, through the URL mentions between university and company official websites. This classic method has been proved to assist in discovering patterns that help gain deeper insights into how institutions engage on the Web, providing evidence about their offline relationships (Minguillo & Thelwall, 2012). Nonetheless, the use of link data on quantifying the interactions between universities and private companies has been treated scarcely, and generally focuses on specific sectors (Stuart & Thelwall 2006) or countries (Garcia-Santiago & Moya-Anegón, 2009; Orduna-Malea & Aytac, 2015). This is probably due to the particular web policies in creating, sharing, disseminating and consuming online information by companies (Vaughan, 2004; Vaughan & Wu 2004), related to legal aspects, market competition, web culture, economic costs, etc.

The main goal of this study is precisely to test the validity of LinkedIn as a source for cybermetric analyses oriented to the quantification of the university-industry interactions. To this end, we propose two different procedures (direct and indirect) based on LinkedIn data, and compare them against a classical procedure widely used in cybermetrics (direct).

The following specific tasks are set out:

1. Determine the feasibility of LinkedIn institutional profiles for cybermetric analyses through the characterization of some web properties (URL syntax, web visibility, number of followers).
2. Measure University-Industry interactions through three different methods (direct based on URL mentions; indirect based on user profiles).
3. Compare the previous methods in revealing interactions between the academia and the market

BACKGROUND

Universities have traditionally had a strong presence in LinkedIn since the platform allows the creation of institutional profiles within the "companies" option. Thus, for example, the Polytechnic University of Valencia (UPV) has an official profile on LinkedIn at the following URL:

<https://www.linkedIn.com/company/upv>

As "companies", universities may include in their institutional profiles (beyond the publication of posts, comments, shares and likes) certain corporate information such as the address, the size of the company (approximate number of employees), industry and type of business ("higher education", and "educational institution" respectively in the case of a university). Likewise, registered users can follow these LinkedIn institutional profiles (for example, as of February 4th 2015, the UPV has 13,494 followers).

These institutional profiles, created as "companies" have certain limitations when connecting universities to students. On the one hand, the ties that bind these two entities can only be made explicit through the "follow" option. Nonetheless, a user can follow a university even if s/he does not have studied or worked in this institution. On the other hand, corporate information depends solely on the profile administrators and their institutional communication policies.

Therefore, limiting institutional information about schools and colleges on these LinkedIn institutional profiles means not taking full advantage of the enormous potential of the LinkedIn database. We should note that this platform is intended, above all, to connect professional users. Consequently, individual users are struggling (hopefully) to complete their profile and résumé in the best way to be as visible and attractive as possible not only to other users (with whom they intend to establish professional relationships or networks) but also with potential companies that may wish to initiate situations vacant. Thereupon users fill their personal profiles with information about their academic training (alma mater) and work experience (companies in which they work or have worked).

If appropriate mined, this huge amount of explicit, comprehensive, structured, and quality data, would yield information about the relationships between universities and companies much richer than merely through the institutional profiles, since students and professionals act as betweenness nodes (users first are students taking courses at the university, and later they become professionals working in a particular company, within a specific sector, and performing specific tasks).

LinkedIn University Pages

With such previous ideas in mind, LinkedIn announced (on August 19th 2013) the creation of a new page feature - labelled as University Pages - supplementary to the already existing university institutional profiles.

These new pages (online available from the menu option Interest/Education), compiled for the first time information on the place where alumni had worked, specific tasks performed or acquired skills, both for current and past companies. Thus, a direct tie between academia and the market was opened out. For example, the URL for the Polytechnic University of Valencia is the following:

<https://www.LinkedIn.com/edu/school?id=12242>

The LinkedIn University Pages provide, among others, the following special features:

- **Search Functionalities:** Exploring universities, schools, colleges, students, and professionals worldwide.
- **Network Information:** How many connections the actual user has with alumni and professionals from the university page consulted, and what the degree of this connection is (first degree, second degree, etc.).
- **News and Conversation Tools:** Receiving regular updates about campus news and activities from any desired school; asking questions and engaging with both the campus community and alumni of schools.
- **Recommendation Panel:** A compilation of texts containing recommendations given by LinkedIn users about engaging with a university, School or academic program.
- **Notable Alumni:** Access to profiles corresponding to important people who studied in one university (for example, in the case of Harvard University, we may find access to the profiles of Bill Gates or Barack Obama, among many others). This feature is not available for all universities.
- **Information about Students & Alumni:** Access to all students and alumni data of each university.

This option is undoubtedly the most important one for the purposes of this manuscript. The user can search for specific people directly using a search box or through some filters: engagement level (people who attended or graduated) and time period.

Figure 1. Example of LinkedIn university page – Stanford University (LinkedIn, 2016)
Source: https://www.LinkedIn.com/edu/school?id=17926

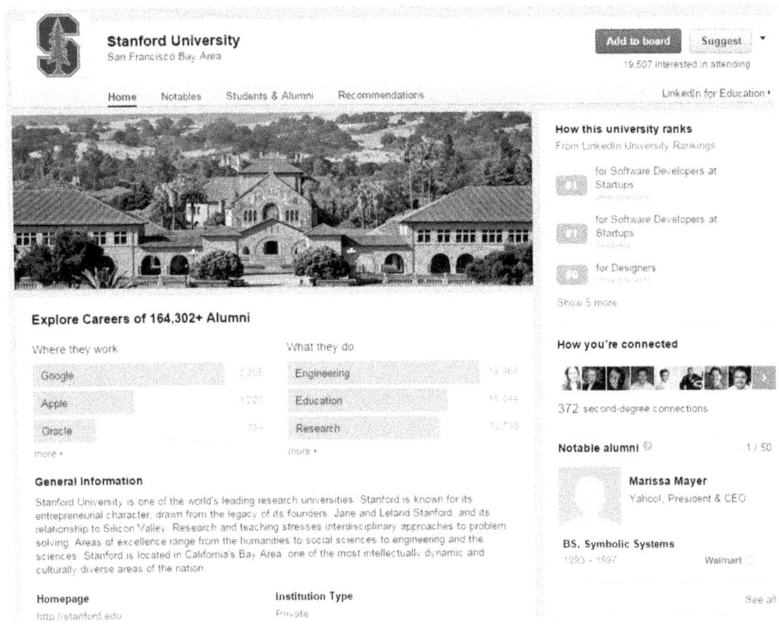

Figure 2. Example of LinkedIn company page – Stanford University (LinkedIn, 2016)
Source: https://www.LinkedIn.com/company/stanford-university

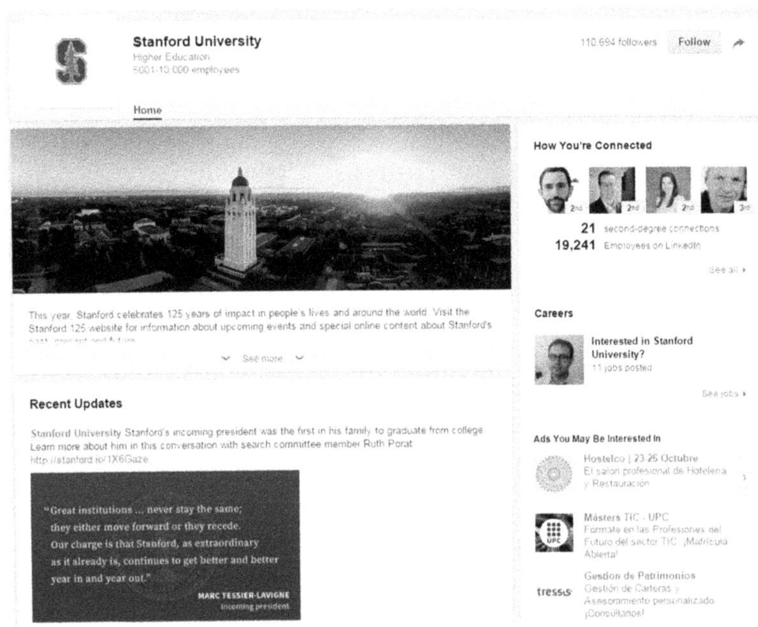

This information about student characteristics is distributed along several filters: where they live, where they work, what they do, what they studied, what they're skilled at, and how the user is connected with these people.

Using these filters, the system provides estimate numbers of LinkedIn profiles that meet the requirements. In the figure 1 we can see an example for Stanford University, where more than 160,000 alumni profiles are analysed. We can find interesting insights, such as the place where students end up working are Google (2,295) and Apple (1,205) in the first and second position respectively, or even the importance of this university for teachers (16,644 alumni work in Education sector) and research (15,736 alumni). Figure 2 shows the information provided by the institutional page as company.

Youniversity

The success and wide acceptance of LinkedIn University Pages (more than 200 universities had adopted these pages at the time of launch) triggered further development for the improvement and expansion of the services offered. Thus, the October 1st 2014, the "Universities Ranking based on career outcomes" was introduced. LinkedIn started thereby moving from aggregated information web pages about universities to a new information system based upon the philosophy of university report cards (Orduna-Malea, 2011) to provide users with a complete and interactive environment aimed at searching universities, colleges, schools, people and academic and professional careers.

The league tables (and University Pages as well) were included as part of a new platform, called Youniversity, consisting of a series of tools (University Rankings, University Finder, Field of Study Explorer, and Create a Board Decision) with the purpose of being a useful information set for both universities and especially for current and future students (LinkedIn, 2014).

University Rankings

To generate interactive rankings, the user first must select the level (Graduate, Undergraduate) and location (currently, only available to United States, Canada, and United Kingdom).

The different university lists appear classified under specialties or fields of knowledge. Currently the following fields are available for undergraduate rankings: Accounting professionals, Designers, Finance Professionals, Investment bankers, Marketers, Media Professionals, Software developers, Software developers at start-ups. In the case of graduate rankings, Management consultants and Sales Professionals are offered as well. Finally, pressing on each field category the user gets access to the corresponding universities' list.

The data offered by Youniversity (and especially the university rankings) have not been prepared from the information of all LinkedIn registered users. For this reason, despite briefly, the procedure followed by LinkedIn to make this academic information service disposable is necessary.

At the time of launch (October 1st, 2014), LinkedIn analysed profiles of about 300 million users; the main ordering criterion was as follows: "the likelihood of relevant graduates from that school landing desirable jobs in that career". To do this, the following steps were performed:

1. Identification of Desirable Jobs
    a. A "desirable job" is considered a job in a "desirable company" for the corresponding job. Therefore, in this first step, the identification of the companies in which professionals of a

certain field/concentration work was undertaken. The capacity of companies for attracting and retaining professionals of certain field determinates a company as a desirable company for this specific field. In order to provide information as current as possible, only transitions and tenures in the last 5 years are considered. Thus, companies with very few relevant employees in one profession over the last five years are excluded for that profession.

2. Relevant Graduates
   a. The second step consists of identifying "relevant graduates" for a desirable job. This is defined by LinkedIn as a member who both attended a graduate program on a specific field and end up working in that career. LinkedIn only considers graduates who obtained their degrees within the last eight years. Once the users are identified, the alma mater is collected from their profiles.
3. Scoring Universities
   a. Finally, for each School, they found the percentage of these alumni (desirable graduates) who found a job at these top companies (desirable companies).

University Finder

Through this option, LinkedIn provides access to a universities' search engine. The academic institutions can be located either directly through a vertical search (specific search box which only attempts to pre-defined topics)[2] or by applying a series of category filters, namely: a) what do you want to study; b) where do you want to work; c) where do you want to live; y d) what do you want to do. Each of the filters offers up to 12 selectable items.

This feature allows the user to select among the most appropriate institutions according to a number of specific needs or tastes, which are essential in university selection criteria. Moreover, the system is able to identify important entities within a university (such as Colleges or Schools) which have created an independent University Page. This is the case for example of the Harvard Law School, Harvard Medical School or Harvard University Graduate School of Education. LinkedIn not only retrieves these internal university entities but also make explicit the hierarchy in the aggregated information offered through a multilevel approach (Orduna-Malea, 2013a).

For example, if a user wants to "study Computer Science", "work on Google" and live in "United States", the system will offer a number of universities, which according to the data collected on thousands of users, are the best suited for get that itinerary.

In the figure 3 we can empirically observe this example. In this case, as expected, the first result corresponds to Stanford University. The system even offers the option to sharpen the location selection breaking down the United States into different geographical areas.

Once a selection from any of the four filters is performed, LinkedIn shows the number of profiles in which the value selected is true for each of the categories filtered. For example, if "Engineering" is selected in the filter "what do you want to do", the system automatically displays the distribution of users per company (in the previous example, IBM occupies the first place with 77,184 users) or that the United States (341,600 users), and India (2,283,520) are the two main countries where LinkedIn users are working as engineers.

Figure 3. Example of LinkedIn university finder (LinkedIn, 2016)
Source: https://www.LinkedIn.com/edu/university-finder

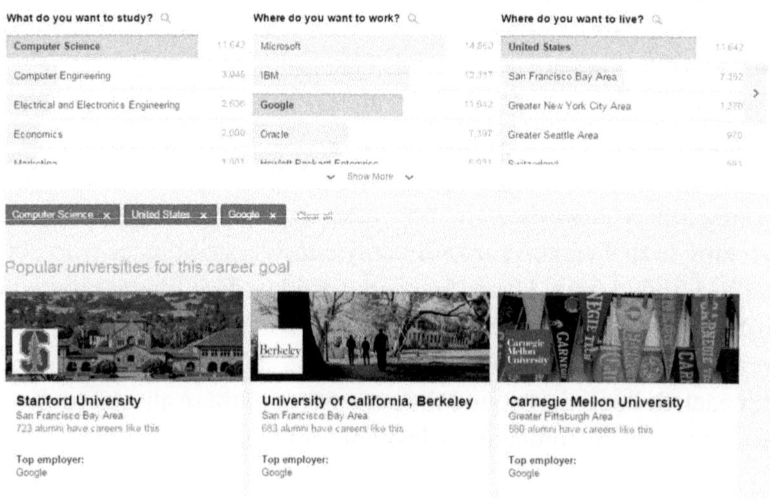

Field of Study Explorer

In this case, the information system allows browsing by selecting from a set of predefined fields of study, available through a drop-down searcher (e.g. acoustics, or Library and Information Science). Once the field is selected, the interface offers the following filters: where they work; what they do; where they went to university; where they live; how you are connected). In this case, each category filter offers up to 25 selectable options.

Create a Decision Board

Finally, the tool allows each user to create a custom panel. From a single interface the user may activate altogether the university finder, university rankings, and field of study explorer, and to design a dashboard with universities that meet all their requirements. This board may further be shared with other users if desired.

In short, with the Youniversity platform service, users can enter their academic interests, desired companies to work with, place or even specific skills, and the system provides back personalized suggestions and social functionalities to create professional networks between alumni and universities. Moreover, companies dispose of valuable data to find good candidates that fit their position requirements.

METHOD

The two methods based on LinkedIn data (Method A and Method B) as well as the classic cybermetric procedure (Method C) are defined below:

Method A: Link analysis, direct through the URL mentions between the institutional LinkedIn profiles of universities and companies.

Method B: Text analysis, indirect through the information contained in the personal LinkedIn profiles of particular users.

Method C: Link analysis, direct through the URL mentions between university and company official websites.

The comparison of methods was performed through network metrics (Barabasi, 2014). To proceed with, the methodology has been divided into three consecutive steps: data gathering (which consisted on selecting and obtaining the samples of universities and companies); data analysis; and finally data visualization (construction of networks).

Data Gathering (I): Universities

The first step consisted on selecting and obtaining the samples of universities. We started thus collecting the whole set of Spanish Universities (77 as of February 2016; 50 public and 27 private). The complete list of institutions was obtained from the official source provided by the Conferencia de Rectores de las Universidades Españolas (CRUE – Conference of Rectors of Spanish Universities).[3]

The Spanish academic system is quite diverse as well as cultural, linguistic and geographically distributed, and it has been also widely analysed under webometric methods (Orduna-Malea, Serrano-Cobos & Lloret-Romero, 2009; Orduna-Malea, 2013b). Therefore, it constitutes an excellent sample test for this research.

For each university, the corresponding official URL was gathered (Site URL). Given that universities tend to possess more than one valid and official URL (using URL aliases and/or redirections), only the main and un-redirected URL was considered.

After this, each university's institutional profile was located in the LinkedIn platform through the direct search box feature. Only the official profile as "company" was taken in this step, avoiding consequently LinkedIn profiles for specific Schools, Faculties, Masters, and so on. Additionally, the number of followers for each institutional profile was gathered as well.

Since LinkedIn provides two different valid URLs for each profile (ID URL and Friendly URL), both versions were taken into account. Next we show an example of both LinkedIn URLs for one company (University of Barcelona):

ID URL: < LinkedIn.com/company/8082>
Friendly URL: < LinkedIn.com/company/ university-of-barcelona>

Data Gathering (II): Private Companies

The second step consisted on the private companies gathering. Since method B determines unintentionally the sample of companies (firms connected with the Spanish universities through LinkedIn users' data), this procedure was utilised as a general basis of obtaining the sample of companies to be used in the remaining methods.

Each of the 77 universities was searched on the "University Finder" feature within the Youniversity platform. The information about companies was located in the "Students & Alumni" section, specifically within the information filter labelled "Where they work", available for each LinkedIn University Page, as explained previously.

The first top 10 companies listed for each university were retrieved as well as the number of LinkedIn users (alumni and professionals) affiliated with or related to each institution. Only private companies were considered, eliminating thus public institutions (such as universities, and regional or local government). This is due to our primarily intention of finding the connection among academia and private labour market, and not public servants within public institutions.

Once all companies were collected, their corresponding number of followers and the different URLs (Site URL, ID URL, and Friendly URL) were also gathered. In the case of international companies, the general web domain was selected unless a Spanish delegation existed under a valid sub-domain form (for example, <telefonica.es> and <vodafone.es> were considered instead of <telefonica.com> or <vodafone.com>).

Data Analysis

This step was exclusively performed to Method A, in order to test the feasibility of URL mentions on LinkedIn institutional profiles.

Despite the number of external inlinks is the key metric for measuring web visibility of websites, this metric cannot be calculated directly in Google (search engine with wider coverage). For this reason, the number of URL mentions targeted to LinkedIn institutional profiles (both for universities and companies) were retrieved from the Spanish version of Google <google.es> by means of hit count estimates, through the launching of the following search query:

<"LinkedIn.com/company/xxxx" -site:LinkedIn.com>

This procedure - despite its well-known limitations - has been tested as a valid alternative procedure to obtain the number of external inlinks to Spanish universities using Google's coverage (Ortega, Orduna-Malea & Aguillo, 2014). In this case, it was applied to both the ID and Friendly LinkedIn URLs, with the purpose of retrieving the whole impact for all existing URL versions.

All data were exported to a spreadsheet in order to be statistically analysed. Spearman correlations ($\alpha < 0.01$) were always used due to the skewed distribution of web data (Barabasi & Albert, 1999).

Data Visualization

Finally, three undirected networks were constructed, reflecting all three alternative methods of measuring the relationships among universities and companies: LinkedIn users (graduate users), LinkedIn profiles (URL mentions), and official websites (URL mentions). All networks were composed by the same 177 nodes: 75 universities and 82 companies.

Two universities (IE University and European University of Madrid) were left out of the networks because neither of them was found at Youniversity platform despite having an institutional profile. In the case of companies, only those providing more than 100 total users to the complete Spanish university system were considered, in order to manage a sample as strong as possible.

1. *URL mention's network for official websites*

The edges for this network represent the URL mentions between a pair University- Company. Since URL mentions are directed in nature, we queried in both directions (mentions from the university website towards the company website; and mentions from company website to the university website), finally adding the two corresponding hit count estimates:

<“university1.tld” site:company1.tld>
<“company1.tld” site:university1.tld>

The total number of queries launched was 12,300 (75*82*2), which ended in 6,150 possible edges. This procedure was automated by a web scraper[4].

2. *URL mention’s network for LinkedIn profiles*

The construction of this network was identical to the previous one, but applied to the institutional LinkedIn profiles. In this case, both LinkedIn URL versions were considered, though only pairs of the same URL type were queried.

<“LinkedIn.com/company/id1-university”
site:LinkedIn.com/company/id2-firm>
<“LinkedIn.com/company/friend1-university”
site:LinkedIn.com/company/friend2-firm>
<“LinkedIn.com/company/id2-university”
site:LinkedIn.com/company/id1-firm>
<“LinkedIn.com/company/friend2-university”
site:LinkedIn.com/company/friend1-firm>

In this case, the total number of queries launched was 24,600 (75*82*2*2).

3. *LinkedIn users’ network:*

The edges for this network represent those LinkedIn users that have studied in one specific university and currently work (or have worked) in one specific company.

For each network, a .net file was developed and exported to Gephi 0.8.2, [5] where both networks were constructed and energised using the Fruchterman & Reingold algorithm (Fruchterman and Reingold, 1991). Some network statistics (such as Diameter, Density, Average cluster coefficient, and Average Degree) were calculated directly from Gephi stats panel. For each node, Degree and Betweenness centrality scores were calculated as well. All web and network metrics used in this study are summarized in Table 1.

All the procedure (data gathering, web data analysis and visualization) was performed along the first week of February 2016. Data about number of users related to companies that have studied in specific universities were gathered on 4th February 2016.

Table 1. Summarize and scope of web and network metrics

Web Impact	SCOPE
URL mention	Times a URL is mentioned from a specific root domain
Node-Level	**SCOPE**
Degree	Number of edges that are adjacent to one node
Betweenness	The average distance from a given node to all other nodes in the network
Clustering coefficient	The degree to which nodes of the neighbourhood of a node "a" are connected to each other
Network-Level	**SCOPE**
Avg. Degree	The average degree over all of the nodes in the network
Avg. clustering coefficient	The average clustering coefficient (degree to which nodes of the neighbourhood of a node "a" are connected to each other) over all of the nodes in the network
Average Path Length	The average graph-distance between all pairs of nodes
Diameter	The maximal distance between all pairs of nodes
Density	How close the network is to complete (density equal to 1)

RESULTS

The two first sections are devoted to the descriptive analysis of institutional LinkedIn profiles whereas the following sections to the comparison of methods through the networks elaborated.

Spanish Universities in LinkedIn

A total of 74 out of 77 Spanish universities have an institutional profile (as company) in LinkedIn (Menendez Pelayo International University, Loyola University Andalusia, and Polytechnic University of Cartagena are the only three institutions without a profile created). The Spanish university with the higher number of followers at the time of writing this manuscript (February 2016) is the Complutense University of Madrid (26,003 followers), followed by the University of Barcelona (20,729).

These are two public, old, general and big universities located at the two main cities in the country (Madrid and Barcelona). The first private university of this list appears at 4th position, corresponding to the Open University of Catalonia (16,022), followed by the European University of Madrid (15,323).

We can observe the top 20 Spanish universities according to the number of followers received in the Table 2. The corresponding total URL mentions (both to ID and friendly URL versions) are too offered. The complete ranking of universities is available in the supplementary material (Appendix I).[6]

As regards the number of URL mentions achieved by each institutional profile, the values obtained are discrete. The National University of Distance Education and the International University of La Rioja are the institutions which receive more URL mentions (40 each one). The Spearman correlation ($\alpha < 0.05$) between the number of followers and the number of URL mentions is statistically significant though low ($r = 0.4$).

Additionally, we should highlight the significant difference in web performance between each of the two existing URL variants. While the 75 friendly URLs provide 517 mentions to the complete Spanish university set, the ID URLs only provide 59, of which 29 just correspond to one university (Camilo Jose Cela University).

Table 2. Top 20 Spanish universities according to the number of followers in LinkedIn

R	UNIVERSITY	TYPE	TOTAL URL MENTION	FOLLOWERS
1	Complutense University of Madrid	Public	4	**26,003**
2	University of Barcelona	Public	10	**20,729**
3	National Univ. of Distance Education	Public	40	**17,960**
4	Open University of Catalonia	Private	15	**16,022**
5	European University of Madrid	Private	3	**15,323**
6	Autonomous University of Barcelona	Public	11	**14,750**
7	Autonomous University of Madrid	Public	5	**14,723**
8	Polytechnic University of Catalonia	Public	5	**14,454**
9	Carlos III University of Madrid	Public	13	**14,169**
10	Polytechnic University of Madrid	Public	4	**14,156**
11	Polytechnic University of Valencia	Public	14	**13,523**
12	University of Seville	Public	4	**13,498**
13	Pompeu Fabra University	Public	18	**12,836**
14	Deusto University	Private	9	**11,401**
15	University of Granada	Public	4	**11,298**
16	International University of La Rioja	Private	40	**10,501**
17	University of Navarre	Private	6	**10,276**
18	University of Valencia	Public	10	**10,066**
19	King Juan Carlos University	Public	4	**8,965**
20	Distance University of Madrid	Private	18	**8,823**

Private Companies Connected with Spanish Universities in LinkedIn

The number of companies which appear at the top 10 ranking where alumni work or have worked and have studied in at least one Spanish university raises to 273. In the Table 3 we show the top 20 companies according to the total number of alumni from Spanish universities who work or have worked in the corresponding firms as well as the number of universities in which each company appears. The number of total URL mentions and followers are provided as well.

The complete list for the 273 companies is available in the supplementary material (Appendix II). [6]

The company that have received more professionals from Spanish universities corresponds to Indra (6,587), which appears at the top 10 ranking for 49 (out of 75) universities. Second and third places are occupied both by banks (BBVA and Santander, with 5,011 and 4,682 users respectively).

As regards the number of followers, IBM occupies the first place (2,727,547), followed at some distance by HP Enterprise (2,217,915). In this case, the Spearman correlation ($\alpha < 0.05$) between the number of followers and the number of URL mentions is significant and high ($r = 0.80$). The correlation between the number of followers and users is smaller though too significant ($r = 0.43$).

The number of URL mentions is also higher in the company environment respect to the university environment. All the 273 companies receive a total of 147,481 URL mentions (142,231 from the friendly

Table 3. Top 20 companies receiving alumni from Spanish universities in LinkedIn

R	COMPANY	TOTAL URL MENTION	FOLLOWERS	ALUMNI	N
1	Indra	310	209,728	**6,587**	49
2	Telefónica	2	6,958	**5,011**	37
3	BBVA	16	13,742	**4,682**	43
4	Banco Santander	436	313,931	**4,669**	35
5	Accenture	36	24,980	**3,525**	31
6	Everis	153	160,609	**2,864**	36
7	Ernst & Young	4,851	1,289,195	**2,444**	19
8	Airbus Group	224	216,210	**2,240**	11
9	Deloitte	65	61,689	**2,213**	21
10	CaixaBank	22	29,586	**1,740**	12
11	Abengoa	846	158,249	**1,356**	13
12	Banco Sabadell	119	41,092	**1,301**	8
13	Repsol	431	168,944	**1,266**	11
14	PwC	3,246	1,052,035	**1,100**	11
15	IBM	58,496	2,727,547	**1,030**	6
16	CatalunyaCaixa	6	4,043	**995**	6
17	Ericsson	1,414	814,321	**970**	5
18	HP Enterprise	565	2,217,915	**944**	7
19	El Corte Inglés	105	65,602	**797**	6
20	Tecnicas Reunidas	83	57,338	**775**	8

version of the URL whereas only 517 from ID URLs). In this case, IBM is the company which receives the great amount of mentions (58,496), followed by the World Bank (9,119).

In Table 4 we can check the number of alumni who work (or have worked) with the corresponding company broken down by university. The top 20 connections between companies and Spanish universities through LinkedIn users are shown. The complete data is available in the supplementary material (Appendix III). [3]

As we can easily observe in the Table 4, the Polytechnic University of Madrid and the Complutense University of Madrid rule the stronger connections (14 out of the 20 stronger connections come from either one of these two institutions), while the stronger connection for a private university corresponds with the Comillas Pontifical University, which provides 449 professionals to Ernst & Young company.

URL Mention' Network for LinkedIn Profiles

The results obtained in the first section of this chapter show an almost inexistent data for total external URL mentions to the universities' institutional profiles (mentions from outside LinkedIn web domain).

As regards the URL mentions from inside LinkedIn domain, specifically between the universities and companies selected, the results are unusable. The two universities receiving more external URL

Table 4. Top 20 connections between companies and Spanish universities in LinkedIn through users

R	COMPANY	ALUMNI	UNIVERSITY	TYPE
1	Indra	1,096	Polytechnic University of Madrid	Public
2	Telefónica	1,081	Polytechnic University of Madrid	Public
3	Airbus Group	893	Polytechnic University of Madrid	Public
4	Banco Santander	859	Complutense University of Madrid	Public
5	BBVA	741	Complutense University of Madrid	Public
6	Indra	723	Complutense University of Madrid	Public
7	Abengoa	712	University of Seville	Public
8	Telefónica	709	Complutense University of Madrid	Public
9	CaixaBank	590	University of Barcelona	Public
10	CaixaBank	555	Pompeu Fabra University	Public
11	Ericsson	550	Polytechnic University of Madrid	Public
12	BBVA	493	Autonomous University of Madrid	Public
13	Ernst & Young	449	Comillas Pontifical University	Private
14	El Corte Inglés	446	Complutense University of Madrid	Public
15	Tecnicas Reunidas	446	Polytechnic University of Madrid	Public
16	Repsol	442	Complutense University of Madrid	Public
17	Bankia	437	Complutense University of Madrid	Public
18	Banco Santander	433	Autonomous University of Madrid	Public
19	INECO	430	Polytechnic University of Madrid	Public
20	Accenture	429	Polytechnic University of Madrid	Public

mentions (National University of Distance Education and International University of La Rioja) receive any URL mention from <linkedin.com>. Likewise, IBM (the most mentioned LinkedIn company profile of the sample) only receives about 1,700 mentions from LinkedIn web space, neither of them from the institutional profile of a Spanish university.

For this reason, this network has not been finally constructed and this method A has been automatically discarded.

LinkedIn Users' Network

The network between Spanish universities and private companies through the number of alumni as connector is offered in the figure 4. As we can observe, the centre of the network is composed by some corporations (mainly BBVA, Indra, Telefónica, Everis, Santander).

This central area is surrounded by most universities. The strong connection of the Polytechnic University of Madrid and the Complutense University of Madrid with these companies is represented as well, which reflects and confirms the data previously shown in the Table 4.

All data corresponding with Degree and Betweenness network scores for all entities are enclosed in the supplementary material (Appendix IV). [3]

Figure 4. Spanish universities and companies connection through LinkedIn users
(n= 157; Fruchterman-Reingold distribution)

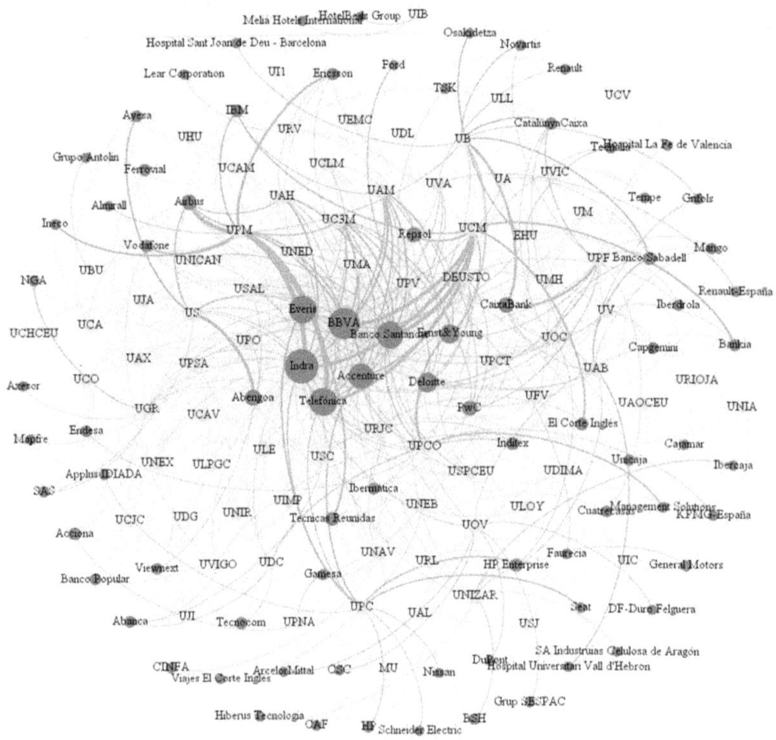

Curiously enough, the universities which achieve the higher betweenness centrality in the network are the University of Oviedo (922,246) and the University of Zaragoza (700,821), both small universities with also relative small number of LinkedIn followers (8,270 and 3,500 respectively), if compared with the big and old traditional Spanish universities. As regards the private companies, Indra achieves the highest Degree (49; this metric obviously corresponds with "N" score in Table 3) and betweenness centrality (2,371.828), followed by the BBVA (Degree= 43; Betweenness= 1,975.154).

Finally, the periphery of the net is composed by low connected entities, among which we may highlight some important companies (such as Mapfre, Acciona, Grifols or Bankia, which all belong to the IBEX35 stock exchange) and some small public (University of Balearic Islands, International University of Andalusia) and private (University of Isabel I, Cardenal Herrera University, International University of Catalonia or Catholic University of Valencia) universities.

URL Mention' Network for Official Websites

The network between Spanish universities and private companies through the number of URL mentions as edge is offered in the figure 5. We can easily detect some important similarities and differences respect to the LinkedIn users' network.

On the one hand, we can find some the same important private companies in the centre of the network (Accenture, Santander, Telefónica) whereas some companies have lost network importance (such as Indra or Everis), and some other have conversely increase its centrality (e.g., Repsol, IBM, Vodafone

Figure 5. Spanish universities and companies network through URL mentions (n= 157; Fruchterman-Reingold distribution)

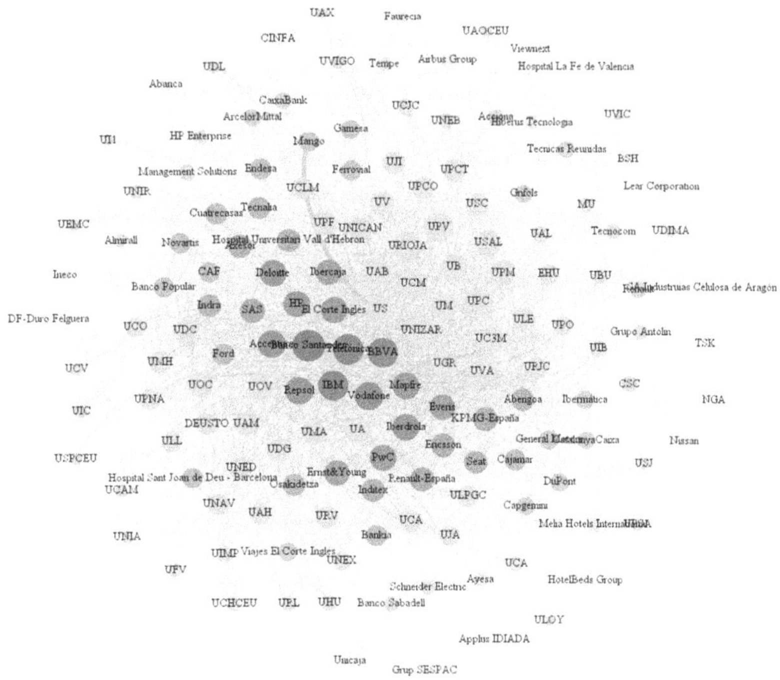

or especially Mapfre). On the other hand, we find that universities are not surrounding the main central private companies.

The URL mentions provide a greater number of companies placed on central positions with elevated InDegree scores. Universities (with higher InDegree scores as well) are placed in two separate clusters divided by the central private companies.

Santander bank is the node with the higher Degree (71) and betweenness centrality (510.112) in this network. Likewise, the Polytechnic University of Valencia becomes the academic institution both with the higher Degree (66) and betweenness centrality (380.199). In the Table 5 we show the top 10 companies and universities according to the Degree score. All data corresponding with network scores for all entities are enclosed in the supplementary material (Appendix V). [3]

We can finally highlight the most connected pair of University-Company through URL mentions, which corresponds to The University of Seville and Mango (133,008 mentions), followed by the pair University of Murcia – Hewlett-Packard (10,023) and the University of Murcia – IBM (4,049).

The main network properties of the two graphs (LinkedIn users and URL mentions) might be compared in the Table 6. As may be expected, URL mention's graph is a denser network, with a lower Diameter and Average Path Lengh. The difference on Average Degree should be normalized because the number of users and URL mentions are on different orders of magnitude. Nonetheless, this metric illustrates the difference nature of each network.

Table 5. Top 10 companies and related Spanish universities with better network degree score (URL mention's network)

COMPANIES			UNIVERSITIES		
NAME	DEGREE	BETWEENNESS	NAME	DEGREE	BETWEENNESS
<santander.com>	**71**	510.112	<upv.es>	**66**	380.199
<telefonica.es>	**68**	382.135	<unizar.es>	**65**	337.982
<ibm.com>	**66**	336.435	<upc.edu>	**64**	314.968
<bbva.es>	**65**	406.807	<uab.cat>	**63**	335.717
<vodafone.es>	**60**	206.982	<unirioja.es>	**63**	238.586
<accenture.com>	**58**	203.888	<ucm.es>	**61**	240.119
<repsol.com>	**58**	174.307	<us.es>	**60**	201.76
<iberdrola.es>	**57**	226.814	<um.es>	**58**	194.625
<mapfre.es>	**56**	179.017	<upm.es>	**56**	185.516
<deloitte.es>	**55**	206.122	<uva.es>	**55**	155.285

DISCUSSION

The results previously presented should be widely discussed and appropriately contextualized due to some limitations that may jeopardize not the general findings but some particular data.

Sample of Universities

The universities under study conforms a whole national academic system (Spain), with its own characteristics and peculiarities so that a possible extrapolation of the results to other countries or academic environments seem difficult.

Regardless of locale, an analysis of word-class universities might probably provide different results due to the different nature (research focused, higher degree of internationalization, private funding), not only related to the web visibility of institutional profiles but also in the nature of relationships with private companies.

In fact, the replication of this study in other environments (either different national systems or specific types of universities) may provide insights about the match and validation of the method proposed.

Table 6. Network properties of LinkedIn users and URL mentions' networks

Metrics	LinkedIn User network	URL mention network
Average Degree	6,611	32,369
Diameter	7	4.00
Density	0.042	0.207
Average Path Length	3.059	2.097

Sample of Private Companies

We should take into account that only top 10 private companies related to each of the 75 Spanish universities (via the number of users connected as alumni) with a LinkedIn University Page activated have been considered.

This decision may bias the sample to some extent. For example, one company may achieve 12th position in its relation to a great university, and providing more LinkedIn users than other company in the first position for a small university. Therefore, results in Table 3 (especially alumni scores) should not be understood as a total ranking since some missing companies may provide a higher total number of alumni from Spanish universities than some of the companies already included.

In any case, the number of companies not appearing in the top ten positions for any University and providing a total number of users from Spanish Universities higher than 100 (threshold used to construct the network) is expected to be limited.

Nonetheless, the expansion of this study to all companies related to universities that LinkedIn provides through the University Page (up to 20, including public entities) is recommended for future work.

Web Data

Despite URL mentions are considered a good proxy of external inlinks to Spanish universities, this metric introduces some important limitations that should be considered. This shortcomings have been previously summarized in the literature (Thelwall and Sud, 2011; Ortega, Orduna-Malea and Aguillo, 2014). Some linguistic limitations of the search engine together with the rounded results provided by Google make a rough estimate results.

The self-mention effect is one of the main pitfalls. In this case, a LinkedIn profile URL belonging to one company/university may be linked to by another website belonging to the same institution. For example, we find that company "Tinsa" (<LinkedIn.com/company/tinsa>) receives a total of 752 URL mentions, though a great quantity come from: <twitter.com/tinsa>, the Twitter profile of this same company.

For this reason, URL mentions must be interpreted in their comparison with the remaining entities under study and never as accurate individual performance.

Another limitation related to web data corresponds to the decision of using just one official URL for each company. The complexity of private firms on the Web has been already studied (Orduna-Malea et al, 2015), finding that the official corporate URL website constitutes in most cases just a small part of the web presence of a company. Namely, the web connection between university website and company website may be produced via alternative web domains (brands, delegations, services, etc.), which may jeopardize the accuracy of the URL mention's network as regards university-company connections via websites.

Notwithstanding, the complexity of discovering and managing all existing web domains related to one company is elevated, requiring expanded technical capacity as well. Moreover, most of these alternative web domains are not devoted or oriented towards universities (or other companies) but for final users. For this reason, the election of the official corporate website is considered a valid solution that not only simplifies the analysis but also serves as a proxy of the real web connectivity between universities and companies in professional and labor terms, since this corporate website is the most important website for a company.

LinkedIn Alumni

Finally, we should rely on the data provided both by LinkedIn (university pages data) and by users (information included in their personal profiles).

Apart of these uncontrollable variables, the coverage of LinkedIn is an important issue since not all real alumni have created a personal profile in this social platform including information related to their alma mater and learning background. Despite this, the interest of users to find a job, of professionals to get professional network, and finally of companies to hire the best professionals, make the information available of quality.

As regards the information on the number of users connecting the universities where they studied and the companies where they work, we should highlight the great volatility of these data.

The statistics available in the LinkedIn University Pages are rapidly changing due to continual updates of the LinkedIn database. For this reason, the number of users from a specific university working on a specific company may change daily, and consequently, the top ten companies displayed. As a result, the date for data collection (in this case 4th February 2016), is necessary.

CONCLUSION

First, the characterization of LinkedIn institutional profiles (both from universities and companies) reveal some important shortcomings if utilized for link analysis. The existence of duplicate URLs (ID and friendly) disperses unnecessarily the web visibility and may affect the search engine optimization as well.

Moreover, the friendly version of the URLs does not appear as the by default URL (only is activated when a user clicks on any menu item of the profile once s/he has accessed to), and it is also configured automatically from the information in the profile's name field. This provokes occasionally (if not treated) the existence of "friend" URLs not as friendly as desirable, including diacritics and other inappropriate symbols in the URL syntax. All the same, the friend URLs achieve on average more external mentions from third parties than ID URLs, and this is particularly evident on companies' profiles.

In this sense, the university LinkedIn institutional profiles are characterized by a very low web visibility (7.8 mentions per university on average; n= 75; seven universities do not even receive any URL mention) as well as a discrete number of followers (6,598 followers on average per university). Although the improvable URL management policy may influence in this issue, an insufficient academic branding from some particular universities in this social platform is thus evidenced.

This contrasts with the values obtained from the main top companies analysed, which obtain higher web visibility (540.2 per university on average; n= 273) and followers (127,117). Nonetheless, these results should be taken cautiously, since the impact of some international companies (such as IBM) biases these descriptive stats and should not be directly compared to some small and domestic universities. Otherwise, some international Spanish companies underperform especially on URL mention scores, such as Telefónica (which receives only two external URL mentions), evidencing the limitation of this method based on the institutional LinkedIn profiles.

This different web behaviour among universities and companies is reflected on the correlation values between URL mentions and the number of followers of each institutional profile (r=0.4 for universities; r= 0.8 for companies). Therefore, these two complementary dimensions of web visibility on LinkedIn profiles (mentions and followers) seem to be strongly correlated once a visibility threshold is achieved.

As a result, the low web visibility of the institutional LinkedIn profiles of Spanish universities invalidates method A (URL mentions among university and company LinkedIn institutional profiles) to constitute one possible method to measure university-industry interactions by link analysis.

The method B (based on the text data contained in user profiles) presents discrete results since the number of professionals connecting universities and companies is low. Nonetheless, the connections obtained are of great interest due to the quality of the information extracted from users' profiles. After analysing the network obtained, we can highlight the following insights:

- Bank, Technology and consultancy firms are those that employ the majority of graduates from Spanish universities.
- Big public universities located in Madrid are those who are better connected with the companies that employ most graduates.
- Detection of special relations between universities and companies mainly due to geographic reasons (such as Abengoa and the University of Seville, or CaixaBank with both the University of Barcelona and Pompeu Fabra university).
- Detection of special relations between specific universities and companies (for example Ford and the Polytechnic University of Valencia) not only due to geographic location but for thematic affinities.

Thus, we conclude that method B is suitable for the purpose of analysing university-industry relations, though understanding its limitations (dependence on LinkedIn University Pages data: low size and high variability).

Finally, the university-industry interactions based on the method C (URL mentions through university and company websites) provide a different picture of relationships in a greater order of magnitude. This procedure confirms the potential of some companies (related to universities regardless the method used) whereas helps showing other hidden relations not based on graduates but on professional collaboration and networks.

Consequently, we conclude that Method C should be used. Nonetheless, the limitations of Google in retrieving web mentions, the web policies of both universities and companies (that may over-emphasize or infra-represent some online relations), and the impossibility of analysing all company websites, make this direct method somewhat incomplete and inaccurate, recommending this only as a complement of other methods.

The results obtained (and especially the methods tested) should be used by users to ascertain how strengthen the interactions between companies and universities are. This may help both of them to adequate marketing strategies, target audience and establish potential partnerships. The network maps generated may additionally guide users to a better understanding of relations among market and academia, serving as a useful complementary information source in decision-making.

REFERENCES

Archambault, A., & Grudin, J. (2012, May). A longitudinal study of facebook, LinkedIn, & twitter use. In *Proceedings of the SIGCHI Conference on Human Factors in Computing Systems* (pp. 2741-2750). New York, NY: ACM. doi:10.1145/2207676.2208671

Barabasi, A. L. (2014). *Linked: the new science of networks*. New York: Perseus Books.

Barabasi, A. L., & Albert, R. (1999). Emergence of Scaling in Random Networks. *Science*, *286*(5439), 509–512. doi:10.1126/science.286.5439.509 PMID:10521342

Bonsón, E., & Bednárová, M. (2013). Corporate LinkedIn practices of Eurozone companies. *Online Information Review*, *37*(6), 969–984. doi:10.1108/OIR-09-2012-0159

Brouer, R. L., Stefanone, M. A., Badawy, R. L., Egnoto, M. J., & Seitz, S. R. (2015). Losing Control of Company Information in the Recruitment Process: The Impact of LinkedIn on Organizational Attraction. In *Proceedings of the 48th Hawaii International Conference on System Sciences* (pp. 1879-1888). Hawaii, HI: IEEE. doi:10.1109/HICSS.2015.226

Dai, K., Nespereira, C. G., Vilas, A. F., & Redondo, R. P. D. (2015). Scraping and Clustering Techniques for the Characterization of LinkedIn Profiles. *Computer Science and Information Technology*, *5*(1), 1–15.

Fruchterman, T. M., & Reingold, E. M. (1991). Graph drawing by force-directed placement. *Software, Practice & Experience*, *21*(11), 1129–1164. doi:10.1002/spe.4380211102

García-Santiago, L., & De Moya-Anegón, F. (2009). Using co-outlinks to mine heterogeneous networks. *Scientometrics*, *79*(3), 681–702. doi:10.1007/s11192-007-1855-0

Guillory, J., & Hancock, J. T. (2012). The effect of LinkedIn on deception in resumes. *Cyberpsychology, Behavior, and Social Networking*, *15*(3), 135–140. doi:10.1089/cyber.2011.0389 PMID:22335544

Kelly, B., & Delasalle, J. (2012). Can LinkedIn and Academia.edu enhance access to open repositories? In *OR2012: The 7th International Conference on Open Repositories*. University of Bath.

Kim, D., Kim, J. H., & Nam, Y. (2014). How does industry use social networking sites? An analysis of corporate dialogic uses of Facebook, Twitter, YouTube, and LinkedIn by industry type. *Quality & Quantity*, *48*(5), 2605–2614. doi:10.1007/s11135-013-9910-9

LinkedIn. (2014). *LinkedIn outcome rankings under the hood*. [Technical sheet]. Retrieved February 15, 2016, from: https://university.linkedin.com/content/dam/university/global/en_US/site/pdf/linkedin_outcome_graduate_rankings.pdf

Mas-Bleda, A., & Aguillo, I. F. (2015). *La web social como nuevo medio de comunicación y evaluación científica*. Barcelona: Editorial UOC.

Minguillo, D., & Thelwall, M. (2012). Mapping the network structure of science parks: An exploratory study of cross-sectoral interactions reflected on the web. *Aslib Proceedings*, *64*(4), 332–357. doi:10.1108/00012531211244716

Orduna-Malea, E. (2011). Personalización e interactividad en los rankings de universidades publicados en la Web. *Anuario ThinkEPI*, *5*, 216–222.

Orduna-Malea, E. (2013a). Aggregation of the web performance of internal university units as a method of quantitative analysis of a university system: The case of Spain. *Journal of the American Society for Information Science and Technology*, *64*(10), 2100–2114. doi:10.1002/asi.22912

Orduna-Malea, E. (2013b). Espacio universitario español en la Web (2010): Estudio descriptivo de instituciones y productos académicos a través del análisis de subdominios y subdirectorios. *Revista Espanola la de Documentacion Cientifica*, *36*(3), e017. doi:10.3989/redc.2013.3.958

Orduna-Malea, E., & Aguillo, I. F. (2015). *Cibermetría. Midiendo el espacio red*. Barcelona: Editorial UOC.

Orduna-Malea, E., & Aytac, S. (2015). Revealing the online network between university and industry: The case of Turkey. *Scientometrics*, *105*(3), 1849–1866. doi:10.1007/s11192-015-1596-4

Orduna-Malea, E., Delgado López-Cózar, E., Serrano-Cobos, J., & Romero, N. L. (2015). Disclosing the network structure of private companies on the web: The case of Spanish IBEX 35 share index. *Online Information Review*, *39*(3), 360–382. doi:10.1108/OIR-11-2014-0282

Orduna-Malea, E., Serrano-Cobos, J., & Lloret-Romero, N. (2009). Las universidades públicas españolas en Google Scholar: presencia y evolución de su publicación académica web. *El profesional de la información, 18*(5), 493-500.

Ortega, J. L., Orduna-Malea, E., & Aguillo, I. F. (2014). Are web mentions accurate substitutes for inlinks for Spanish universities? *Online Information Review*, *38*(1), 59–77. doi:10.1108/OIR-10-2012-0189

Skeels, M. M., & Grudin, J. (2009). When social networks cross boundaries: a case study of workplace use of facebook and LinkedIn. In *Proceedings of the ACM 2009 International Conference On Supporting Group Work* (pp. 95-104). New York: ACM. doi:10.1145/1531674.1531689

Stuart, D., & Thelwall, M. (2006). Investigating triple helix relationships using URL citations: A case study of the UK West Midlands automobile industry. *Research Evaluation*, *15*(2), 97–106. doi:10.3152/147154406781775968

Thelwall, M. (2004). *Link analysis: an information science approach*. San Diego: Academic Press.

Thelwall, M., & Sud, P. (2011). A comparison of methods for collecting web citation data for academic organisations. *Journal of the American Society for Information Science and Technology*, *62*(8), 1488–1497. doi:10.1002/asi.21571

Unkelos-Shpigel, N., Sherman, S., & Hadar, I. (2015). Finding the missing link to industry: LinkedIn professional groups as facilitators of empirical research. In *Proceedings of the Third International Workshop on Conducting Empirical Studies in Industry* (pp. 43-46). Piscataway, NJ: IEEE Press. doi:10.1109/CESI.2015.14

Van Dijck, J. (2013). You have one identity: Performing the self on Facebook and LinkedIn. *Media Culture & Society*, *35*(2), 199–215. doi:10.1177/0163443712468605

Vaughan, L. (2004). Exploring website features for business information. *Scientometrics*, *61*(3), 467–477. doi:10.1023/B:SCIE.0000045122.93018.2a

Vaughan, L., & Wu, G. (2004). Links to commercial web sites as a source of business information. *Scientometrics*, *60*(3), 487–496. doi:10.1023/B:SCIE.0000034389.14825.bc

ENDNOTES

1 http://blog.linkedin.com/2015/10/29/400-million-members
2 https://en.wikipedia.org/wiki/Vertical_search
3 http://www.crue.org
4 https://en.wikipedia.org/wiki/Web_scraping
5 https://gephi.org/
6 http://doi.org/10.13140/RG.2.1.4470.5043

Chapter 10
Good Practices for Universities on Twitter

Susana Herrera Damas
Universidad Carlos III de Madrid, Spain

ABSTRACT

According to Klamm (2011), some of the best practices for universities seeking to build effective social media engagement are: 1) to develop a strategy and set goals; 2) to pick and choose the platforms; 3) to empower and support individual departments; 4) to put guidelines in place; 5) to develop a consistent voice across platforms; and 6) to communicate across campus. In the following pages, we will delve more deeply into Twitter for universities. After checking 10 reasons why universities should take Twitter seriously, we will offer some good practices for universities willing to use Twitter in a way that optimizes the tool; that is, not only for disseminating information, but also for talking with the community and taking advantage of its multimedia potential. We will illustrate these good practices with screenshots to better show what we mean and also to reinforce their utility with contributions from some of the early theorists on what the proper use of social media should be.

INTRODUCTION

According to Klamm (2011), some of the best practices for universities seeking to build effective social media engagement are the following:

1. To develop a strategy and set goals
2. To pick and choose the platforms
3. To empower and support individual departments
4. To put guidelines in place
5. To develop a consistent voice across platforms
6. To communicate across campus

Similarly, Ahlquist (2013) provides these 10 best social media practices in higher education:

DOI: 10.4018/978-1-5225-0917-2.ch010

1. To implement a social media strategy
2. To produce quality and accurate content
3. To manage platforms with social media managers and student leaders
4. To use an authentic and transparent voice
5. To represent the university/department brand and university resources
6. To collaborate and support other university social media pages
7. To respect the community
8. To dive into data
9. To empower influencers and engage audience
10. To get internal buy-in

In the following pages, we will delve more deeply into Twitter for universities.

Why Universities Should Take Twitter Seriously

1. **It is a Free Service:**Anyone can access this space; the only requirement is opening an account in the service. In addition, the free access allows users to connect with the rest of the world. For universities, which often face financial problems, it is an affordable new channel for promoting its brand and strengthening contact with their community.
2. **It is Where People Live:**Until 2012, Twitter had 500 million users registered worldwide. Over 340 million tweets are posted daily. Along with Facebook, Google+ and Instagram, Twitter is one of the most popular platforms. Since its launch, it has won a large number of users. In the case of universities, it is also interesting to note that Twitter attracts many young people:

For universities, deciding to use social media is a no-brainer. The 18- to 24-year-old college student demographic is all over the social web, and its younger counterpart (the high school crowd) is equally immersed. Alumni, recent and far-removed, use social networks to engage and stay connected with the world. Community members, parents of students, potential donors, faculty and staff and other constituents are just a tweet or 'like' away. With so many key populations embracing social media, universities almost have no choice but to integrate these platforms into their marketing and communications plans (Klamm, 2011).

3. **It is Easier to Cross Borders:** Twitter is accessible from different platforms and through a variety of mobile devices such as smartphones, tablets, etc. Specifically, mobile phones facilitate users to interact freely, regardless of where they are.
4. **It Helps in Getting Feedback:** The immediacy of this tool allows messages to be spread instantly. Through Twitter, universities can get feedback from their teachers or students without time or space limits.
5. **It Modernizes the Image and the Skills of the University:** With the arrival of technology, the world has changed a lot, especially the way we communicate. The use of Twitter allows universities to modernize their image and skills. This might not be their top priority but it is an unavoidable consequence that brings indisputable advantages for the brand.

6. **It Promotes the Connections with Other Universities:** Since more and more universities around the world have accounts on Twitter, they can interact among each other in a quick and simple way without significant differences of time and space. According to Guzmán, del Moral and Fernández (2012: 30), these connections can also take place directly among scholars:

Teachers share information and resources related to their practice (Mateos & Labra, 2008), asking for help, offering suggestions and maintaining relationships with other teachers that are geographically dispersed (Katz, 2010; Veletsianos, 2011). It also allows them to collaborate taking advantage of the synergy of crowdsourcing and intelligence application in higher education (Gowers and Nielsen, 2009). It also promotes the participation of research groups and the integration of professionals from different areas with complementary and enriching perspectives (Usluel, 2009, Forkosh & Hershokovitz, 2011).

7. **It Contributes to an Enthusiastic Community of Followers:** Twitter allows universities to create communities through the exchange of information, views and experiences. This platform makes it easier for universities to provide information of interest to their community, to encourage them and to thank them.
8. **It Facilitates the Spread of News Affecting the University:** Because of the immediacy of the service, universities can also spread new information quickly and easily.
9. **It Allows Collaborative Work with Other Universities:** Universities can also collaborate in a collective way, without the traditional problems of physical distance.
10. **It Fosters Creativity:** After the launch of Twitter in March 2006 and following the release of its API in September 2006, the tool improves its performance every day. This is also possible thanks to the proposals made by users. At the same time, the brevity required by the service together with the need to highlight challenge universities to be very creative in their approaches to spread their messages in an attractive way and serve their communities of followers.

From a more specific perspective, Guzmán, del Moral and González (2012) add that Twitter provides very useful features for education processes:
the key is in its:

a. agile interaction between teachers and students (Reed, et al., 2010; Stepanyan, 2010 et al.);
b. its permanent access to messages online, where followers can read, comment and participate from different places (Stieger and Burger, 2010);
c. the generation of immediate responses (Ebner et al., 2010),
d. the multimedia format that facilitates quick learning and communication (Grosseck & Holotescu, 2010),
e. the facility to create a community interested in a subject supporting the participation and comments at conferences and events.

It also allows direct contact with educators and experts from different disciplines (Holotescu & Grosseck, 2010).

According to these scholars (Guzmán, del Moral and González, 2012), two different uses of Twitter can be observed at universities:

1. The teaching one, as a way of supporting different subjects
2. As a tool for providing institutional information, promoting communication between the institution and its public

Without underestimating Twitter's efficiency as a teaching tool (Guzmán, del Moral and González, 2012: 28-30), we will focus here on the use of Twitter for corporate communication of the activities carried out by universities.

GOOD PRACTICES FOR UNIVERSITIES ON TWITTER

In the following pages, we will offer some good practices for universities willing to use Twitter in a way that optimizes the tool; that is, not only for disseminating information, but also for talking with the community and taking advantage of its multimedia potential (Dai, 2014). We will illustrate these good practices with screenshots to better show what we mean and also to reinforce their utility with contributions from some of the early theorists on what the proper use of social media should be.

To Have a Human Voice

One of the first recommendations on which social media theorists insist is that the institutions have to be genuine, showing a human voice:

Be Real. Authenticity is the secret ingredient behind any good and valuable social media marketing campaign. If you know your audience, locate them online, listen, add value, respond, refrain from spamming and just be yourself, you'll have far better and more long-lasting positive results than if you try to be someone -or something- you're not (Sherman, 2009).

Similarly, Johnson (2009) asserts that "the social warmth of all those stray details shouldn't be taken lightly" while Roy insists on the need to provide personality to the messages:

The automated feed simply culls headlines from the front page and links users to the related articles, which can lead to a pretty dry feed. Human management of a Twitter feed means wit can shine, making tweets more interesting and users more likely to interact with them. Fascinating and poignant details that weren't included in the headline can be applied to make a link more clickable. Personality is a major driver for popular Twitter feeds; without it, your account is little more than a RSS feed (Roy, 2011).

In Ahlquist's advice:

Be a real person (not a robot or marketing machine). Social Media needs to be authentic. Especially if your target is students, this audience can tell when posts are automated or not personal (Ahlquist, 2013).

Similarly, the University of Leicester's social media guideline regarding Twitter states:

Planning, training, coordination and integration with social tools is imperative but Twitter use can appear disingenuous and inhuman if the approach is too structured, to the point where your community may be turned off. Treat your Twitter relationships the same way you would any other relationship. Familiarise yourself with the rest of the Social Media Handbooks to understand the parameters and rules for engagement.

Klamm (2011) also recalls the need to show a consistent voice across the different platforms despite the nuances of adjusting it to the singularities of each type of platform:

A university's social media presence is an extension of the school's brand. What is your brand all about? Is it playful and joking or conservative and buttoned-up? It's important that a consistent voice is implemented across all of the school's major social media platforms —a school with an ultra-serious Facebook page and an offbeat and sarcastic Twitter account will just look like it's having an identity crisis (Klamm, 2011).

We can find some examples of universities using a human voice on Twitter in the following messages:

Figure 1. Tweets showing a human voice

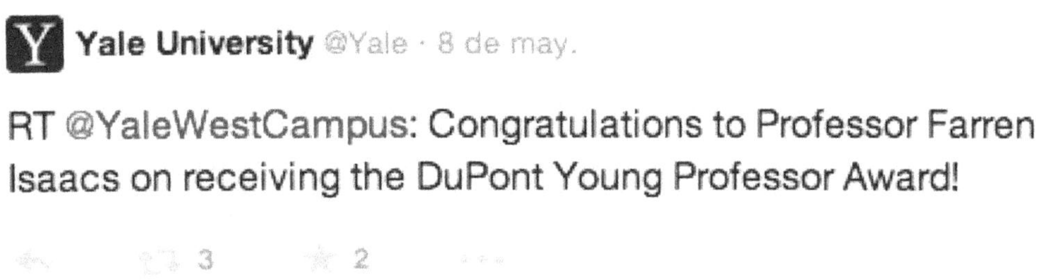

Figure 2. Tweets showing a human voice

Figure 3. Tweets showing a human voice

Figure 4. Tweets showing a human voice

Figure 5. Tweets showing a human voice

To Add Multimedia Value to Its Updates

As said at the beginning of this chapter, one of the attractions that Twitter provides is the ability to add value to the updates so that these are composed not only of text, but also of photos, audio, video, graphs, surveys, etc. This is of special interest in a culture as predominantly visual as ours is. In the following screenshots, we can find good examples of messages that make the most of the multimedia potential that Twitter provides:

To Offer Useful and Timely Information

Given the amount of information we are exposed to everyday, social media scholars also remind us that messages must add value:

Add Value. Enter any online conversation with the aim of adding value. Before posting a message as a new participant in a forum, ask yourself: How is this providing value to the conversation? To the com-

Figure 6. Tweets adding multimedia value to its updates

Figure 7. Tweets adding multimedia value to its updates

Figure 8. Tweets adding multimedia value to its updates

Figure 9. Tweets adding multimedia value to its updates

Figure 10. Tweets adding multimedia value to its updates

munity? In some circles, talking about your product or service can be considered valuable, but in most, it is unwelcome and intrusive (Sherman, 2009).

Figure 11. Tweets adding multimedia value to its updates

Figure 12. Tweets offering useful and timely information

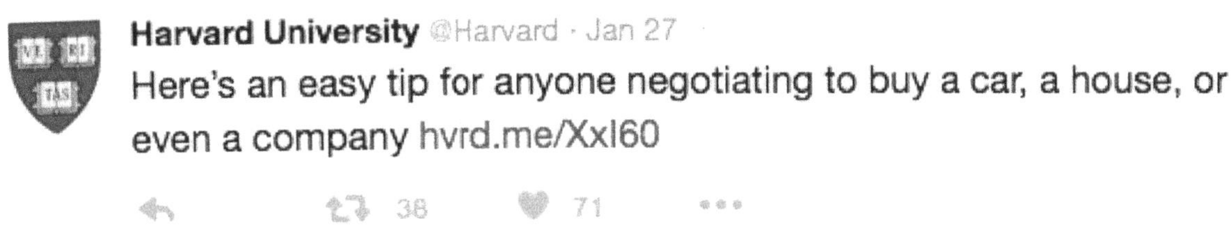

Figure 13. Tweets offering useful and timely information

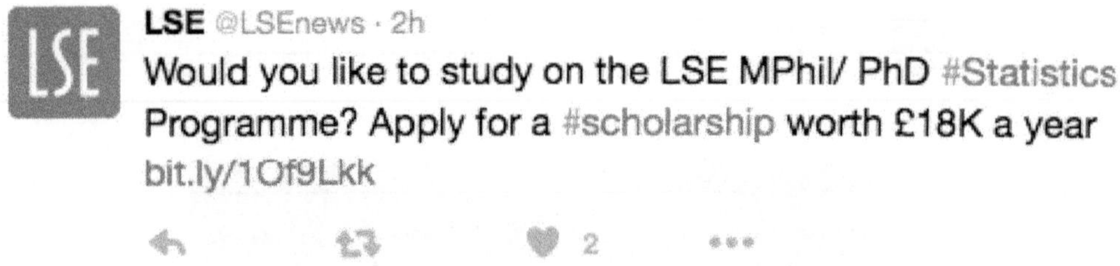

Figure 14. Tweets offering useful and timely information

Figure 15. Tweets offering useful and timely information

In this regard, universities can use Twitter to spread useful information to their audiences, in forms of news, benefits, scholarships, discounts or job opportunities.

Figure 16. Tweets offering useful and timely information

Figure 17. Tweets offering useful and timely information

Figure 18. Tweets offering useful and timely information

To Provide Live Coverages

Twitter also allows universities to broadcast live events, serving as a backchannel for the dissemination of events so that both the community and other interested people can follow them, even if they don't have the opportunity to physically witness them.

Following Guzmán, del Moral and Fernández (2012: 30), this coverage can also be provided through hashtags, which allow the transmission of real-time events such as conferences, forums, etc. with the option of posing questions to the speakers and surveys to the followers:

To use Creative Hashtags

Hashtags allow universities to extend the dissemination of their messages beyond their own networks. In their study on how Latin American universities were using Twitter, Guzmán, del Moral and Fernández (2012: 34) analyze the use of hashtags for:

Figure 19. Tweets providing live coverages

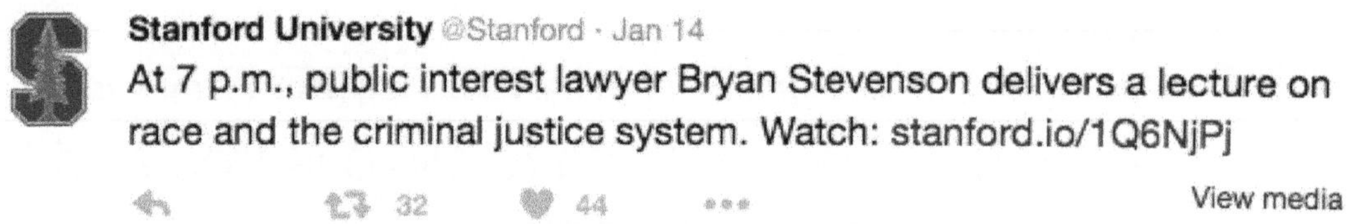

Figure 20. Tweets providing live coverages

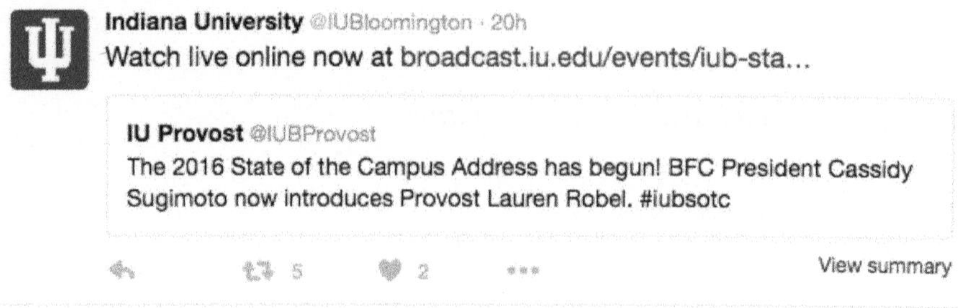

Figure 21. Tweets providing live coverages

Figure 22. Tweets providing live coverages

1. Promoting services
2. Spreading cultural programs
3. Promoting interaction betweeen students and faculty
4. Communicating research and projects
5. Promoting library services
6. Publicizing calls to participate in activities of social interest
7. Inviting and participating in forums, seminars and conferences

When possible, this should be achieved in an attractive manner, as can be seen in the following screenshots:

Figure 23. Tweets using creative hashtags

Figure 24. Tweets using creative hashtags

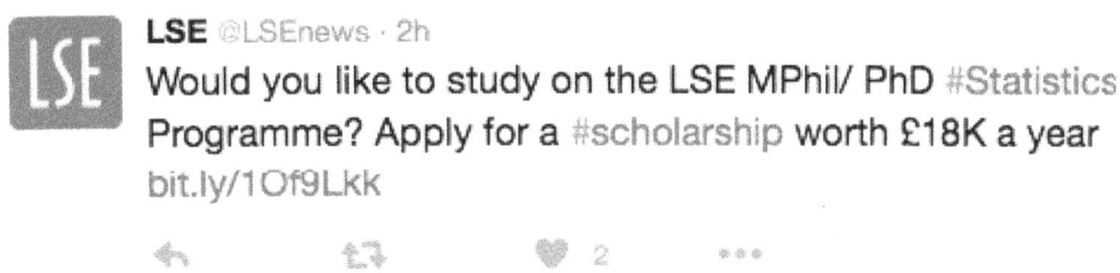

Figure 25. Tweets using creative hashtags

Figure 26. Tweets using creative hashtags

Figure 27. Tweets using creative hashtags

To Retweet and Link External Material to Enrich One's Contribution

Universities can also retweet messages that might be of interest to their audience although they originally come from another institution. Retweets allow original authors to be told that their content is relevant while followers are told that there is some stuff that is worth paying attention to. For these two reasons, retweeting is also an effective way to create communities.

In addition, in the absence of context, universities may also include links pointing to the source from which the information has been taken. Linking to external material is another good practice when creating communities:

Link to others (...) social media is a tremendous tool for finding interesting content, and you should share it when you find it, not just keep it to yourself. If you do this, others are more likely to share your links when you post them (Ingram, 2011).

Besides, linking to external material allows one to get around the brevity restriction imposed by Twitter with its 140-character limit. A good external link can be found in the following screenshot:

To Give Credit to Other Users

Several social media scholars insist on the obligation to give others credit, something that appears as a central part of social media "ethos":

Figure 28. Tweet linking external material

Figure 29. Tweet giving credit to other users

Social media works when you are generous. There is nothing wrong with self-promotion, but things really take off when you give others praise or a moment in the spotlight. The rise of retweeting -real retweeting, not spammy retweeting- shows how far giving credit to others can go in social spaces (Sherman, 2009).

A very effective way to give others credit and strengthen with it the relations with the community is by making external references. We can find external mentions, not directly linked to the university, in the following example:

To Pose Questions

Furthermore, as an additional way to create community, institutions should not be limited to using social media as one-way media, but should strive to find ways of talking to it:

Being in conversation with our community means listening as well as talking, and adjusting what we do and cover based on what we hear. It involves: hosting discussions in person and online on topics that matter to the community. Participating in conversations we're not hosting, both in person and online

(...) Using web analytics to better understand what people are showing us they value in what we do, and basing at least some of our decisions about content and staff resources based on what we're seeing (Mayer, 2011).

To justify this, it is important to bear in mind that these are called social media for a reason:

Talk to people: this has nothing to do with promoting your own content. It means engaging in conversation about issues, and responding to and/or asking questions of others. It's called a conversational medium for a reason (Ingram, 2011).

Similarly, Ahlquist (2013) recalls the need to engage and interact:

Engagement and interaction is more important than the number of subscribers, followers or fans. Communications cues such as likes, comments, RT, shares and mentions are stronger than total community members (Ahlquist, 2013).

In this regard, a certain way of talking with followers is posing them questions:

Don't be a 'me me me' tweeterer. The best friends in real life are the ones that listen more than they speak. It's the same on Twitter. Ask questions about the other person, be genuinely interested, be a good listener... err... reader of tuits. Be empathetic (Adams, 2009).

We find proper examples of tweets that pose questions to followers in the following updates:

Figure 30. Tweets posing questions

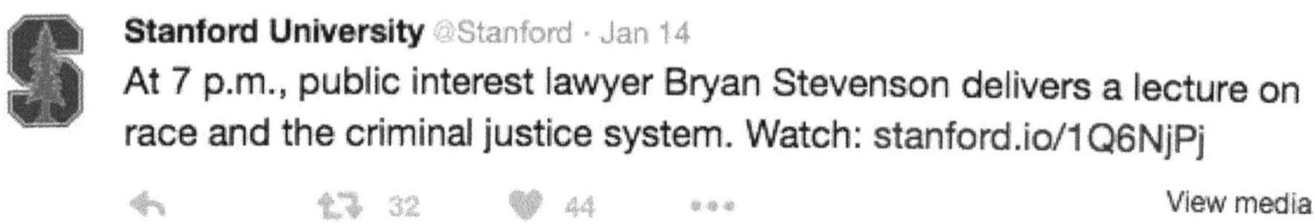

Figure 31. Tweets posing questions

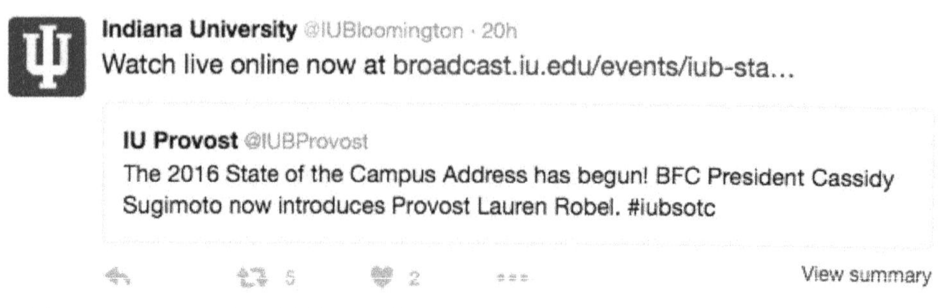

To Reply to the Mentions

In addition, many scholars reiterate the utility of these spaces as a way to keep abreast of the concerns, needs and interests of some members of the community

Listening thoughtfully gives you a better sense of not only what people are saying but also how they are feeling. In virtual spaces where there are no visual cues, good listening skills become a powerful asset. Listening also helps you map out your current social media footprint and measure your marketing campaigns over time. The key to successful social media marketing is listening (Sherman, 2009).

An effective way to verify that the university is listening is to respond to the mentions, a very good practice in social media:

Reply when you are spoken to: *if you don't respond when someone asks you a direct question or makes a point in reference to you, it's like ignoring someone who is standing right beside you and talking to you. That doesn't mean responding to every troll or flame (Ingram, 2011).*

According to Sherman, there are several ways to lessen this task:

The burden of response can be great, but it can be lessened by using the right tools and crowdsourcing answers. A quick response is more important than ever, and thanks to search tools, alert apps and other services, it is possible to achieve. Don't be a dam in a conversation flow (Sherman, 2009).

Similarly, Buttry states that replying to followers is a very convenient way to interact with them:

If your Twitter feed is a one-way stream of self-promoting headlines and links, people are not likely to engage with it. But if you answer their questions, ask them how they know newsworthy information that they tweet, retweet their observations about community life, they will engage with you (Buttry, 2011).

We find some examples of fast and efficient replies in the following messages:

Figure 32. Tweets replying to the mentions

Figure 33. Tweets replying to the mentions

To Link the Profile with the One on Other Social Networks

Twitter also allows an institution to synchronize its profile with the one on other social networks such as Facebook, Instagram, Pinterest, etc. This increases the visibility and productivity of the institution in social media and the community can continue growing in these other places. However, at this point, it is very necessary to remember the singularity of each platform:

Keep in mind that each platform has nuances, so the voice will never be exactly the same across all major university accounts. Content needs to be re-worked to fit with the feel of each platform (Klamm, 2011).

Here are some examples:

Figure 34. Tweets linking with profiles on other social networks

Figure 35. Tweets linking with profiles on other social networks

Figure 36. Tweets linking with profiles on other social networks

Figure 37. Tweets linking with profiles on other social networks

Figure 38. Tweets linking with profiles on other social networks

Figure 39. Tweets linking with profiles on other social networks

Figure 40. Tweets linking with profiles on other social networks

Figure 41. Tweets linking with profiles on other social networks

REFERENCES

Ahlquist, J. (2013). 10 Social Media Best Practices in Higher Education. *Josie Ahlquist*. Retrieved February, 15th, 2016, from http://www.josieahlquist.com/2013/10/07/10smbp/

Dai, Y. (2014). *Análisis comunicacional del uso que las universidades españolas están haciendo de Twitter: evaluación e implicaciones para la práctica*. (Unpublished Master's thesis). Universidad Carlos III de Madrid, Madrid, Spain.

Ebner, M., Lienhardt, C., Rohs, M., & Meyer, I. (2010). Microblogs in Higher Education. A chance to facilitate informal and process-oriented learning? *Computers & Education*, *55*(1), 92–100. doi:10.1016/j.compedu.2009.12.006

Forkosh, A., & Hershkovitz, A. (2011). The use of social networks by higher education institutes in Israel. In *Proceedings of the Chais Conference on instructional technologies research 2011: Learning in the technological era* (pp. 14-20). Raanana: The Open University of Israel.

Gowers, T., & Nielsen, M. (2009). Massively collaborative mathematics. *Nature*, *461*(7266), 879–881. doi:10.1038/461879a PMID:19829354

Grosseck, G., & Holotescu, C. (2010). Microblogging multimedia-based teaching methods best practices with Cirip. Eu. *Procedia: Social and Behavioral Sciences*, *1*, 495–501.

Guzmán, A. P., Del Moral, M. E., & González, F. (2012). Usos de Twitter en las universidades iberoamericanas. *Revista Latinoamericana de Tecnología Educativa, 11*(1), 27–39.

Holotescu, C., & Grosseck, G. (2010). Learning to microblogging to learn. A case study on learning scenarios in a microblogging context. In *Proceedings of the 6th international scientific conference e learning and software for educational Bucharest* (pp. 15-16). Bucharest: RomaniaUniversity Press.

Junco, R., Heiberger, G., & Loken, E. (2010). The effect of Twitter on college student engagement and grades. *Journal of Computer Assisted Learning, 27*(2), 119–132. doi:10.1111/j.1365-2729.2010.00387.x

Katz, R. N. (2010). Scholars, scholarship and scholarly enterprise in the digital age. *EDUCAUSE Review, 45*(2), 44–56.

Klamm, D. (2011). 6 Best Practices for Universities Embracing Social Media. *Mashable*. Retrieved February, 15th, 2016, from http://mashable.com/2011/10/10/universities-social-media/#lyqXRDcSp5qa

Mateos, C. V., & Labra, J. P. (2008). Evaluación del uso y manejo de las tecnologías de la información y la comunicación en los docentes universitarios. Un caso mexicano. *Revista Latinoamericana de Tecnología Educativa, 7*(1), 13–32.

Stepanyan, K., Borau, K., & Ullrich, C. (2010). A social network analysis perspective on student interaction within the Twitter microblogging environment. In *Proceedings of the Advanced Learning Technologies (ICALT), 2010 IEEE 10th International Conference* (pp. 70-72), Sousse. doi:10.1109/ICALT.2010.27

Stieger, S., & Burger, C. (2010). Let's go informative: Continuous student ratings with web 2.0 application Twitter. *Cyberpsychology, Behavior, and Social Networking, 13*(2), 163–167. doi:10.1089/cyber.2009.0128 PMID:20528272

Tufts University. (n.d.). *Social Media Best Practices*. Retrieved February, 15th, 2016, from http://webcomm.tufts.edu/social-media-best-practices13/

University of Leicester. (n.d.). *Twitter best practice*. Retrieved February, 15th, 2016, from http://www2.le.ac.uk/offices/cap/marcomms/communications/social/handbook/twitter/twitter-best-practise

Usluel, Y. K., & Mazman, S. G. (2009). Adoption of web 2.0 tools in distance education. *Procedia: Social and Behavioral Sciences, 1*(1), 818–823. doi:10.1016/j.sbspro.2009.01.146

Veletsianos, G. (2011). Higher education scholars' participation and practices on Twitter. *Journal of Computer Assisted Learning*. Retrieved February, 15th, 2016, from http://onlinelibrary.wiley.com/doi/10.1111/j.1365-2729.2011.00449.x/full

Chapter 11
The Use of Geomarketing to Promote Personal Branding:
Make Your Mark as a Teacher in the Territory of Internet

Gersón Beltrán López
Universitat de València, Spain

ABSTRACT

This document aims to show how teachers can make their mark in the territory through the appropriate use of their personal brand and geomarketing techniques, with the example of the author, and how this methodology can be transferred to students in their connection with the professional world. It is a process of structuring and decentralization of personal branding in the territory, linking physical space with online space and where geography acquires a new integral dimension through SoLoMo methodology (Social, Local, and Mobile).

INTRODUCTION

The Internet has revolutionized our lives; all aspects of our daily lives have been incorporated into this information and communication space. Education is one of the most sensitive sectors to these changes: on the one hand for the possibilities offered to teachers on an educational level and on the other hand because students are continuously connected and have access to information like never before. Teachers no longer issue unidirectional information and their dominant position in the classroom must involve teaching the students to search for information and to use critical judgement to select that which is relevant, thus finding themselves closer to the students, working bidirectionally, in a collaborative, accompanying position.

On the other hand, the territory has transcended borders and is no longer merely physical, but unites the physical and the online in one through the SoLoMo concept: we live in a world connected by social networks, we communicate from a specific place thanks to our mobile devices. Geo-positioning is of vital importance, since it indicates the location from where we communicate

DOI: 10.4018/978-1-5225-0917-2.ch011

and unites this physical space with online space, becoming a communication tool between both. Therefore, geo-positioning will help us to use localized data as a key strategic variable to work with large volumes of data, which has become known as "big data".

Finally, the possibilities each individual has to develop actions and communicate on the internet has led to the development of "Personal Branding"; a step beyond the Curriculum Vitae, where each person's work leaves a mark which serves to show their skills, both dynamically and in real time, as if it were a living, breathing résumé. To quote Risto Mejide "Any proper name is a brand. The products are your works. And what you say, your campaigns" (Mejide, 2014).

Therefore this document aims to show how teachers can make their mark in the territory through appropriate use of their personal brand and geomarketing techniques, with the example of the author himself, through a six-step approach and diverse online geolocalization tools, in addition to indicators which allow us to measure the results.

For them it works on two levels linked to technology and where the teacher's geopositioning becomes the nexus of union between them both, generating information in the territory and communicating it through various social geopositioning tools such as Facebook Places, Google My Business, Foursquare and Yelp: New Technologies of Information and Communication (TICs), people, who thanks to social networks generate conversations on the internet and creative collaborative projects and New Technologies of Geographical Information (TIGs), space, which generate geopositioned information on the internet.

The text does not make reference to academic websites such as Academia.edu, Research Gate or Mendeley, since it is understood that they are part of "classic" university communication, transferred to a digital environment, and are therefore complementary tools but different to those analysed here.

In the example, the author's own activity transcends the offices of the University to leave their mark or their personal brand on other areas: whether giving a lecture, a course or going on a field trip outing and recording it through check-ins or posting it implies that they are making all the territory their communication space and their workspace, making their mark, generating a subjective map of their activity which allows them to be interrelated both on a physical and digital level with the surrounding environment and to pass on their knowledge in various settings.

We must bear in mind that information distribution nowadays is part of the "long tail" (Anderson, 2008) and therefore communication tools cater to segmented groups of demand: the use of each tool is aimed at a very specific group of users. The same event, such as a conference, reported in various tools (Facebook, Twitter, Instagram, Youtube, etc) has a huge scope but is highly targeted, so efficiency is optimal. All this, while taking into account that there must be a specific strategy which defines targets, customer segments, communication tools and measurement indicators.

Therefore, this article talks about a new education in a new liquid environment, to use the words of Z.Bauman (2003), where teachers have an obligation to be at the forefront, to communicate with future professionals learning with us, in this global classroom which is the interconnected world of the Internet.

PREVIOUS CONCEPTS

Internet, Education, and Learning

The Internet has revolutionized our lives and without doubt has revolutionized education. Over 25 years ago, Isaac Asimov (Asimov, 1988) warned, in a completely visionary way, that computers would be connected to each other and would bring about a change in the way we learn. We come from a traditional background where a teacher offers students a range of content, often unidirectionally, where the teacher was the source of knowledge and the students had to learn from them. Knowledge was not compared but protected, and all this generated élites who had control over it.

However, everyone is different, and therefore all students are different. The theory of multiple intelligences indicates that there are different types of intelligence and we cannot pass judgement on only one of them, or, in other words: "if you judge a fish by its ability to climb trees, it will live its whole life believing it is useless" (Megias, 2014). The emergence of the Internet allows a complete change in the way knowledge is transmitted; going from understanding it as a structured product to its concept as a managed service, as indicated by Genís Roca (2014): to go from an analogue product to a digital product and from there to a digital service.

Indeed, there has been a change in society which is reaching education: this, as Z. Bauman (2003) says, has become liquid and solid old structures are being surpassed by an increasingly rapidly-changing world, with a high dose of uncertainty. With this scenario, when in the classroom, students are able to compare the information given them on their laptops and mobile phones. If they need information, they have access to it in real time and from anywhere, but, in addition, when they leave the classroom they receive so many impacts are that they return with much more information.

In addition, education is approaching the "just in time" industrial concept (Yasuhiro, 1996), in as such that contents cannot be produced in a Fordist way, but each content must be adapted to an increasingly segmented market, that is to say consumers grouped by common interests. MOOCs or free university courses appear, Youtube becomes a channel for collaborative learning, Google Plus Hangouts are in real time, etc... In other words, the distribution of educational content has diversified into the "long tail" and nowadays the student has many ways to learn away from the traditional.

Therefore the teacher has the obligation to train themselves in order to better comprehend this new situation and adapt to it, so that they may offer their knowledge but with new tools and new approaches, bringing everyone closer to this new reality. It is no longer about accumulating knowledge, but rather of how to manage it, to know where this information may be found and to provide students with the tools to find what they are searching for, and to use their critical capacity to evaluate whether this provides them with what they were looking for.

This fact is not without problems that must be addressed: the loss of privacy, which involves having a very clear separation (or not) between the personal and professional environment; lack of time, which may be compensated with proper social media management, incorporating it into daily routines as part of one's daily work; lack of abilities and skills, involving continuous training by teachers; and the possibility that teachers refuse to perform these tasks, implying that mechanisms must be established governing their use, such as, for example, rewarding it, as is the case with indexation in specialized journals.

We are shifting from so-called formal learning to social learning, where the key is based on the connectivity of people and the capacity to share information, creating a collective global intelligence (Plasencia, 2015). One of the best examples is Wikipedia, a global collaborative encyclopaedia, made in an open, participative way (Plasencia, 2015) or Openstreetmap, a world map created by users with cartographic information and open data which is shared and changed (Moore, E & Garzón, C; 2010). To quote Enrique Dans, "You are what you share" (Dans, 2014); we are social beings and sharing becomes the key to the transmission of knowledge and development today.

A good example of this new way to collaborate and train is the Training Department at Alicante Provincial Council (Spain), where they are developing very interesting innovation projects in public administration (Latorre, 2015).

Social, Local, and Mobile: Geomarketing and Big Data

The SoLoMo Manifesto indicates that for some years we have found ourselves in a social environment, where we share local information on social networks through mobile devices (Reed, 2011). However, in addition, this concept is evolving towards SoCoMo, where the local element is surpassed by the contextual element, that is to say, that the significance is given not only due to the location where things happen, but for the special context of where they happen (Buhalis, 2015).

It was Tim Bernerss Lee who said "Data is the raw material of the twenty-first century" (Beltrán, 2014), that is to say, the significance of data which invades everything in volumen, variety, speed and value, in what has become known as "big data" (Invattur, 2015). The importance is not as much in obtaining data, but in the capacity to select, organize and, above all, interpret it.

Moreover, these data are generated in a specific space and therefore the location and information are linked. Geomarketing (Chasco, 2003) is the way to manage these spatial data to take decisions and to better manage projects.

These data are generated by people, and they are generated in space, and the tools which allow them to do so are the new technologies. Therefore we find two independent aspects which are at the same time interdependent:

1. Technologies of Geographical Information (TIG), spatial analysis tools which show the territory in which we move, where information is generated objectively and quantitatively.
2. Technologies of Information and Communication (TIC), communication tools which show us the people who generate objective and quantitative information.

The use of new technologies by teachers is the basis of this document, as by using it they are able to make their mark on the Internet and share their knowledge, wherever they may be, with whoever is interested. However, this requires you to have a strong personal branding which sets you apart from the rest and which gives added value.

PERSONAL BRANDING

In this new era, anyone can become a personal brand. It is said that "there is no profession of the future, but there is a professional of the future" (Chicheri, 2012). Personal branding implies that people are very important and provides added value to organizations in which they work or with which they collaborate (Pérez Ortega, 2012).

The teacher must not only be an educational employee, but must also be recognized as a specialist in the matters they dominate. The qualitative change is that this acknowledgement surpasses the limits of the scientific community in which they moved and must go on to become a social recognition through the work they share on the internet on social networks (Alcántara, 2015).

Therefore, when we talk about personal branding on the internet, we are speaking about positioning and visibility, at this moment. This is identified by appearing among the first results in Google - not only with the name of the teaching professional, but also with key terms or words in which they specialize.

In the case of the author of this article, he has worked on his personal brand by moving his profession as a geographer to the Internet, that is to say, trying to shift the analysis and methodology of this science to a new environment such as the Internet.

A fundamental aspect is to generate quality original knowledge. That is to say, in the medium term what works on the Internet is to generate knowledge and to transmit it. Therefore we may speak about two aspects: the format in which knowledge is available - which may be in the form of text, but also photography, video, audio, documents or maps; and secondly how it is transmitted by networks, distinguishing whether they are general or specific networks, or whether they are networks that are specialized depending on the type of format.

An example can be had through a conference: a teacher specialized in geopositioning and tourism offers a lecture on it, in a specific space, such as a city congress centre. From here we can use a time criterion to analyze the use of personal branding:

Before the conference, the author can generate content by way of a summary of the conference and this content can be created in video format, with a brief statement of how it will be, accompanied by an article on their blog where they will discuss this invitation. This content is transmitted to their followers on their social networks such as Facebook chat, Twitter, Google Plus and LinkedIn, as well as specific groups who may be interested in attending the conference or obtaining information from it.

During the conference the author will perform a check-in on Swarm, (formerly Foursquare), indicating that they are at the Congress Hall to give a lecture and sharing their location on Twitter and Facebook. They could also prepare a PDF with the conference and hang it on Slideshare, scheduling a few tweets through Hootsuite to be tweeted every 15 minutes, announcing that it is now available with the links. People viewing this content know that they can consume this knowledge in real time and, most importantly, viralize the content by sharing it in turn with their followers.

Furthermore, this conference could be transmitted by streaming or even with a Google Plus Hangout, so it could be viewed in real time anywhere in the world. Nowadays it could even be transmitted via Twitter using the Periscope tool.

Equally interesting it is to generate a specific hashtag on the event to work as a structuring and unifying element for the whole conference so that people can learn about the event through it. This would be available for consultation via the Google search engine and Facebook, Twitter, Google Plus, Linkedin and Instagram.

In this way, what the speaker is doing is diversifying the sources of knowledge distribution. It is no longer a question of using only attendance, but to offer their event through all possible channels, openly, to reach a wider audience, segmented in turn depending on their use of social tools.

After the conference, the speaker may share knowledge through another article on their blog, uploading a video of the conference to Youtube (in the case of Google Hangouts this is done automatically), converting the audio to an mp3 format in order to upload it to podcast platforms such as Ivoox or iTunes, etc. All this work should be duly labelled using keywords (in this case geo-positioning, social networks, tourism, etc) and shared in common conversation channels and in specialized groups for each topic area.

Therefore we can show how this new liquid reality actually works and how teachers and speakers can adapt to it and become integrated into its environment through their personal brand which, well managed, may lead them to leave their mark on the Internet by using the promotion possibilities which geomarketing has to offer.

THE CENTRAL ISSUE

The problem is the adaptation to a new medium in which the teacher must be the first to know in order to later transmit it to their students and must do so under these premises:

The importance of sharing: "You are what you share": the old formula of knowledge accumulation is no longer useful today - knowledge must be shared, because by doing so it improves with contributions from others and becomes larger and more alive. To do this, we must transform the concept of "copyright" to that of "copy is right", provided the original source of that knowledge is cited and contributes added value.

Generate conversations: As the Cluetrain Manifesto said "The market is conversations" (Levine, 2009) and in this global era, conversations occur at any time and through very different media. The fact that we converse in social media enables us to communicate with other people and share this previously cited knowledge. We are social beings and conversation is one of the ways we communicate.

Measure everything: One of the advantages of the Internet is that everything can be measured. We have tools which allow us find out who has seen or shared our publications and their social profile. "What cannot be measured cannot be improved" (Herce, 2012), but in order to do this, we will also need to know which indicators and metrics to use to obtain these measurements.

Continuous assessment: In a phase of continuous beta, the speed with which information is generated and shared today has led us to speak of "infoxication" (Cornella, 1999), a sort of poisoning brought about by information overload. To manage it, we must keep abreast of what is happening, but above all be able to select the most relevant information in order to keep up with everything happening in our area. Everything changes and we will therefore miss the boat if we are not updated continuously.

SOLUTIONS

The Aims and the Methodology

The solution lies in generating a methodology which is transferable to other cases where geomarketing tools allow the teacher to promote their personal branding and leave their mark on the territory of the internet.

After years of research and experimentation, the author has developed their own methodology which allows any professional to use geomarketing over six chronological steps based on the proposed objectives.

1. **Positioning:** The first objective proposed is positioning, that is to say to indicate the position in space of the company or person. Here there are two proposed scenarios:
   a. In the event of having a physical location where an occupation is performed, (University training centre, research centre, etc.), it is to be registered on different platforms, so that it appears in the most widely used online map. At this time it should be registered in three major platforms:
      i. **Google Maps:** The most widely used map at this time, owned by communications giant Google and which also acts as a search engine for geo-positioning information.
      ii. **Bing Maps:** Very similar to the previous but dependent on Microsoft, which has its own search engine (Bing) and its own maps.
      iii. **Here:** As a mapping provider, it has acquired importance by being the basis of the maps for the Facebook social network.

There is another platform which is currently under review: Apple Maps. Although a resounding failure on its initial release, it must not be forgotten that it is dependent on a leading company in other areas of communication and must therefore be taken into account.

For all these platforms there is a way to register places through the use of the business version or "places" with a very similar method: it is looked for on the map, if it appears then it may be claimed in order to manage it, and if it does not appear, it may be registered for management.

The result is that we have control of our information centre on the Internet and from there we can manage all the objectives which follow.

   b. In the event of not having a physical location or the possibility of managing the location, for example, such as in the author's case, who works for the University of Valencia but moves all over the country giving courses and conferences, the work methodology is different.

Social geo-positioning tools allow you to check-in or to say where in the territory a person is to be located at all times, and to share this information with their followers and on other social networks.

It is therefore a good way to "make your mark" on the territory, that is to say, to indicate each time you visit a location where specific training is to be given and even to use a specific hashtag to serve as the fundamental brand structure, sharing it on social networks, making for an even greater scope.

It is a strategy of territorial structuring strategy through personal branding. It does not have a single physical location but as many locations as the teacher chooses to visit, and so the effect is not only greater in quantity but also in quality, multiplying the impact exponentially every time a teaching project is developed at a location.

2. **Information:** All social tools offer the possibility to contribute information about a business or our person, but not all possible information is completed in all cases, missing the opportunity to make ourselves known. There are several basic elements:
   a. **Name:** The name of the organization or the person, which, if possible, must be the same in all social networks.
   b. **Photograph:** If an organization, the logo, and if a person, their face or similar picture, which shows their real image.

c. **Description:** You must be able to describe in approximately 140 characters what the business or professional does and what objective they have.
d. **Contact:** If offered the possibility of including the email address, telephone number or website, it is important to do so, so that you may be contacted.

3. **Participation:** Communication is the key, and this is based on conversations. Therefore, every time an activity takes place, it must be shared in addition to offering valuable original content on specific aspects. This will get people involved in this content, who will then in turn share and comment on this, eventually creating a user community around a specific topic.

Participation is achieved with a continuous dynamization of content and day to day consistency that makes the professional be present in the particular field of knowledge.

4. **Size:** This refers to the number of followers a professional has on social media. The fact they encourage participation on specific topics of interest makes people want to follow them and therefore their number of followers increases.

It is important to achieve an "engagement", namely an emotional closeness with the user, something that generates an interest in them following you. To have many followers is pointless and detrimental to your personal branding if you do not interact with them.

5. **Reputation:** This refers to the opinion followers have of the brand and has two similar variables on all social media:
   a. **Scoring:** Often expressed graphically by five stars where the user values the brand from 1 to 5 in order of importance. This is a quantitative and somewhat objective assessment as it allows comparison with others and shows the value given it by users.
   b. **Opinion:** Often expressed in text form, this is where the user explains what they think of the brand using comments. This is a qualitative and subjective assessment which may be basically positive, negative or neutral.

Online reputation is a core element nowadays, taking the traditional reputation of the scientific community over the years from being more closed and segmented to a reputation open to everyone on the internet, generating much more quickly and therefore having a much faster and much more incisive impact on teaching

Moreover, reputation directly affects the sales process, namely social comments are very much taken into account by users and condition them when making a decision on the consumption of a particular product or service. A positive reputation on a course may increase the demand for a teacher, whereas a negative reputation may mean the opposite.

6. **Positioning:** It refers to the presence of teacher information on the first page of Google search results.

In the past, for a business to be considered important it had to be on a main street in a big city, or a teacher had to be in a recognized department at a prestigious university. Nowadays things have changed and this main street is called Google and the most important thing is to be well positioned. This positioning is given in three ways:

a. **Naturally:** Generating quality original content which users value by linking them or commenting on them.
b. **Artificially:** Paying for advertising which allows us to appear prominently as an "advertisement".
c. **In a geolocated way:** showing where the activity is performed, which appears with a symbol of a marker or on a map.

The use of these strategies must receive proper consideration. Advertising can be paid for in order to appear on Google in the short term, but in the medium term the best strategy is to be positioned and geolocated by brand content (Maciá, 2011).

One very important aspect is that the use of diverse social media allows the brand to have an internet presence with multiple entries, with all entries even appearing on the first results page, giving us our own control over brand information online.

The result of using geomarketing tools with this methodology allows the promotion and marketing of teaching services on the Internet and therefore enhances their status.

Social Tools

Social media are tools we use to communicate with each other on the internet. There are many types and these may be grouped in many different ways, but in this case I will make a personal and practical division according to the use which is given to them:

Conversation Tools: These are the tools which we use to speak to each other. The main ones are currently Facebook, Twitter, Google Plus and Linkedin, each different from the other but with one common denominator: the possibility of generating a content which has a short duration in time and where we may follow people or people may follow us in order to consume that content.

Tools contents: These are tools which generate content in a specific format and can last longer in time, as these tools serve in turn as content stores. In the case of audiovisual content (Youtube and Vimeo), photographs (Instagram and Pinterest), podcasts (Ivoox and Soundcloud) in pdf or similar (Slideshare and Issuu) or blogs in text format (Wordpress, Tumblr and Blogspot), although these final tools have the power to allow the above formats.

Geopositioning tools: These are tools for viewing maps (Google Maps, Bing maps or Nokia Here) and tools which allow you to share information from a location (Facebook Places, Google Local, Foursquare y Yelp).

In the case of this document, the main tools which allow for the use of social geomarketing are these last ones, as they have some similar components which make them comparable:

1. There is a personal profile and a page and/or place, although the latter is the relevant criterion.
2. The possibility to create a place socially on the internet, namely without being the manager or administrator of the business.
3. The possibility to manage the business account reclaiming the business.
4. There are internal usage statistics to analyze the behaviour of users based on their location.
5. Location of a business or resources at a particular address.

6. Possibility to generate comments about a location, opinions which involve sharing a subjective perception of the location.
7. Possibility to generate scores about a location which involves sharing an objective perception of the location based on a star ratings system.
8. Possibility to perform online advertising campaigns based on location.

Social Dynamization

One of the main aspects when using these tools is the ability to dynamize them correctly and this requires four steps in chronological order:

Generate information: There must be something to say about wherever. For example, a conference a guest speaker is to offer at a university or the day to day events from the workplace itself.

Post it: The speaker will complete a check-in at the time of giving the conference and can share it on social media, in addition to sharing the conference so that everyone may have access to it. It is important to note the possibility offered by social media to post the same information on different platforms, depending on the different formats worked with.

- **Dynamize It:** Posting content is not enough; the key is to dynamize it. One has to know where to post this information and to be proactive when it comes to engaging users. In this regard, the use of segmented groups on social networks offers very good results as they generate considerable engagement.
- **Measure It:** Once these actions have been performed, the teacher should be able to measure the consequences of their actions and assess whether they have been useful in achieving proposed objectives.

From this point, feedback is made on the process and it returns to the beginning. Performance measurement provides data to improve future projects or to consolidate actions which have worked well.

MEASURING THE TOOLS

"What cannot be measured cannot be improved". Under the previously expressed maxim, it is essential to reiterate the need to take the importance of obtaining and interpreting data into account. The advantage of using these tools is that all of them have their own statistical usage data, to a greater or lesser extent, so that we can analyze how, when and where they are used, continuously evaluating the outcome in order to optimize online communication and these channels, thereby improving efficiency and meeting proposed objectives.

However, it is not enough to have only the data for each tool, because what is actually adding value is the analysis as a whole based on the methodology we have outlined above. The realization of a matrix as a control panel is therefore proposed, where both sets of variables may be analyzed:

On the one hand the objectives to be fulfilled, namely location, information, size, participation, reputation and positioning.

On the other hand the tools to analyze: Google Local, Facebook Places, Foursquare and Yelp.

The data to be measured are:

- **Location:** measured depending on whether it is located on the map (1) or if it is not (0). Information measured depending on whether it has complete information (2) or incomplete information (1).
- **Size:** measured by the number of "likes" or followers.
- **Participation:** measured by interactions with followers.
- **Reputation:** measured by scoring (five stars) and opinions (also called reviews, tips or comments)
- **Positioning:** measured by the position of the brand on the first Google page.

From here on, a column should be added indicating the objectives to be achieved and then as many columns as there are measurements you are to perform periodically.

In this case we are talking about social geopositioning tools, although this control panel enables the addition of as many measuring tools as you desire (Twitter, Pinterest, Instagram, Youtube, etc).

Faced with the problem of infoxication or attending to too many social media at the same time, there is no perfect method beyond common sense. That is to say, one should only use those that we are able to effectively maintain, and that the model is as flexible or as "liquid" as possible, i.e., to effectively change depending on the needs of each moment.

FUTURE TREND

The evolution of all these aspects is going to be increasingly faster and requires a different analysis effort and understanding. We can no longer base ourselves on strategic methodologies such as business plans, because they do not address current problems; we are unable to predict the medium term and therefore we need agile methodologies. In this sense, Micro-innovation, the Canvas Model, Empathy Model, Design Thinking and Lean Startup (Sobejano, 2015) offer a much more appropriate way to confront reality and must be incorporated into a teacher's day to day activity.

Moreover, new gadgets will appear, such as augmented reality glasses, smart watches, interconnected bracelets, flexible screens, the use of the body as a tool, nanotechnology, etc, but it must not be forgotten that these are tools and therefore simply require a computer adaptation or product. The human element continues to be the key.

The human element becomes increasingly important, not only in the concept of engagement but also in the vision of teaching from the demand point of view, to understand the student as a customer and put ourselves in their place (empathize) and use it as the core of the marketing strategy. It is essential to avoid technocratic and marketer-related trends which turn us into mere algorithm data and we must put the human and anthropocentric element at the heart of any actions.

Finally, we must understand the territory as a communication platform where each and every person acts as a communication node and the edges are messages posted and shared on the internet. Collaborative elements are key factors in its development. Now more than ever, we operate holistically in a global network. The Internet of things is directly related to the territory through smart cities which manage this data on a global scale (big data).

CONCLUSION

One of the most important aspects is the speed at which change takes place in the social, local and mobile world. “It’s not about being the best, but about being different” and we must provide some added value. In the case of teaching, I consider there to be three fundamental elements:

Real time: To be constantly informed and updated on what is happening in our specific areas of knowledge to be a reference in our field by mastering material in terms of updating knowledge. It is no longer about writing extensive articles for scientific magazines but rather a steady flow of shorter articles to be shared on our blog.

Local element: We are informed of what happens globally by news agencies, but information which happens near us is generated locally and then shared with the world. When it comes to providing training, it is important to have local data with examples to empathize with users and show them that we have made the effort to learn about local reality.

Sharing: Although we have already mentioned this aspect before, the importance of sharing knowledge without fear and, in turn, of using other shared knowledge for learning, always quoting the original source, improves collective intelligence as the basis of internet content.

REFERENCES

Alcántara, A. (2014). *Súper profesional: Tómate tu vida profesional como algo personal.* Alienta.

Anderson, C. H. (2008), *The Long Tail: Why the Future of Business is Selling Less of More. Hyperon Books.*

Beltrán López, G. (2012). Geolocalización y redes sociales: Un munso social, local y móvil (1a edición ed.). Valencia, Spain: Bubok.

Beltrán López, G. (2014). Geomarketing: geolocalización, redes sociales y turismo (1a edición ed.). Valencia, Spain: Bubok.

Buhalis, D. (2015). SoCoMo marketing for travel and tourism: Empowering co-creation of value. *Journal of Destination Marketing & Management.*

Chasco, C. (2003). El geomárketing y la distribución comercial. *Investigación y Marketing*, *79*, 6–13.

Chicheri, J. (2012). *31 historias de branding personal contadas en primera persona.* Bubok.

Cornellá, A. (1999). *Cómo sobrevivir a la infoxicación.* Academic Press.

Dans, E. (2014). *Eres lo que compartes.* Retrieved 06/25, 2014, from http://www.enriquedans.com/2014/06/eres-lo-que-compartes.html

Herce, D. (2012). *Lo que no se mide no se mejora.* Retrieved 06/25, 2012, from http://dherce.es/2012/12/23/lo-que-no-se-mide-no-se-mejora/

Invat.tur. (2015). *Big Data: retos y oportunidades para el turismo.* Retrieved from http://www.slideshare.net/invattur/estudio-big-data-retos-y-oportunidades-para-el-turismo

Latorre, J. (2015). *Innovando en formación pública.* Retrieved 06/26, 2015, from https://josanlatorre.wordpress.com/2015/06/24/innovando-en-formacion-publica/

Levine, R., & Locke, C. (2009). *El manifiesto cluetrain: El ocaso de la empresa.* Deusto SA ediciones.

Maciá, F. (2011). *Técnicas avanzadas de posicionamiento en buscadores*. Anaya Multimedia.

Megias, J. (2014). *Manifiesto contra la mediocridad.* Retrieved 06/26, 2015, from http://javiermegias.com/blog/2014/07/manifiesto-contra-la-mediocridad/

Mejide, R. (2014). *Urbrands*. Espasa Calpe.

Moore, E., & Garzón, C. (2010). Social cartography: The art of using maps to build community power. Academic Press.

Moyers, B. (2008). *Bill moyers rewind: Isaac Asimov (1988).* Retrieved 06/17, 2015, from http://www.pbs.org/moyers/journal/blog/2008/03/bill_moyers_rewind_isaac_asimo_1.html

Pérez Ortega, A. (2012). Marca personal, cómo convertirse en la opción preferente (2nd ed.) ESIC.

Plasencia, A. (2015). *Recordar el future.* Retrieved from http://adolfoplasencia.es/blog/recordar-el-futuro/

Plasencia, A. (2015). *Wikipedia: Un diálogo con Jimmy Wales.* Retrieved from http://adolfoplasencia.es/blog/wikipedia-un-dialogo-con-jimmy-wales/

Reed, R. (2011). *SOLOMO manifesto: Just about everything marketers need to know about de convergence of social, local, and mobile (SoLoMo).* Academic Press.

Roca, G. (2014). *Estrategias digitales para hacer crecer el negocio.* Congreso de hoteleros españoles, Sta Cruz de Tenerife.

Sobejano, J. (2012). *Microinnovación a la vuelta de la esquina.* capítulo del libro "Nunca te fies de un economista que no dude", Deusto ediciones.

Yasuhiro, M. (1996). *El Just in time hoy en Toyota*. Deusto SA ediciones.

Chapter 12
Audiobranding and Its Importance to Your Personal Brand:
New Ways to Communicate Our Messages through Sound

Raúl Terol
Polytechnic University of Valencia, Spain

ABSTRACT

In this chapter we will try to discover the importance of audio and voice as key elements in building a personal brand. Podcast, audioblogs and web tools that reference the sound will be analyzed based on their degree of influence on personal branding. By using the voice we can differentiate ourselves from our competitors, so the audio branding, or sonic branding as it is also known, shows a verbal identity of his own personality. The sound has always been known for being a great tool to convey memorable messages to consumers, in fact we are able to consume the sound since we are in the womb. Having an audio brand is one way to ensure that no other company uses a similar audio and make our brand is consistent. The radio language, consisting of voice, music, silence and the special effects are responsible for generating emotions in the receiving public. The audio conveys information, entertains consumers and in the long term, helps build a positive image that reinforces the brand values. The design of an audio brand goes beyond creating a catchy jingle or a piece of melody.

INTRODUCTION

Nowadays the importance of audio gains points in the field of personal branding. Internet favors largely interactivity between transmitter and receiver, generating a feedback once nonexistent or very difficult to take place. In his radio theory, the German writer and poet Bertolt Brecht commented that radio, media-based audio transmission, should be able not only to send a message but also receive messages from listeners to know the effect that this had occurred. A radio that not only broadcasted but also was

DOI: 10.4018/978-1-5225-0917-2.ch012

able to receive, something unusual for the time when he proclaimed his theory, between 1927 and 1932, but today takes a relatively large force through social media and the high degree of connection users (Brecht, 1927). In 1995, another visionary Nicholas Negroponte in this case, he dared to venture that we were moving into a world in which people would be permanently connected. It was undoubtedly the digital world well recounted in his text (Negroponte, 1995). Twenty years later, most people have a high technological dependence. A study by *Nokia* in 2012 revealed that each person looking at his smartphone 150 times a day, approximately every six minutes. Tomi Ahonen, a former manager of the company's mobile terminals, exposed in one of his lectures an update of this study, where it was shown that by 2014 smartphone users looked at him 221 times a day, about every 4 minutes and 18 seconds approximately. This technological dependence means that companies must be present in the mobile market, where customers will access more easily, so you can interact with them.

Traditionally it is said that a picture is worth a thousand words, but not only of image lives the Personnel Branding. The voice is a communication tool that transmits non-verbal communication. "Studies also show that although someone we like it at first, the attraction can disappear right away, how they communicate, especially because of his voice, because the most important is not what someone says but how he says" (Glass, 1994: 205-206). For this reason, the voice becomes very important in radio, so it is one of the four pillars of which Balsebre listed as elements of radio language, along with music, sound effects, all without forgetting the silence.

The information and communications technology (ICT) make it possible to communicate is at the reach of our hand and, further, to communicate through audio, or video, there is something available to media technologies high cost and inaccessible licenses. The live audio transmission, which by streaming or podcasting, radio programs or parts of prerecorded audio, hosted on the network and synchronized with users who have previously subscribed, known are some of the tools based on the audio will help to project our brand image on the Internet. There are specialized podcasts in areas or sections, in most cases, are related to the tastes of the creator of the same, but also include the existence of online radio stations in which the value of the brand is praised. It is specialized radios whose emission is done via the Internet and that allow us to strengthen our self-image or our brand or product. In this chapter we will take care to publicize some success stories, and review some of the platforms where our pieces to accommodate audio or podcasts, so that our potential customers can go to them.

INTERACTIVITY IN ONLINE RADIO

Technological user today has become a combination between the producer-consumer, which has been called *prosumer*. At your fingertips you have tools with which to produce their own sound files, upload them to a web page and become a transmitter. A *prosumer*, like any user who consume online radio, follow the rule of three A: Anywhere, anytime, anyway (Roel, 2010). The user becomes a programmer, selecting as desired the content found on the different pages (Roel, 2010). Thanks to this selective capacity of the new interactive listeners, the *prosumer* can develop their own agenda setting (Rufí Pérez, 2013). Before the emergence of the term, Jean Cloutier (1973) coined the expression *EMEREC*, an acronym combining the *émetteur* and *récepteur* words. This term was intended to describe a process Cloutier through which transmitter and receiver were producers and receivers of messages, which Martinez Zarandona defined as a mutual exchange between them. A communication model exposes Palma Peña, who serve for sending and receiving an active role, enabling them to establish patterns of dialogue for the "*emirecs*" the spanish version, build relationships in the horizontal plane and in a more or less democratized, where you can freely express their tastes and preferences. For Garcia Matilla, EMEREC theory "transgresses the conventions of classical communication schemes in which it comes to giving a chain of transmission and response between sender and receiver".

This is exactly what Bertold Brecht asked in the early 1930s, when formulating the Theories of Radio. The German poet and playwright stressed the communicative character that defined the radio media and postulated that radio would be the best means of communication if their function is not limited to distributing a message, but also could receive, instead of isolating the listener achieved do enter into a relationship with the environment and participation in, the contents:

I remember how you first heard about the radio. (...) He felt that it was a matter not only fashionable, but really modern. (...) A colossal triumph of technique finally able to put within reach of the whole world a Viennese waltz and a recipe. (...) In my opinion, you should (addressing the artistic directors of the radio) Radio trying to make a democratic thing really. (...) The radio has a face, which should have two. It's a simple dispenser simply dealt.

And to be positive, that is, to discover the positive aspects of broadcasting, a proposal to change the operation of the radio: we must transform the radio, turn distribution device communication device. The radio would be the most fabulous communication device imaginable from public life, a system of fantastic pipeline, that is, if he knew it would not only transmit but also receive, therefore the listener not only hear but also make him talk, not isolate, but to communicate with him.

(...) If you consider this utopian, I beg you to reflect on why it is utopian (Brecht, 1927).

The *prosumer* and *emirec* concepts connect clearly and effectively with the proposals formulated by Brecht more than eighty years ago and that "the radio should leave the business of supplying and organizing his listeners as suppliers" (Brecht, 1927).

In his treatise Dialogues, Plato, three centuries B.C., showed their disagreement with the new invention of the book, as in his opinion it was "not interlocutive entity with which one can not discuss" (Del Rio Pereda, 1996). To Palma Peña what the Greek philosopher missed, albeit unknowingly, was a medium like radio and, thanks to its analysis of the shortcomings of the book defines and anticipates its creation (Peña Jimenez & Ventero Velasco, 2011: 235).

Interactivity is responsible for establishing a different relationship as far as participation is concerned. In the traditional radio model participation is a piece that is part of the show and the audience involvement in program content, perceived as a whole and not as two different actions, argues Mariano Cebrián Herreros. The audience follows the plot laid out by the station, as well as the different interventions of the participants, a link to another so that may arise from what has been shown by the broadcaster or expressed in the interventions. Moreover, as Cebrián Herreros classified as the "*ciberradio*" interactivity is comprised of a rather different way, "is a dialogue between the parties do not have to be followed by the others but is established individually between what is offered by the station and the involvement of every Internet" (Cebrián Herreros, 2008: 46).

We are facing a change that impacts the full radio media. The traditional radio, found with the phone the best ally to achieve a high degree of audience participation in its programs. Successful song requests, calls to see a specialist, to vote any of the proposed options or to give his testimony about what he was talking about. It was along the traditional letter by mail, one of the few chances to get feedback from the recipient of your message.

The appearance of text messages from mobile phones, known by SMS, or multimedia messages, MMS, would start producing a change in the radio medium in terms of interactivity is concerned. Email, chat applications like *Windows Live Messenger* or *Google Talk*, are very valuable input in this regard. *IPVoice* services such as *Skype*, one of the most widely deployed today. *Skype* allows calls over the Internet for free, or pay if what you want is to contact any telephone line. Cebrián Herreros postulated that the sound created great difficulties in network development, especially as it related to speech recognition. Therefore, he claimed to be followed carefully the development of that system and observing attentively those specific applications that can take place in the *ciberradio* (Cebrián Herreros, 2008: 72). Nowadays, stations like *Europa FM*, in the weekend show where listeners are the protagonists through musical dedications, have included new ways of participation among which voice messages through the application *Whatsapp*.

In the radio media, the concept of interactivity is somewhat higher than the social interaction between individuals. Computer and telematics are responsible for making possible human-computer connection to access the information in the network, or connect to two or more alternating the roles of sender and receiver traditional users to become interactive through the sounds, either using a service *IPVoice* or sharing audio files.

One of the main characteristics of radio interactivity is the peculiarity of all communication components involved, as well as by the differences in respect of variants that can be seen in other fields. The items of note are (Cebrián Herreros, 2008: 41):

- An exchange in the roles of users. The issuer ceases to be one to become participants in various radio communication processes.
- Technical mediation produced by the computer and telematics, through the vast fabric of wire line, wireless or mobile networks.
- The contents are set in an open way so that they can be expanded and modified by other users or willing to exchange. Content developed at the same time or stored waiting for someone to interact with them. Synchronous interactivity and other asynchronous appears as Cebrián Herreros aspect that makes the difference with the radio broadcast by radio, dialogic and participatory above.
- The messages are reflected in oral, musical or environmental sounds, but also secondarily are recorded through writing or iconic visual components.

Using content modifies the spatial components. The barriers that limited space and time are demolished and now the space fits all the coverage of the networks through a local outreach to other global. The temporal dimension expands its traditional instant issuance, we knew that the fugacity of the message, now it becomes timeless. With asynchrony added value, it is time to raise participation and interactivity (Cebrián Herreros, 2008: 41).

THE IMPORTANCE OF SOUND IN PERSONAL BRANDING

Audio branding is also known by the name of sonic branding and tries to give an identity through sound (music, voice...). It's a verbal identity that promotes the creation of our personal brand, through the use of voice mainly. The audio logo is one of the tools with which we define our brand. An audio logo is short, a characteristic tune, also known as it jingles, or another sequence of sound. Often position them, usually at the beginning or at the end of the announcement. Through sound we reinforce personal identity or company and brand. Thus, it should be noted that the sound has always been a great tool for the transmission of messages that they are easily memorable by the receiving public, we all remember examples of wedges radio or TV spots in which has been us the catchy chorus of the jingle. Since we are an embryo already we are potential consumers of audio. With only twelve weeks of gestation already the auditory capacity develops in the fetus, a long before the ears appear, across every cell, we listen to the beatings of the heart of our mother. It is a question of the first sense that we develop and it will be the last one that is still operative before our death.

Unlike the sight, the ears remain active 24 hours of the day. While we dream our ears they are analyzing the sounds that are received. Our auditory range is ten times major that our visual scope, in terms of relative frequency. In intensity, our audible range is 10,000 times higher than the scope of our vision. We will be able to understand a good display, even using discordant colors; nevertheless, it can be that we never manage to understand a good presentation in a space with a terrible sound.

But the integration of the sound in the construction of the personal image was not an easy task, makes clear Maria Elena Durón in his article. Although the image and video have greater acceptance among users and has a better web positioning, the audio is gaining strength. Having an audio will make us more powerful identify our brand image. The mind has an amazing ability to memorize a sound. In Alzheimer's patients it has shown that they can remember and sing songs long after they have ceased to recognize faces and names. There is growing evidence that listening to music can help stimulate lost memories and even to take part in the restoration of some cognitive function.

Through sound emotions are transmitted, which will allow us to believe an emotional connection with the recipients of our message. The audio transmits information and allows the positive association with the clients that we have and favors the consolidation of the values of brand. The way more common of communicating is by means of the sound, for what it is one of the principal motives to bearing in mind at the moment of starting a plan of communication and defining which will be the communicative strategy to continuing in order to start a consistent brand image. We will be able to create a good or bad sound, in the first case we will observe as the process of communication it develops in an effective way and that exists, on the part of the clients, a feedback who gives us to understand that our strategy had successful. Across the sound we transmit the meaning and the ideas. As Durón appears, in previous studies one could have verified that an audio brand is so much or more striking than a visual brand. In the case in which our creation does not meet with the expectations, we will realize that the meaning or the energy differ from the values of the brand and there is not transmitted really what is wished. An audio that lacks consistency, or rather be confused, it can generate a disconnection in the recipient, which will meet reflected with the loss of a client.

To design an audio brand hides a work of major density that the one that could be create a short melody or a sticky jingle that hammers in the heads of our users. First, they should be heard large number of tracks to know that you are choosing the best option for it to be transmitted through the audio. In case of the personal image, he is you the one that must be at the head of all this process, you know what values you would like to convey in every aspect of your brand. You have to take with time and research; interact with the producer and the writer to see first-hand that is what they arrange to do before beginning his work. Hereby you will turn into the strategist and leader of his own personal brand, and he makes sure himself that everything what expresses in his communicative strategy is that one that really wanted to transmit across his brand.

The sound is very effective, even to subliminal level. For Lars Deutsch, the companies that they do not invest in the audio branding are stopping to spend a great opportunity of marketing. On the contrary, the companies that yes do it, enrich his brands and enjoy a clear advantage on his competitors. The audio logo is the acoustic equivalent to the visual logo; a combination of both will use to generate the recognition of a brand. A logo is the face of a brand; in the audio the logo is his voice.

THE USE OF THE VOICE AS A DIFFERENTIATOR

The voice, which Armando Balsebre names as the word, is one of the elements of the radio language with major importance together with the music, the effects of sound or the silence. For the construction of our brand image across the audio we can use any of these elements, but the one that undoubtedly is going to receive a singular protagonism is going to be the word and for the person who is going to be pronounced. Hereby, for Balsebre, the words were characterized by his color, his melody, and his harmony and for his pace (Balsebre, 1994). These characteristics are those that will come to define our message and will be the persons in charge of the image that the consumer recreates in his mind. Across these characteristics we are going to transmit the values that we have decided to grant to our brand.

The appearance of the color of the voice is based on the existing interrelationship between the dimensions that define in an acoustic way the sound of the word: the ringer, the tone and the intensity. Hereby, for Balsebre, the "color" treats itself about the result of the above-mentioned interrelationship in the form in which we perceive it. The ringer, for your part, is defined by the spectral composition, the form of assault and extinction of the transitory ones and the number and distribution of the "formantic zones". The tone, which measurement is realized in cycles for second, is a basic parameter to be able to recognize a masculine or feminine voice. On the other hand, the intensity of the voice will depend on the force in the one that breathes. When there is realized a voluntary control of the pace of breathing will be exercised a decisive function regarding the control of the intensity of the voice (Balsebre, 1994: 46). The melody of the word enunciates the dramatization of the reality that is transmitted. Across her they remain registered the intention of reporting and the affective movement. Hereby, the audio, it is capable of transmitting a great affectability. The melody also is capable of expressing the notion of continuity of the reported message. The ringer, the tone and the intensity of the voice are those who are going to grant him a color or other one to our message, and thanks to them we will be capable of transmitting in an effective way those aspects that we have described in our communicative strategy. The harmony of the word is defined, while musically, for being a psychoacoustic dimension of the sound that proves on having produced notes of a simultaneous way. And, finally, the pace of the word, it refers to the dimension that fixes the nature of the periodicity in which the different sonorous sequences are perceived.

The qualities of our voice are going to be extremely important and, into some cases they will turn even decisive, regarding the social and public relations. They are going to determine what the others think about us. And it is that not only we judge the individuals for his image, but also for his voice. If all these characteristics that we have enumerated can determine the relation between the persons, they are going to be decisive at the moment of communicating our message and creating, in the mind of our clients, the brand image that we want to transmit.

NEW TOOLS FOR COMMUNICATING WITH THE AUDIO

As we were saying, the Internet arrival has made possible the appearance of new tools that favor the communication between the persons. Hereby, the use of the programs of instantaneous messenger company for the interpersonal communication, or video audio transmission or to users' multitude, is what does that we could understand the power of communication that has the web 2.0. Other one of the characteristics of this web is the creation of the social networks, where every day more users are connected and communicate between networks of contacts. Networks across which the persons can share videos, images, texts and, though in minor measure, audio. If we were speaking about the step of the consumer to the *prosumer*, there has come the moment to observe with what tools the user can communicate his image across the audio.

The streaming is the principal tool to issue live sounds and that the others could connect to listening to our product. It is a question of the radio online, if there exists a daily and normalized emission, though it can serve us to broadcast some event of our company of a sporadic way. One of the principal characteristics that possess the radio online, according to the graph that we estimate in the part of low of this text, is that it favors the brand image, the branding. And clear, if we speak about brand and personal brand, transmitting across the radio online, it would be an audio example branding. No doubt it is one of most economic ways of being able to create a brand image, across the sound, in the mind of our user or consumer, in such a way that it connects to issuer online from our web page or from the application for smartphone where it wants that he meets.

Figure 1. Graph on the radio online
(IAB Spain, 2014)
Source: IAB Spain, 2014

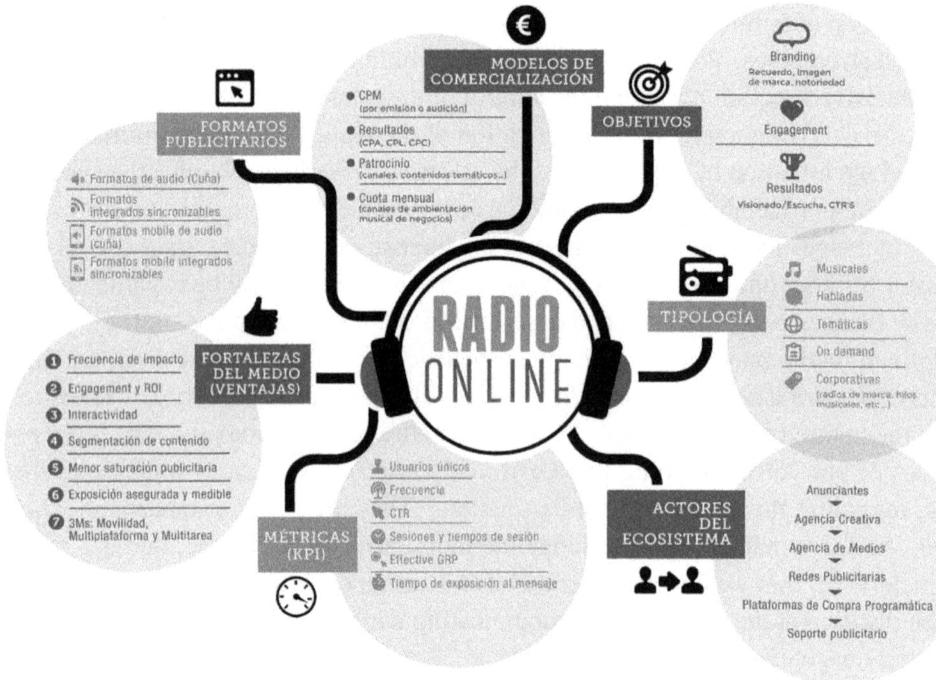

An audio example branding offered across the creation of a radio online is the fact that Francisco Sanchez and Cristina Mosquera offers the company *Expressa*, founded in the year 2011 by Nacho Montero, with the initial capital of 100.000 Euros. His mission is to create, to manage and to develop radios across Internet for companies or commercial brands. Radio stations online for companies, organisms and institutions in order that they could develop his strategy of marketing and be able like that to spread his products, promotions and all kinds of information that could be of interest both for his own employees and for the final client. Between its services is of offering announcers, music, programming besides the whole putting in March to technological level, with the creation of the application for mobiles, the listen of the issuer in his web or intranet...

Types of Corporate Radio

When we talk about corporate radius we refer to four type's radio: Brand radio, In store radio, Staff radio and Event radio. Let's see what the characteristics that define each it.

First, Brand radio is an online radio station that has the main objective of enhancing the recognition and own brand identity or company. This type of corporate radio allows us to develop personalized radio design and high quality, with music and information that reinforce the values that you wish to convey. We can also create additional channels in order to promote a product or a segment of our production in particular, so we connect with different customer segments, and could also be used to reinforce specific advertising campaigns. *Universia* has a radio channel for users of its portal, *Banco Santander*, which have one million registered users and a monthly audience of 50,000 listeners. But not only music, ads and corporate messages feed the Brand radios, but its content goes far beyond. For *Galp FM*, online radio station GALP Energy, its production combines contemporary music and of all time, meeting these successes in space "*Lo esencial de Galp FM*", but also offers its listeners programs Radio as "*Pinchos, parches y tachuelas*" are by the musician Armando Soria, or "*Centrifugado*" by the presenter, writer and director of musical programs within Jose Luis Casado.

Meanwhile, *In store Radio* is an innovative marketing and communication tool that unifies music and messages at all points of sale. The music of a store, sometimes tune in large national chains or a local radio station can help create our brand audio. Obviously if you only have a shop our efforts are focused in other areas, but if it is a franchise with stores throughout the state would be interesting to unify the messages that are thrown at them. A clear example of this type of radio is used in supermarket chains, which have listings of their products and store messages always to corporate voice. In store Radios are completely flexible and customizable. One example is the company Verdecora, which provides this service to its hypermarkets in Spain, with an estimated 200.000-listeners/month hearing.

The following types of corporate radius we find the Staff Radio systems, which are designed with the definite purpose of providing workplaces for a custom musical and informative solution. So, allow content adaptation employee profile and issuing specific messages that are addressed to workers. An example of such messages can range from congratulations on your birthday or anniversary, to call for meetings, events, and new training courses... It is an innovative element of internal communication of the company that is very effective as a motivational tool and support to the various work teams. Companies like *ThePhoneHouse* or *Imaginarium* stores have this service.

Finally we reached the Event radio, this type of corporate Radio is broadcast live any event, whether corporate or public, for brands or companies, using a specific online radio channel for the event. It relies on social networks to win in impact on your audience and take advantage to become viral. It is based on interactivity, something that we pointed out earlier in this chapter, as well as the involvement of the followers of the event, with the aim of obtaining a large number of followers and achieves customer loyalty to the brand. This format increasingly companies are using it, including trade fairs to promote their content and their brands. So, there are companies, such as *Sonido on Line*, in charge of assembling the online radio for your company or event, so they do not have to worry at all about the technical specifications. The company slogan is "do the radio your event, we sound to your brand". *Sonido on Line* bases its sound on online business to promote friendly communication, direct, emotional and reliable basis; propose the creation of a community and transforming customers into fans by identifying with the values of the brand; the issue can be followed from anywhere and anytime; create content to further disseminate and add value in social networks; own means is created to control the final message; is selected accurately a target audience and analytical data are achieved. They claim from the company "radio is a simple way to go to the public to present your products, provide added value to your customers and create community around your company".

Podcasting or Goodbye to Fleetingness Message

One of the primordial characteristics of the radio was the transience of his message. Listening to the radio the maximum of the "here and now" is established. Another problem was that the topography of the area where the transmitter and the antenna was going to limit our scope of coverage. All this disappeared with the advent of ICT and Web 2.0. The streaming that we talked breaks down geographical barriers and policies and will be the podcasting the cause that we can go to an audio as often as we want. And more to go, which is the audio which is downloaded to your device when upgrading, as discussed below.

Podcasting has its direct antecedent in the service of "Radio on demand", in fact it is the same service but improved through Really Simple Syndication (RSS) technology. In this way, the creator of the audio file will handle hang on the digital platform its new program and the users/listeners receive it on their computer or mobile phone and can download the latest edition of the program or podcast, which have been subscribed previously. This service might think it's exclusively for radio, but it's not like that, anyone with a computer, a microphone, a small mixer and, what is more important, something to tell, can have a podcast to the Internet, according to your interests, you can go subscribing.

To Audio branding this format can be very useful for communicating with customers on a regular basis, the latest news we have in our company and thus loyalty and transmit audio with our brand image. It is accessible technology and easy to use. The company *Thyssen Krupp* performed monthly basis, a podcast in Spanish and English, full of internal company information and addressed to all employees a twenty offices throughout Europe, obtaining an estimated audience of 20.000 downloads a month.

One of the most important aspects in producing a podcast are its contents, almost as much as the technology that might be available. Xavier Sellas says that the way the listener breaks the traditional way to consume and radio programming, so you decide what, when, how and where to listen to the radio. If the break between space and time is one of the most important consequences of integration between radio and the Internet, what will become critical will be the relationship that is established between the user Internet and audio, either as a consumer or creator thereof (Sellas, 2008).

The sociologist Manuel Castells states those applications such as podcasting and everything that is related to Internet communication is closely linked to freedom of expression in all its forms, based on free and interactive expression (Castells, 2001: 256).

MAIN PLATFORMS FOR INTERNET AUDIO BROADCATSING

There are companies, as we have seen, they offer their services from installation and set-up of our corporate radio, but what degree of difficulty takes this process? To get the issue of what will be our online through streaming station, we will need a computer connected to the Internet and you will enter through the audio input, what we will issue. You can use different applications for which broadcast via streaming, some of them free as are the *Shoutcast*, *Radionomy*, the same company that acquired and currently distributes the *Winamp* player. This program will launch our broadcast directly to a specialized server, so it is advisable to hire the services of companies offering online radio servers at a low monthly cost, a rate that will depend on listeners have the radio and compression and quality of sound in kilobytes per second (Kbps) we want to have our broadcast. The listener just going to have to go to our homepage to click where it tells can listen to online radio, and he does not need to know all the technical framework that we outlined.

Another option available to the user is download the application for your smartphone in which the station is heard directly through streaming. If we have not created your own application, you can always resort to applications acting as databases online radios and where we can listen to virtually all who are in the network.

Each time the audio streaming service becomes more important and that is why many large companies in the ICT sector want to be present in this market. A market that dawned with the endless controversies that have always aroused copyright and which they were led by strong companies like *Spotify*, with more than 20 million subscribers among its 75 million users, *Rdio*, *Pandora*, *Deezer*. The emergence this summer 2015 *Apple* audio streaming has broken all the schemes and it's open season. The *Google* Company also wants to have a presence in this market and announced that soon will bring one service with which to compete with *Apple Music* and existing. Returning to the Cupertino Company, CEO Tim Cook announced that the service will be available from autumn to users of Android, and its cost will be about $ 9.99 per month. One of the innovations is the launch of *Beats 1*, a global radio broadcasts 24 hours a day, with songs of new artists, albums and renowned presenters in the US.

In Spain the audio streaming services lead us to the example of the innovative online radio station that created the announcer José Antonio Abellán. It is *Radio4G*, which offers several themed stations in addition to the general where he performs his morning show every day with that achieved fame, "*La Jungla*". It also serves as a database for hear other online radio stations or create your own radio based on your musical tastes, which would be a musical prescriber. An example of the latter would be launching *RadioGym*, an online radio station dedicated to the people who are doing sport. The radio tunes into your rhythm as well like to be reminded, offers a reproducer with five music channels classified according to their pulsations or beats per minute, the athlete can hear the rhythm music coaching. Besides this novelty, the streaming service has four thematic channels that have been created specifically for the practice of running, cycling, walking and yoga.

Moreover, we find podcasting, a much more simple technology at the user level if we use the platforms, free in most cases, which offer this service. So there are different companies that are going to provide services to hang the podcast on the net. In Spain we *Ivoox*, a service which describes itself as the audio kiosk, which allows us to raise our program using an MP3 audio file format and through RSS technology, users to organize Web content, as well as share in different social networks that we have user account. The service offers *Ivoox* is completely free and without any restriction on length of the content. Created in 2008 by Juan Ignacio Solera, the *Ivoox* platform about gathering podcasts and audios of all sorts of topics providing interaction and classify as. His initial idea was to facilitate access, creation, classification and dissemination of podcast.

Another service that can be found on the Internet is *Audioboo*, which allows us to record content anytime, anywhere, generating its own channel of content so that any listener can subscribe and follow on the Web. *Streamcaster* of *Gunpmedia*, we will facilitate access to different services of podcasting through a simple interface that will allow us to move through the different podcast to which we subscribe and be able to hear.

CONCLUSION

The sound through the creation of an online radio or by posting a series of podcasts with a certain periodicity, may favor the consolidation of our personal brand or our company. The sound has lain dormant throughout these years of the XXI century, but with the arrival of Beats1 by Apple and all the work that has been developing during this time, will change the current landscape communication through the network. Both streaming and podcast are two of the tricks with playing audio on the Web 2.0 and can make a big impact on our target audience and achieve what we want, create a brand image in the consumer's mind and transmit a series of values. The video did not kill the radio star in the early 80's, and YouTube has removed this medium centenary but emerges more strongly. Apple and Google have announced their commitment in this 2015 for the streaming of music, something magical has radio, and in general, the audio on the Internet that we can't forget him.

REFERENCES

Ahonen, T. (2011). *Main Trends in the Telecommunications Market.* Retrieved April 15, 2015 from http://www.citia.co.uk/content/files/50_44-887.pdf

Balsebre, A. (1994). *El lenguaje radiofónico.* Madrid: Cátedra.

Brecht, B. (1927). Teoría de la radio. In *De las ondas rojas a las radios libres.* Barcelona: Gustavo Gili.

Castells, M. (2001). *La Galaxia Internet.* Barcelona: Random House Mondadori.

Cebrián Herreros, M. (2008). *La radio en Internet: De la ciberradio a las redes sociales y la radio móvil.* Buenos Aires: La Crujía.

Cloutier, J. (1973). *La communication audio-scripto-visuelle à heure des self-media, ou l'ère d'Emerec.* Montreal: Presses de l'Université de Montreal.

Del Río Pereda, P. (1996). *Psicología de los medios de comunicación: hacia el diseño sociocultural en comunicación audiovisual.* Madrid: Síntesis.

Deutsch, L. (2012). *Can you hear your Brand.* Retrieved April 13, 2015, from http://www.larsdeutsch.net/doc/CanYouHearYourBrand.pdf

Durón, M. E. (2013). *Audiobranding and its importance to your personal Brand.* Retrieved April 13, 2015, from http://www.personalbrandingblog.com/audio-branding-and-its-importance-to-your-personal-brand/

Gallego Pérez, J.I. (2010). *Podcasting. Nuevos modelos de distribución para los contenidos sonoros.* Barcelona: UOC Press.

García Matilla, A. (2009). *El concepto de Emirec. Los medios para la comunicación educativa.* Retrieved August 5, 2014, from http://ayura.udea.edu.co/medios/documentos/EL%20CONCEPTO%20DE%20EMIREC

Glass, L. (1994). *Cómo expresarse correctamente.* Barcelona: Paidós.

Martínez Zarandona, I. (2005). *Tres pilares de la educación para los medios.* Retrieved August 2, 2014, from http://www.uned.es/ntedu/espanol/master/primero/modulos/teorias-del-aprendizaje-y-comunicacion-educativa/artirene.htm

Negroponte, N. (2000). *El mundo digital.* Barcelona: Ediciones B.

Peña Jiménez, P. (2012). Interactividad y redes sociales en la radio española, nuevas formas de participación. *Telos*, (92), 105–117.

Peña Jiménez, P., & Ventero Velasco, M. (2011). Nuevas formas de participación en radio. In *Radio 3.0: Una nueva radio para una nueva era.* Madrid: Fragua.

Pérez Rufí, J. P. (Ed.). (2013). *Nuevos tiempos para la industria radiofónica en España.* Málaga: EUMED - Universidad de Málaga.

Roel, M. (2010). Desafíos de la televisión ante la consolidación del ecosistema digital. *Ámbitos, 19*, 25-42.

Sellas Güell, T. (2011). *El podcasting. La (r)evolución sonora.* Barcelona: UOC.

Woollaston, V. (2014). *How often do you look at your phone? The average user now picks up their device more than 1,500 times a week.* Retrieved April 15, 2015, from http://www.dailymail.co.uk/sciencetech/article-2783677/How-YOU-look-phone-The-average-user-picks-device-1-500-times-day.html

KEY TERMS AND DEFINITIONS

Audiobrand: Also known as Sonic branding, this term we mean when we use the sound in order to strengthen the brand identity, whether personal or business.

Event Radio: It is a type of corporate radio handles enhances the brand image through the live broadcast by streaming events.

In Store Radio: A tool that allows customization of shopping centers through musical, informative and advertising solution homogeneous and integrated manner.

Interactivity: The level of relationship established between user/s and other user/s (social relationships), and/or between a user and machine (computer, video game, traditional media like radio, television, press...), with the claim that active and feel part of the communication process, protagonist able to decide on the future of a number of future events, although this implementation capacity is limited by the options provided by the machine.

Podcasting: is the term resulting from the union of the English term pod, capsule (which also gives its name to the popular iPod portable audio player), and Casting, short for the word broadcasting, mass distribution of content. The origin is also associated with the acronym POD: Portable On Demand (portable on demand).

Radio Brand: A model of online radio whose main objective is to promote recognition and brand identity through voice and sound.

Radio Staff: These systems have been designed to provide workplaces with a personalized musical and informative solution.

Streaming: Is a technology for transmitting and receiving audio and video on the Internet alleging that the sound or visual document plays as it downloads to your computer.

Related References

To continue our tradition of advancing academic research, we have compiled a list of recommended IGI Global readings. These references will provide additional information and guidance to further enrich your knowledge and assist you with your own research and future publications.

Adeyemo, O. (2013). The nationwide health information network: A biometric approach to prevent medical identity theft. In *User-driven healthcare: Concepts, methodologies, tools, and applications* (pp. 1636–1649). Hershey, PA: Medical Information Science Reference; doi:10.4018/978-1-4666-2770-3.ch081

Adler, M., & Henman, P. (2009). Justice beyond the courts: The implications of computerisation for procedural justice in social security. In A. Martínez & P. Abat (Eds.), *E-justice: Using information communication technologies in the court system* (pp. 65–86). Hershey, PA: Information Science Reference; doi:10.4018/978-1-59904-998-4.ch005

Aflalo, E., & Gabay, E. (2013). An information system for coping with student dropout. In L. Tomei (Ed.), *Learning tools and teaching approaches through ICT advancements* (pp. 176–187). Hershey, PA: Information Science Reference; doi:10.4018/978-1-4666-2017-9.ch016

Ahmed, M. A., Janssen, M., & van den Hoven, J. (2012). Value sensitive transfer (VST) of systems among countries: Towards a framework. *International Journal of Electronic Government Research, 8*(1), 26–42. doi:10.4018/jegr.2012010102

Aikins, S. K. (2008). Issues and trends in internet-based citizen participation. In G. Garson & M. Khosrow-Pour (Eds.), *Handbook of research on public information technology* (pp. 31–40). Hershey, PA: Information Science Reference; doi:10.4018/978-1-59904-857-4.ch004

Aikins, S. K. (2009). A comparative study of municipal adoption of internet-based citizen participation. In C. Reddick (Ed.), *Handbook of research on strategies for local e-government adoption and implementation: Comparative studies* (pp. 206–230). Hershey, PA: Information Science Reference; doi:10.4018/978-1-60566-282-4.ch011

Aikins, S. K. (2012). Improving e-government project management: Best practices and critical success factors. In *Digital democracy: Concepts, methodologies, tools, and applications* (pp. 1314–1332). Hershey, PA: Information Science Reference; doi:10.4018/978-1-4666-1740-7.ch065

Akabawi, M. S. (2011). Ghabbour group ERP deployment: Learning from past technology failures. In E. Business Research and Case Center (Ed.), Cases on business and management in the MENA region: New trends and opportunities (pp. 177-203). Hershey, PA: Business Science Reference. doi:10.4018/978-1-60960-583-4.ch012

Akabawi, M. S. (2013). Ghabbour group ERP deployment: Learning from past technology failures. In *Industrial engineering: Concepts, methodologies, tools, and applications* (pp. 933–958). Hershey, PA: Engineering Science Reference; doi:10.4018/978-1-4666-1945-6.ch051

Akbulut, A. Y., & Motwani, J. (2008). Integration and information sharing in e-government. In G. Putnik & M. Cruz-Cunha (Eds.), *Encyclopedia of networked and virtual organizations* (pp. 729–734). Hershey, PA: Information Science Reference; doi:10.4018/978-1-59904-885-7.ch096

Akers, E. J. (2008). Technology diffusion in public administration. In G. Garson & M. Khosrow-Pour (Eds.), *Handbook of research on public information technology* (pp. 339–348). Hershey, PA: Information Science Reference; doi:10.4018/978-1-59904-857-4.ch033

Al-Shafi, S. (2008). Free wireless internet park services: An investigation of technology adoption in Qatar from a citizens' perspective. *Journal of Cases on Information Technology*, *10*(3), 21–34. doi:10.4018/jcit.2008070103

Al-Shafi, S., & Weerakkody, V. (2009). Implementing free wi-fi in public parks: An empirical study in Qatar. *International Journal of Electronic Government Research*, *5*(3), 21–35. doi:10.4018/jegr.2009070102

Aladwani, A. M. (2002). Organizational actions, computer attitudes and end-user satisfaction in public organizations: An empirical study. In C. Snodgrass & E. Szewczak (Eds.), *Human factors in information systems* (pp. 153–168). Hershey, PA: IRM Press; doi:10.4018/978-1-931777-10-0.ch012

Aladwani, A. M. (2002). Organizational actions, computer attitudes, and end-user satisfaction in public organizations: An empirical study. *Journal of Organizational and End User Computing*, *14*(1), 42–49. doi:10.4018/joeuc.2002010104

Allen, B., Juillet, L., Paquet, G., & Roy, J. (2005). E-government and private-public partnerships: Relational challenges and strategic directions. In M. Khosrow-Pour (Ed.), *Practicing e-government: A global perspective* (pp. 364–382). Hershey, PA: Idea Group Publishing; doi:10.4018/978-1-59140-637-2.ch016

Alshawaf, A., & Knalil, O. E. (2008). IS success factors and IS organizational impact: Does ownership type matter in Kuwait? *International Journal of Enterprise Information Systems*, *4*(2), 13–33. doi:10.4018/jeis.2008040102

Ambali, A. R. (2009). Digital divide and its implication on Malaysian e-government: Policy initiatives. In H. Rahman (Ed.), *Social and political implications of data mining: Knowledge management in e-government* (pp. 267–287). Hershey, PA: Information Science Reference; doi:10.4018/978-1-60566-230-5.ch016

Amoretti, F. (2007). Digital international governance. In A. Anttiroiko & M. Malkia (Eds.), *Encyclopedia of digital government* (pp. 365–370). Hershey, PA: Information Science Reference; doi:10.4018/978-1-59140-789-8.ch056

Amoretti, F. (2008). Digital international governance. In A. Anttiroiko (Ed.), *Electronic government: Concepts, methodologies, tools, and applications* (pp. 688–696). Hershey, PA: Information Science Reference; doi:10.4018/978-1-59904-947-2.ch058

Amoretti, F. (2008). E-government at supranational level in the European Union. In A. Anttiroiko (Ed.), *Electronic government: Concepts, methodologies, tools, and applications* (pp. 1047–1055). Hershey, PA: Information Science Reference; doi:10.4018/978-1-59904-947-2.ch079

Amoretti, F. (2008). E-government regimes. In A. Anttiroiko (Ed.), *Electronic government: Concepts, methodologies, tools, and applications* (pp. 3846–3856). Hershey, PA: Information Science Reference; doi:10.4018/978-1-59904-947-2.ch280

Amoretti, F. (2009). Electronic constitution: A Braudelian perspective. In F. Amoretti (Ed.), *Electronic constitution: Social, cultural, and political implications* (pp. 1–19). Hershey, PA: Information Science Reference; doi:10.4018/978-1-60566-254-1.ch001

Amoretti, F., & Musella, F. (2009). Institutional isomorphism and new technologies. In M. Khosrow-Pour (Ed.), *Encyclopedia of information science and technology* (2nd ed., pp. 2066–2071). Hershey, PA: Information Science Reference; doi:10.4018/978-1-60566-026-4.ch325

Andersen, K. V., & Henriksen, H. Z. (2007). E-government research: Capabilities, interaction, orientation, and values. In D. Norris (Ed.), *Current issues and trends in e-government research* (pp. 269–288). Hershey, PA: CyberTech Publishing; doi:10.4018/978-1-59904-283-1.ch013

Anderson, K. V., & Henriksen, H. Z. (2005). The first leg of e-government research: Domains and application areas 1998-2003. *International Journal of Electronic Government Research, 1*(4), 26–44. doi:10.4018/jegr.2005100102

Anttiroiko, A. (2009). Democratic e-governance. In M. Khosrow-Pour (Ed.), *Encyclopedia of information science and technology* (2nd ed., pp. 990–995). Hershey, PA: Information Science Reference; doi:10.4018/978-1-60566-026-4.ch158

Association, I. R. (2010). Networking and telecommunications: Concepts, methodologies, tools and applications (Vols. 1–3). Hershey, PA: IGI Global; doi:10.4018/978-1-60566-986-1

Association, I. R. (2010). Web-based education: Concepts, methodologies, tools and applications (Vols. 1–3). Hershey, PA: IGI Global; doi:10.4018/978-1-61520-963-7

Baker, P. M., Bell, A., & Moon, N. W. (2009). Accessibility issues in municipal wireless networks. In C. Reddick (Ed.), *Handbook of research on strategies for local e-government adoption and implementation: Comparative studies* (pp. 569–588). Hershey, PA: Information Science Reference; doi:10.4018/978-1-60566-282-4.ch030

Becker, S. A., Keimer, R., & Muth, T. (2010). A case on university and community collaboration: The sci-tech entrepreneurial training services (ETS) program. In S. Becker & R. Niebuhr (Eds.), *Cases on technology innovation: Entrepreneurial successes and pitfalls* (pp. 68–90). Hershey, PA: Business Science Reference; doi:10.4018/978-1-61520-609-4.ch003

Becker, S. A., Keimer, R., & Muth, T. (2012). A case on university and community collaboration: The sci-tech entrepreneurial training services (ETS) program. In Regional development: Concepts, methodologies, tools, and applications (pp. 947-969). Hershey, PA: Information Science Reference. doi:10.4018/978-1-4666-0882-5.ch507

Bernardi, R. (2012). Information technology and resistance to public sector reforms: A case study in Kenya. In T. Papadopoulos & P. Kanellis (Eds.), *Public sector reform using information technologies: Transforming policy into practice* (pp. 59–78). Hershey, PA: Information Science Reference; doi:10.4018/978-1-60960-839-2.ch004

Bernardi, R. (2013). Information technology and resistance to public sector reforms: A case study in Kenya. In *User-driven healthcare: Concepts, methodologies, tools, and applications* (pp. 14–33). Hershey, PA: Medical Information Science Reference; doi:10.4018/978-1-4666-2770-3.ch002

Bolívar, M. P., Pérez, M. D., & Hernández, A. M. (2012). Municipal e-government services in emerging economies: The Latin-American and Caribbean experiences. In Y. Chen & P. Chu (Eds.), *Electronic governance and cross-boundary collaboration: Innovations and advancing tools* (pp. 198–226). Hershey, PA: Information Science Reference; doi:10.4018/978-1-60960-753-1.ch011

Borycki, E. M., & Kushniruk, A. W. (2010). Use of clinical simulations to evaluate the impact of health information systems and ubiquitous computing devices upon health professional work. In S. Mohammed & J. Fiaidhi (Eds.), *Ubiquitous health and medical informatics: The ubiquity 2.0 trend and beyond* (pp. 552–573). Hershey, PA: Medical Information Science Reference; doi:10.4018/978-1-61520-777-0.ch026

Borycki, E. M., & Kushniruk, A. W. (2011). Use of clinical simulations to evaluate the impact of health information systems and ubiquitous computing devices upon health professional work. In *Clinical technologies: Concepts, methodologies, tools and applications* (pp. 532–553). Hershey, PA: Medical Information Science Reference; doi:10.4018/978-1-60960-561-2.ch220

Buchan, J. (2011). Developing a dynamic and responsive online learning environment: A case study of a large Australian university. In B. Czerkawski (Ed.), *Free and open source software for e-learning: Issues, successes and challenges* (pp. 92–109). Hershey, PA: Information Science Reference; doi:10.4018/978-1-61520-917-0.ch006

Buenger, A. W. (2008). Digital convergence and cybersecurity policy. In G. Garson & M. Khosrow-Pour (Eds.), *Handbook of research on public information technology* (pp. 395–405). Hershey, PA: Information Science Reference; doi:10.4018/978-1-59904-857-4.ch038

Burn, J. M., & Loch, K. D. (2002). The societal impact of world wide web - Key challenges for the 21st century. In A. Salehnia (Ed.), *Ethical issues of information systems* (pp. 88–106). Hershey, PA: IRM Press; doi:10.4018/978-1-931777-15-5.ch007

Burn, J. M., & Loch, K. D. (2003). The societal impact of the world wide web-Key challenges for the 21st century. In M. Khosrow-Pour (Ed.), *Advanced topics in information resources management* (Vol. 2, pp. 32–51). Hershey, PA: Idea Group Publishing; doi:10.4018/978-1-59140-062-2.ch002

Bwalya, K. J., Du Plessis, T., & Rensleigh, C. (2012). The "quicksilver initiatives" as a framework for e-government strategy design in developing economies. In K. Bwalya & S. Zulu (Eds.), *Handbook of research on e-government in emerging economies: Adoption, e-participation, and legal frameworks* (pp. 605–623). Hershey, PA: Information Science Reference; doi:10.4018/978-1-4666-0324-0.ch031

Cabotaje, C. E., & Alampay, E. A. (2013). Social media and citizen engagement: Two cases from the Philippines. In S. Saeed & C. Reddick (Eds.), *Human-centered system design for electronic governance* (pp. 225–238). Hershey, PA: Information Science Reference; doi:10.4018/978-1-4666-3640-8.ch013

Camillo, A., Di Pietro, L., Di Virgilio, F., & Franco, M. (2013). Work-groups conflict at PetroTech-Italy, S.R.L.: The influence of culture on conflict dynamics. In B. Christiansen, E. Turkina, & N. Williams (Eds.), *Cultural and technological influences on global business* (pp. 272–289). Hershey, PA: Business Science Reference; doi:10.4018/978-1-4666-3966-9.ch015

Capra, E., Francalanci, C., & Marinoni, C. (2008). Soft success factors for m-government. In A. Anttiroiko (Ed.), *Electronic government: Concepts, methodologies, tools, and applications* (pp. 1213–1233). Hershey, PA: Information Science Reference; doi:10.4018/978-1-59904-947-2.ch089

Cartelli, A. (2009). The implementation of practices with ICT as a new teaching-learning paradigm. In A. Cartelli & M. Palma (Eds.), *Encyclopedia of information communication technology* (pp. 413–417). Hershey, PA: Information Science Reference; doi:10.4018/978-1-59904-845-1.ch055

Charalabidis, Y., Lampathaki, F., & Askounis, D. (2010). Investigating the landscape in national interoperability frameworks. *International Journal of E-Services and Mobile Applications*, *2*(4), 28–41. doi:10.4018/jesma.2010100103

Charalabidis, Y., Lampathaki, F., & Askounis, D. (2012). Investigating the landscape in national interoperability frameworks. In A. Scupola (Ed.), *Innovative mobile platform developments for electronic services design and delivery* (pp. 218–231). Hershey, PA: Business Science Reference; doi:10.4018/978-1-4666-1568-7.ch013

Chen, I. (2005). Distance education associations. In C. Howard, J. Boettcher, L. Justice, K. Schenk, P. Rogers, & G. Berg (Eds.), *Encyclopedia of distance learning* (pp. 599–612). Hershey, PA: Information Science Reference; doi:10.4018/978-1-59140-555-9.ch087

Chen, I. (2008). Distance education associations. In L. Tomei (Ed.), *Online and distance learning: Concepts, methodologies, tools, and applications* (pp. 562–579). Hershey, PA: Information Science Reference; doi:10.4018/978-1-59904-935-9.ch048

Chen, Y. (2008). Managing IT outsourcing for digital government. In A. Anttiroiko (Ed.), *Electronic government: Concepts, methodologies, tools, and applications* (pp. 3107–3114). Hershey, PA: Information Science Reference; doi:10.4018/978-1-59904-947-2.ch229

Chen, Y., & Dimitrova, D. V. (2006). Electronic government and online engagement: Citizen interaction with government via web portals. *International Journal of Electronic Government Research*, *2*(1), 54–76. doi:10.4018/jegr.2006010104

Chen, Y., & Knepper, R. (2005). Digital government development strategies: Lessons for policy makers from a comparative perspective. In W. Huang, K. Siau, & K. Wei (Eds.), *Electronic government strategies and implementation* (pp. 394–420). Hershey, PA: Idea Group Publishing; doi:10.4018/978-1-59140-348-7.ch017

Chen, Y., & Knepper, R. (2008). Digital government development strategies: Lessons for policy makers from a comparative perspective. In H. Rahman (Ed.), *Developing successful ICT strategies: Competitive advantages in a global knowledge-driven society* (pp. 334–356). Hershey, PA: Information Science Reference; doi:10.4018/978-1-59904-654-9.ch017

Cherian, E. J., & Ryan, T. W. (2014). Incongruent needs: Why differences in the iron-triangle of priorities make health information technology adoption and use difficult. In C. El Morr (Ed.), *Research perspectives on the role of informatics in health policy and management* (pp. 209–221). Hershey, PA: Medical Information Science Reference; doi:10.4018/978-1-4666-4321-5.ch012

Cho, H. J., & Hwang, S. (2010). Government 2.0 in Korea: Focusing on e-participation services. In C. Reddick (Ed.), *Politics, democracy and e-government: Participation and service delivery* (pp. 94–114). Hershey, PA: Information Science Reference; doi:10.4018/978-1-61520-933-0.ch006

Chorus, C., & Timmermans, H. (2010). Ubiquitous travel environments and travel control strategies: Prospects and challenges. In M. Wachowicz (Ed.), *Movement-aware applications for sustainable mobility: Technologies and approaches* (pp. 30–51). Hershey, PA: Information Science Reference; doi:10.4018/978-1-61520-769-5.ch003

Chuanshen, R. (2007). E-government construction and China's administrative litigation act. In A. Anttiroiko & M. Malkia (Eds.), *Encyclopedia of digital government* (pp. 507–510). Hershey, PA: Information Science Reference; doi:10.4018/978-1-59140-789-8.ch077

Ciaghi, A., & Villafiorita, A. (2012). Law modeling and BPR for public administration improvement. In K. Bwalya & S. Zulu (Eds.), *Handbook of research on e-government in emerging economies: Adoption, e-participation, and legal frameworks* (pp. 391–410). Hershey, PA: Information Science Reference; doi:10.4018/978-1-4666-0324-0.ch019

Ciaramitaro, B. L., & Skrocki, M. (2012). mHealth: Mobile healthcare. In B. Ciaramitaro (Ed.), Mobile technology consumption: Opportunities and challenges (pp. 99-109). Hershey, PA: Information Science Reference. doi:10.4018/978-1-61350-150-4.ch007

Comite, U. (2012). Innovative processes and managerial effectiveness of e-procurement in healthcare. In A. Manoharan & M. Holzer (Eds.), *Active citizen participation in e-government: A global perspective* (pp. 206–229). Hershey, PA: Information Science Reference; doi:10.4018/978-1-4666-0116-1.ch011

Cordella, A. (2013). E-government success: How to account for ICT, administrative rationalization, and institutional change. In J. Gil-Garcia (Ed.), *E-government success factors and measures: Theories, concepts, and methodologies* (pp. 40–51). Hershey, PA: Information Science Reference; doi:10.4018/978-1-4666-4058-0.ch003

Cropf, R. A. (2009). ICT and e-democracy. In M. Khosrow-Pour (Ed.), *Encyclopedia of information science and technology* (2nd ed., pp. 1789–1793). Hershey, PA: Information Science Reference; doi:10.4018/978-1-60566-026-4.ch281

Cropf, R. A. (2009). The virtual public sphere. In M. Pagani (Ed.), *Encyclopedia of multimedia technology and networking* (2nd ed., pp. 1525–1530). Hershey, PA: Information Science Reference; doi:10.4018/978-1-60566-014-1.ch206

D'Abundo, M. L. (2013). Electronic health record implementation in the United States healthcare industry: Making the process of change manageable. In V. Wang (Ed.), *Handbook of research on technologies for improving the 21st century workforce: Tools for lifelong learning* (pp. 272–286). Hershey, PA: Information Science Publishing; doi:10.4018/978-1-4666-2181-7.ch018

Damurski, L. (2012). E-participation in urban planning: Online tools for citizen engagement in Poland and in Germany. *International Journal of E-Planning Research*, *1*(3), 40–67. doi:10.4018/ijepr.2012070103

de Almeida, M. O. (2007). E-government strategy in Brazil: Increasing transparency and efficiency through e-government procurement. In M. Gascó-Hernandez (Ed.), *Latin America online: Cases, successes and pitfalls* (pp. 34–82). Hershey, PA: IRM Press; doi:10.4018/978-1-59140-974-8.ch002

de Juana Espinosa, S. (2008). Empirical study of the municipalitites' motivations for adopting online presence. In A. Anttiroiko (Ed.), *Electronic government: Concepts, methodologies, tools, and applications* (pp. 3593–3608). Hershey, PA: Information Science Reference; doi:10.4018/978-1-59904-947-2.ch262

de Souza Dias, D. (2002). Motivation for using information technology. In C. Snodgrass & E. Szewczak (Eds.), *Human factors in information systems* (pp. 55–60). Hershey, PA: IRM Press; doi:10.4018/978-1-931777-10-0.ch005

Demediuk, P. (2006). Government procurement ICT's impact on the sustainability of SMEs and regional communities. In S. Marshall, W. Taylor, & X. Yu (Eds.), *Encyclopedia of developing regional communities with information and communication technology* (pp. 321–324). Hershey, PA: Information Science Reference; doi:10.4018/978-1-59140-575-7.ch056

Devonshire, E., Forsyth, H., Reid, S., & Simpson, J. M. (2013). The challenges and opportunities of online postgraduate coursework programs in a traditional university context. In B. Tynan, J. Willems, & R. James (Eds.), *Outlooks and opportunities in blended and distance learning* (pp. 353–368). Hershey, PA: Information Science Reference; doi:10.4018/978-1-4666-4205-8.ch026

Di Cerbo, F., Scotto, M., Sillitti, A., Succi, G., & Vernazza, T. (2007). Toward a GNU/Linux distribution for corporate environments. In S. Sowe, I. Stamelos, & I. Samoladas (Eds.), *Emerging free and open source software practices* (pp. 215–236). Hershey, PA: Idea Group Publishing; doi:10.4018/978-1-59904-210-7.ch010

Diesner, J., & Carley, K. M. (2005). Revealing social structure from texts: Meta-matrix text analysis as a novel method for network text analysis. In V. Narayanan & D. Armstrong (Eds.), *Causal mapping for research in information technology* (pp. 81–108). Hershey, PA: Idea Group Publishing; doi:10.4018/978-1-59140-396-8.ch004

Dologite, D. G., Mockler, R. J., Bai, Q., & Viszhanyo, P. F. (2006). IS change agents in practice in a US-Chinese joint venture. In M. Hunter & F. Tan (Eds.), *Advanced topics in global information management* (Vol. 5, pp. 331–352). Hershey, PA: Idea Group Publishing; doi:10.4018/978-1-59140-923-6.ch015

Drnevich, P., Brush, T. H., & Luckock, G. T. (2011). Process and structural implications for IT-enabled outsourcing. *International Journal of Strategic Information Technology and Applications*, *2*(4), 30–43. doi:10.4018/jsita.2011100103

Dwivedi, A. N. (2009). Handbook of research on information technology management and clinical data administration in healthcare (Vols. 1–2). Hershey, PA: IGI Global; doi:10.4018/978-1-60566-356-2

Elbeltagi, I., McBride, N., & Hardaker, G. (2006). Evaluating the factors affecting DSS usage by senior managers in local authorities in Egypt. In M. Hunter & F. Tan (Eds.), *Advanced topics in global information management* (Vol. 5, pp. 283–307). Hershey, PA: Idea Group Publishing; doi:10.4018/978-1-59140-923-6.ch013

Eom, S., & Fountain, J. E. (2013). Enhancing information services through public-private partnerships: Information technology knowledge transfer underlying structures to develop shared services in the U.S. and Korea. In J. Gil-Garcia (Ed.), *E-government success around the world: Cases, empirical studies, and practical recommendations* (pp. 15–40). Hershey, PA: Information Science Reference; doi:10.4018/978-1-4666-4173-0.ch002

Esteves, T., Leuenberger, D., & Van Leuven, N. (2012). Reaching citizen 2.0: How government uses social media to send public messages during times of calm and times of crisis. In K. Kloby & M. D'Agostino (Eds.), *Citizen 2.0: Public and governmental interaction through web 2.0 technologies* (pp. 250–268). Hershey, PA: Information Science Reference; doi:10.4018/978-1-4666-0318-9.ch013

Estevez, E., Fillottrani, P., Janowski, T., & Ojo, A. (2012). Government information sharing: A framework for policy formulation. In Y. Chen & P. Chu (Eds.), *Electronic governance and cross-boundary collaboration: Innovations and advancing tools* (pp. 23–55). Hershey, PA: Information Science Reference; doi:10.4018/978-1-60960-753-1.ch002

Ezz, I. E. (2008). E-governement emerging trends: Organizational challenges. In A. Anttiroiko (Ed.), *Electronic government: Concepts, methodologies, tools, and applications* (pp. 3721–3737). Hershey, PA: Information Science Reference; doi:10.4018/978-1-59904-947-2.ch269

Fabri, M. (2009). The Italian style of e-justice in a comparative perspective. In A. Martínez & P. Abat (Eds.), *E-justice: Using information communication technologies in the court system* (pp. 1–19). Hershey, PA: Information Science Reference; doi:10.4018/978-1-59904-998-4.ch001

Fagbe, T., & Adekola, O. D. (2010). Workplace safety and personnel well-being: The impact of information technology. *International Journal of Green Computing*, *1*(1), 28–33. doi:10.4018/jgc.2010010103

Fagbe, T., & Adekola, O. D. (2011). Workplace safety and personnel well-being: The impact of information technology. In *Global business: Concepts, methodologies, tools and applications* (pp. 1438–1444). Hershey, PA: Business Science Reference; doi:10.4018/978-1-60960-587-2.ch509

Farmer, L. (2008). Affective collaborative instruction with librarians. In S. Kelsey & K. St.Amant (Eds.), *Handbook of research on computer mediated communication* (pp. 15–24). Hershey, PA: Information Science Reference; doi:10.4018/978-1-59904-863-5.ch002

Favier, L., & Mekhantar, J. (2007). Use of OSS by local e-administration: The French situation. In K. St.Amant & B. Still (Eds.), *Handbook of research on open source software: Technological, economic, and social perspectives* (pp. 428–444). Hershey, PA: Information Science Reference; doi:10.4018/978-1-59140-999-1.ch033

Fernando, S. (2009). Issues of e-learning in third world countries. In M. Khosrow-Pour (Ed.), *Encyclopedia of information science and technology* (2nd ed., pp. 2273–2277). Hershey, PA: Information Science Reference; doi:10.4018/978-1-60566-026-4.ch360

Filho, J. R., & dos Santos Junior, J. R. (2009). Local e-government in Brazil: Poor interaction and local politics as usual. In C. Reddick (Ed.), *Handbook of research on strategies for local e-government adoption and implementation: Comparative studies* (pp. 863–878). Hershey, PA: Information Science Reference; doi:10.4018/978-1-60566-282-4.ch045

Fletcher, P. D. (2004). Portals and policy: Implications of electronic access to U.S. federal government information services. In A. Pavlichev & G. Garson (Eds.), *Digital government: Principles and best practices* (pp. 52–62). Hershey, PA: Idea Group Publishing; doi:10.4018/978-1-59140-122-3.ch004

Fletcher, P. D. (2008). Portals and policy: Implications of electronic access to U.S. federal government information services. In A. Anttiroiko (Ed.), *Electronic government: Concepts, methodologies, tools, and applications* (pp. 3970–3979). Hershey, PA: Information Science Reference; doi:10.4018/978-1-59904-947-2.ch289

Forlano, L. (2004). The emergence of digital government: International perspectives. In A. Pavlichev & G. Garson (Eds.), *Digital government: Principles and best practices* (pp. 34–51). Hershey, PA: Idea Group Publishing; doi:10.4018/978-1-59140-122-3.ch003

Franzel, J. M., & Coursey, D. H. (2004). Government web portals: Management issues and the approaches of five states. In A. Pavlichev & G. Garson (Eds.), *Digital government: Principles and best practices* (pp. 63–77). Hershey, PA: Idea Group Publishing; doi:10.4018/978-1-59140-122-3.ch005

Gaivéo, J. M. (2013). Security of ICTs supporting healthcare activities. In M. Cruz-Cunha, I. Miranda, & P. Gonçalves (Eds.), *Handbook of research on ICTs for human-centered healthcare and social care services* (pp. 208–228). Hershey, PA: Medical Information Science Reference; doi:10.4018/978-1-4666-3986-7.ch011

Garson, G. D. (1999). *Information technology and computer applications in public administration: Issues and trends*. Hershey, PA: IGI Global; doi:10.4018/978-1-87828-952-0

Garson, G. D. (2003). Toward an information technology research agenda for public administration. In G. Garson (Ed.), *Public information technology: Policy and management issues* (pp. 331–357). Hershey, PA: Idea Group Publishing; doi:10.4018/978-1-59140-060-8.ch014

Garson, G. D. (2004). The promise of digital government. In A. Pavlichev & G. Garson (Eds.), *Digital government: Principles and best practices* (pp. 2–15). Hershey, PA: Idea Group Publishing; doi:10.4018/978-1-59140-122-3.ch001

Garson, G. D. (2007). An information technology research agenda for public administration. In G. Garson (Ed.), *Modern public information technology systems: Issues and challenges* (pp. 365–392). Hershey, PA: Idea Group Publishing; doi:10.4018/978-1-59904-051-6.ch018

Gasco, M. (2007). Civil servants' resistance towards e-government development. In A. Anttiroiko & M. Malkia (Eds.), *Encyclopedia of digital government* (pp. 190–195). Hershey, PA: Information Science Reference; doi:10.4018/978-1-59140-789-8.ch028

Gasco, M. (2008). Civil servants' resistance towards e-government development. In A. Anttiroiko (Ed.), *Electronic government: Concepts, methodologies, tools, and applications* (pp. 2580–2588). Hershey, PA: Information Science Reference; doi:10.4018/978-1-59904-947-2.ch190

Ghere, R. K. (2010). Accountability and information technology enactment: Implications for social empowerment. In E. Ferro, Y. Dwivedi, J. Gil-Garcia, & M. Williams (Eds.), *Handbook of research on overcoming digital divides: Constructing an equitable and competitive information society* (pp. 515–532). Hershey, PA: Information Science Reference; doi:10.4018/978-1-60566-699-0.ch028

Gibson, I. W. (2012). Simulation modeling of healthcare delivery. In A. Kolker & P. Story (Eds.), *Management engineering for effective healthcare delivery: Principles and applications* (pp. 69–89). Hershey, PA: Medical Information Science Reference; doi:10.4018/978-1-60960-872-9.ch003

Gil-Garcia, J. R. (2007). Exploring e-government benefits and success factors. In A. Anttiroiko & M. Malkia (Eds.), *Encyclopedia of digital government* (pp. 803–811). Hershey, PA: Information Science Reference; doi:10.4018/978-1-59140-789-8.ch122

Gil-Garcia, J. R., & González Miranda, F. (2010). E-government and opportunities for participation: The case of the Mexican state web portals. In C. Reddick (Ed.), *Politics, democracy and e-government: Participation and service delivery* (pp. 56–74). Hershey, PA: Information Science Reference; doi:10.4018/978-1-61520-933-0.ch004

Goldfinch, S. (2012). Public trust in government, trust in e-government, and use of e-government. In Z. Yan (Ed.), *Encyclopedia of cyber behavior* (pp. 987–995). Hershey, PA: Information Science Reference; doi:10.4018/978-1-4666-0315-8.ch081

Goodyear, M. (2012). Organizational change contributions to e-government project transitions. In S. Aikins (Ed.), *Managing e-government projects: Concepts, issues, and best practices* (pp. 1–21). Hershey, PA: Information Science Reference; doi:10.4018/978-1-4666-0086-7.ch001

Gordon, S., & Mulligan, P. (2003). Strategic models for the delivery of personal financial services: The role of infocracy. In S. Gordon (Ed.), *Computing information technology: The human side* (pp. 220–232). Hershey, PA: IRM Press; doi:10.4018/978-1-93177-752-0.ch014

Gordon, T. F. (2007). Legal knowledge systems. In A. Anttiroiko & M. Malkia (Eds.), *Encyclopedia of digital government* (pp. 1161–1166). Hershey, PA: Information Science Reference; doi:10.4018/978-1-59140-789-8.ch175

Graham, J. E., & Semich, G. W. (2008). Integrating technology to transform pedagogy: Revisiting the progress of the three phase TUI model for faculty development. In L. Tomei (Ed.), *Adapting information and communication technologies for effective education* (pp. 1–12). Hershey, PA: Information Science Reference; doi:10.4018/978-1-59904-922-9.ch001

Grandinetti, L., & Pisacane, O. (2012). Web services for healthcare management. In D. Prakash Vidyarthi (Ed.), *Technologies and protocols for the future of internet design: Reinventing the web* (pp. 60–94). Hershey, PA: Information Science Reference; doi:10.4018/978-1-4666-0203-8.ch004

Groenewegen, P., & Wagenaar, F. P. (2008). VO as an alternative to hierarchy in the Dutch police sector. In G. Putnik & M. Cruz-Cunha (Eds.), *Encyclopedia of networked and virtual organizations* (pp. 1851–1857). Hershey, PA: Information Science Reference; doi:10.4018/978-1-59904-885-7.ch245

Gronlund, A. (2001). Building an infrastructure to manage electronic services. In S. Dasgupta (Ed.), *Managing internet and intranet technologies in organizations: Challenges and opportunities* (pp. 71–103). Hershey, PA: Idea Group Publishing; doi:10.4018/978-1-878289-95-7.ch006

Gronlund, A. (2002). Introduction to electronic government: Design, applications and management. In Å. Grönlund (Ed.), *Electronic government: Design, applications and management* (pp. 1–21). Hershey, PA: Idea Group Publishing; doi:10.4018/978-1-930708-19-8.ch001

Gupta, A., Woosley, R., Crk, I., & Sarnikar, S. (2009). An information technology architecture for drug effectiveness reporting and post-marketing surveillance. In J. Tan (Ed.), *Medical informatics: Concepts, methodologies, tools, and applications* (pp. 631–646). Hershey, PA: Medical Information Science Reference; doi:10.4018/978-1-60566-050-9.ch047

Hallin, A., & Lundevall, K. (2007). mCity: User focused development of mobile services within the city of Stockholm. In I. Kushchu (Ed.), Mobile government: An emerging direction in e-government (pp. 12-29). Hershey, PA: Idea Group Publishing. doi:10.4018/978-1-59140-884-0.ch002

Hallin, A., & Lundevall, K. (2009). mCity: User focused development of mobile services within the city of Stockholm. In S. Clarke (Ed.), Evolutionary concepts in end user productivity and performance: Applications for organizational progress (pp. 268-280). Hershey, PA: Information Science Reference. doi:10.4018/978-1-60566-136-0.ch017

Hallin, A., & Lundevall, K. (2009). mCity: User focused development of mobile services within the city of Stockholm. In D. Taniar (Ed.), Mobile computing: Concepts, methodologies, tools, and applications (pp. 3455-3467). Hershey, PA: Information Science Reference. doi:10.4018/978-1-60566-054-7.ch253

Hanson, A. (2005). Overcoming barriers in the planning of a virtual library. In M. Khosrow-Pour (Ed.), *Encyclopedia of information science and technology* (pp. 2255–2259). Hershey, PA: Information Science Reference; doi:10.4018/978-1-59140-553-5.ch397

Haque, A. (2008). Information technology and surveillance: Implications for public administration in a new word order. In T. Loendorf & G. Garson (Eds.), *Patriotic information systems* (pp. 177–185). Hershey, PA: IGI Publishing; doi:10.4018/978-1-59904-594-8.ch008

Hauck, R. V., Thatcher, S. M., & Weisband, S. P. (2012). Temporal aspects of information technology use: Increasing shift work effectiveness. In J. Wang (Ed.), *Advancing the service sector with evolving technologies: Techniques and principles* (pp. 87–104). Hershey, PA: Business Science Reference; doi:10.4018/978-1-4666-0044-7.ch006

Hawk, S., & Witt, T. (2006). Telecommunications courses in information systems programs. *International Journal of Information and Communication Technology Education*, *2*(1), 79–92. doi:10.4018/jicte.2006010107

Helms, M. M., Moore, R., & Ahmadi, M. (2009). Information technology (IT) and the healthcare industry: A SWOT analysis. In J. Tan (Ed.), *Medical informatics: Concepts, methodologies, tools, and applications* (pp. 134–152). Hershey, PA: Medical Information Science Reference; doi:10.4018/978-1-60566-050-9.ch012

Hendrickson, S. M., & Young, M. E. (2014). Electronic records management at a federally funded research and development center. In J. Krueger (Ed.), *Cases on electronic records and resource management implementation in diverse environments* (pp. 334–350). Hershey, PA: Information Science Reference; doi:10.4018/978-1-4666-4466-3.ch020

Henman, P. (2010). Social policy and information communication technologies. In J. Martin & L. Hawkins (Eds.), *Information communication technologies for human services education and delivery: Concepts and cases* (pp. 215–229). Hershey, PA: Information Science Reference; doi:10.4018/978-1-60566-735-5.ch014

Hismanoglu, M. (2011). Important issues in online education: E-pedagogy and marketing. In U. Demiray & S. Sever (Eds.), *Marketing online education programs: Frameworks for promotion and communication* (pp. 184–209). Hershey, PA: Information Science Reference; doi:10.4018/978-1-60960-074-7.ch012

Ho, K. K. (2008). The e-government development, IT strategies, and portals of the Hong Kong SAR government. In A. Anttiroiko (Ed.), *Electronic government: Concepts, methodologies, tools, and applications* (pp. 715–733). Hershey, PA: Information Science Reference; doi:10.4018/978-1-59904-947-2.ch060

Holden, S. H. (2003). The evolution of information technology management at the federal level: Implications for public administration. In G. Garson (Ed.), *Public information technology: Policy and management issues* (pp. 53–73). Hershey, PA: Idea Group Publishing; doi:10.4018/978-1-59140-060-8.ch003

Holden, S. H. (2007). The evolution of federal information technology management literature: Does IT finally matter? In G. Garson (Ed.), *Modern public information technology systems: Issues and challenges* (pp. 17–34). Hershey, PA: Idea Group Publishing; doi:10.4018/978-1-59904-051-6.ch002

Holland, J. W. (2009). Automation of American criminal justice. In M. Khosrow-Pour (Ed.), *Encyclopedia of information science and technology* (2nd ed., pp. 300–302). Hershey, PA: Information Science Reference; doi:10.4018/978-1-60566-026-4.ch051

Holloway, K. (2013). Fair use, copyright, and academic integrity in an online academic environment. In *Digital rights management: Concepts, methodologies, tools, and applications* (pp. 917–928). Hershey, PA: Information Science Reference; doi:10.4018/978-1-4666-2136-7.ch044

Horiuchi, C. (2005). E-government databases. In L. Rivero, J. Doorn, & V. Ferraggine (Eds.), *Encyclopedia of database technologies and applications* (pp. 206–210). Hershey, PA: Information Science Reference; doi:10.4018/978-1-59140-560-3.ch035

Horiuchi, C. (2006). Creating IS quality in government settings. In E. Duggan & J. Reichgelt (Eds.), *Measuring information systems delivery quality* (pp. 311–327). Hershey, PA: Idea Group Publishing; doi:10.4018/978-1-59140-857-4.ch014

Hsiao, N., Chu, P., & Lee, C. (2012). Impact of e-governance on businesses: Model development and case study. In *Digital democracy: Concepts, methodologies, tools, and applications* (pp. 1407–1425). Hershey, PA: Information Science Reference; doi:10.4018/978-1-4666-1740-7.ch070

Huang, T., & Lee, C. (2010). Evaluating the impact of e-government on citizens: Cost-benefit analysis. In C. Reddick (Ed.), *Citizens and e-government: Evaluating policy and management* (pp. 37–52). Hershey, PA: Information Science Reference; doi:10.4018/978-1-61520-931-6.ch003

Hunter, M. G., Diochon, M., Pugsley, D., & Wright, B. (2002). Unique challenges for small business adoption of information technology: The case of the Nova Scotia ten. In S. Burgess (Ed.), *Managing information technology in small business: Challenges and solutions* (pp. 98–117). Hershey, PA: Idea Group Publishing; doi:10.4018/978-1-930708-35-8.ch006

Hurskainen, J. (2003). Integration of business systems and applications in merger and alliance: Case metso automation. In T. Reponen (Ed.), *Information technology enabled global customer service* (pp. 207–225). Hershey, PA: Idea Group Publishing; doi:10.4018/978-1-59140-048-6.ch012

Iazzolino, G., & Pietrantonio, R. (2011). The soveria.it project: A best practice of e-government in southern Italy. In D. Piaggesi, K. Sund, & W. Castelnovo (Eds.), *Global strategy and practice of e-governance: Examples from around the world* (pp. 34–56). Hershey, PA: Information Science Reference; doi:10.4018/978-1-60960-489-9.ch003

Imran, A., & Gregor, S. (2012). A process model for successful e-government adoption in the least developed countries: A case of Bangladesh. In F. Tan (Ed.), *International comparisons of information communication technologies: Advancing applications* (pp. 321–350). Hershey, PA: Information Science Reference; doi:10.4018/978-1-61350-480-2.ch014

Inoue, Y., & Bell, S. T. (2005). Electronic/digital government innovation, and publishing trends with IT. In M. Khosrow-Pour (Ed.), *Encyclopedia of information science and technology* (pp. 1018–1023). Hershey, PA: Information Science Reference; doi:10.4018/978-1-59140-553-5.ch180

Islam, M. M., & Ehsan, M. (2013). Understanding e-governance: A theoretical approach. In M. Islam & M. Ehsan (Eds.), *From government to e-governance: Public administration in the digital age* (pp. 38–49). Hershey, PA: Information Science Reference; doi:10.4018/978-1-4666-1909-8.ch003

Jaeger, B. (2009). E-government and e-democracy in the making. In M. Khosrow-Pour (Ed.), *Encyclopedia of information science and technology* (2nd ed., pp. 1318–1322). Hershey, PA: Information Science Reference; doi:10.4018/978-1-60566-026-4.ch208

Jain, R. B. (2007). Revamping the administrative structure and processes in India for online diplomacy. In A. Anttiroiko & M. Malkia (Eds.), *Encyclopedia of digital government* (pp. 1418–1423). Hershey, PA: Information Science Reference; doi:10.4018/978-1-59140-789-8.ch217

Jain, R. B. (2008). Revamping the administrative structure and processes in India for online diplomacy. In A. Anttiroiko (Ed.), *Electronic government: Concepts, methodologies, tools, and applications* (pp. 3142–3149). Hershey, PA: Information Science Reference; doi:10.4018/978-1-59904-947-2.ch233

Jauhiainen, J. S., & Inkinen, T. (2009). E-governance and the information society in periphery. In C. Reddick (Ed.), *Handbook of research on strategies for local e-government adoption and implementation: Comparative studies* (pp. 497–514). Hershey, PA: Information Science Reference; doi:10.4018/978-1-60566-282-4.ch026

Jensen, M. J. (2009). Electronic democracy and citizen influence in government. In C. Reddick (Ed.), *Handbook of research on strategies for local e-government adoption and implementation: Comparative studies* (pp. 288–305). Hershey, PA: Information Science Reference; doi:10.4018/978-1-60566-282-4.ch015

Jiao, Y., Hurson, A. R., Potok, T. E., & Beckerman, B. G. (2009). Integrating mobile-based systems with healthcare databases. In J. Erickson (Ed.), *Database technologies: Concepts, methodologies, tools, and applications* (pp. 484–504). Hershey, PA: Information Science Reference; doi:10.4018/978-1-60566-058-5.ch031

Joia, L. A. (2002). A systematic model to integrate information technology into metabusinesses: A case study in the engineering realms. In F. Tan (Ed.), *Advanced topics in global information management* (Vol. 1, pp. 250–267). Hershey, PA: Idea Group Publishing; doi:10.4018/978-1-930708-43-3.ch016

Jones, T. H., & Song, I. (2000). Binary equivalents of ternary relationships in entity-relationship modeling: A logical decomposition approach. *Journal of Database Management*, *11*(2), 12–19. doi:10.4018/jdm.2000040102

Juana-Espinosa, S. D. (2007). Empirical study of the municipalitites' motivations for adopting online presence. In L. Al-Hakim (Ed.), *Global e-government: Theory, applications and benchmarking* (pp. 261–279). Hershey, PA: Idea Group Publishing; doi:10.4018/978-1-59904-027-1.ch015

Jun, K., & Weare, C. (2012). Bridging from e-government practice to e-government research: Past trends and future directions. In K. Bwalya & S. Zulu (Eds.), *Handbook of research on e-government in emerging economies: Adoption, e-participation, and legal frameworks* (pp. 263–289). Hershey, PA: Information Science Reference; doi:10.4018/978-1-4666-0324-0.ch013

Junqueira, A., Diniz, E. H., & Fernandez, M. (2010). Electronic government implementation projects with multiple agencies: Analysis of the electronic invoice project under PMBOK framework. In J. Cordoba-Pachon & A. Ochoa-Arias (Eds.), *Systems thinking and e-participation: ICT in the governance of society* (pp. 135–153). Hershey, PA: Information Science Reference; doi:10.4018/978-1-60566-860-4.ch009

Juntunen, A. (2009). Joint service development with the local authorities. In C. Reddick (Ed.), *Handbook of research on strategies for local e-government adoption and implementation: Comparative studies* (pp. 902–920). Hershey, PA: Information Science Reference; doi:10.4018/978-1-60566-282-4.ch047

Kamel, S. (2001). *Using DSS for crisis management.* Hershey, PA: IGI Global; doi:10.4018/978-1-87828-961-2.ch020

Kamel, S. (2006). DSS for strategic decision making. In M. Khosrow-Pour (Ed.), *Cases on information technology and organizational politics & culture* (pp. 230–246). Hershey, PA: Idea Group Publishing; doi:10.4018/978-1-59904-411-8.ch015

Kamel, S. (2009). The software industry in Egypt as a potential contributor to economic growth. In M. Khosrow-Pour (Ed.), *Encyclopedia of information science and technology* (2nd ed., pp. 3531–3537). Hershey, PA: Information Science Reference; doi:10.4018/978-1-60566-026-4.ch562

Kamel, S., & Hussein, M. (2008). Xceed: Pioneering the contact center industry in Egypt. *Journal of Cases on Information Technology, 10*(1), 67–91. doi:10.4018/jcit.2008010105

Kamel, S., & Wahba, K. (2003). The use of a hybrid model in web-based education: "The Global campus project. In A. Aggarwal (Ed.), *Web-based education: Learning from experience* (pp. 331–346). Hershey, PA: Information Science Publishing; doi:10.4018/978-1-59140-102-5.ch020

Kardaras, D. K., & Papathanassiou, E. A. (2008). An exploratory study of the e-government services in Greece. In G. Garson & M. Khosrow-Pour (Eds.), *Handbook of research on public information technology* (pp. 162–174). Hershey, PA: Information Science Reference; doi:10.4018/978-1-59904-857-4.ch016

Kassahun, A. E., Molla, A., & Sarkar, P. (2012). Government process reengineering: What we know and what we need to know. In *Digital democracy: Concepts, methodologies, tools, and applications* (pp. 1730–1752). Hershey, PA: Information Science Reference; doi:10.4018/978-1-4666-1740-7.ch086

Khan, B. (2005). Technological issues. In B. Khan (Ed.), *Managing e-learning strategies: Design, delivery, implementation and evaluation* (pp. 154–180). Hershey, PA: Information Science Publishing; doi:10.4018/978-1-59140-634-1.ch004

Khasawneh, A., Bsoul, M., Obeidat, I., & Al Azzam, I. (2012). Technology fears: A study of e-commerce loyalty perception by Jordanian customers. In J. Wang (Ed.), *Advancing the service sector with evolving technologies: Techniques and principles* (pp. 158–165). Hershey, PA: Business Science Reference; doi:10.4018/978-1-4666-0044-7.ch010

Khatibi, V., & Montazer, G. A. (2012). E-research methodology. In A. Juan, T. Daradoumis, M. Roca, S. Grasman, & J. Faulin (Eds.), *Collaborative and distributed e-research: Innovations in technologies, strategies and applications* (pp. 62–81). Hershey, PA: Information Science Reference; doi:10.4018/978-1-4666-0125-3.ch003

Kidd, T. (2011). The dragon in the school's backyard: A review of literature on the uses of technology in urban schools. In L. Tomei (Ed.), *Online courses and ICT in education: Emerging practices and applications* (pp. 242–257). Hershey, PA: Information Science Reference; doi:10.4018/978-1-60960-150-8.ch019

Kidd, T. T. (2010). My experience tells the story: Exploring technology adoption from a qualitative perspective - A pilot study. In H. Song & T. Kidd (Eds.), *Handbook of research on human performance and instructional technology* (pp. 247–262). Hershey, PA: Information Science Reference; doi:10.4018/978-1-60566-782-9.ch015

Kieley, B., Lane, G., Paquet, G., & Roy, J. (2002). e-Government in Canada: Services online or public service renewal? In Å. Grönlund (Ed.), Electronic government: Design, applications and management (pp. 340-355). Hershey, PA: Idea Group Publishing. doi:10.4018/978-1-930708-19-8.ch016

Kim, P. (2012). "Stay out of the way! My kid is video blogging through a phone!": A lesson learned from math tutoring social media for children in underserved communities. In *Wireless technologies: Concepts, methodologies, tools and applications* (pp. 1415–1428). Hershey, PA: Information Science Reference; doi:10.4018/978-1-61350-101-6.ch517

Kirlidog, M. (2010). Financial aspects of national ICT strategies. In S. Kamel (Ed.), *E-strategies for technological diffusion and adoption: National ICT approaches for socioeconomic development* (pp. 277–292). Hershey, PA: Information Science Reference; doi:10.4018/978-1-60566-388-3.ch016

Kisielnicki, J. (2006). Transfer of information and knowledge in the project management. In E. Coakes & S. Clarke (Eds.), *Encyclopedia of communities of practice in information and knowledge management* (pp. 544–551). Hershey, PA: Information Science Reference; doi:10.4018/978-1-59140-556-6.ch091

Kittner, M., & Van Slyke, C. (2006). Reorganizing information technology services in an academic environment. In M. Khosrow-Pour (Ed.), *Cases on the human side of information technology* (pp. 49–66). Hershey, PA: Idea Group Publishing; doi:10.4018/978-1-59904-405-7.ch004

Knoell, H. D. (2008). Semi virtual workplaces in German financial service enterprises. In P. Zemliansky & K. St.Amant (Eds.), *Handbook of research on virtual workplaces and the new nature of business practices* (pp. 570–581). Hershey, PA: Information Science Reference; doi:10.4018/978-1-59904-893-2.ch041

Koh, S. L., & Maguire, S. (2009). Competing in the age of information technology in a developing economy: Experiences of an Indian bank. In S. Koh & S. Maguire (Eds.), *Information and communication technologies management in turbulent business environments* (pp. 326–350). Hershey, PA: Information Science Reference; doi:10.4018/978-1-60566-424-8.ch018

Kollmann, T., & Häsel, M. (2009). Competence of information technology professionals in internet-based ventures. In I. Lee (Ed.), *Electronic business: Concepts, methodologies, tools, and applications* (pp. 1905–1919). Hershey, PA: Information Science Reference; doi:10.4018/978-1-60566-056-1.ch118

Kollmann, T., & Häsel, M. (2009). Competence of information technology professionals in internet-based ventures. In A. Cater-Steel (Ed.), *Information technology governance and service management: Frameworks and adaptations* (pp. 239–253). Hershey, PA: Information Science Reference; doi:10.4018/978-1-60566-008-0.ch013

Kollmann, T., & Häsel, M. (2010). Competence of information technology professionals in internet-based ventures. In *Electronic services: Concepts, methodologies, tools and applications* (pp. 1551–1565). Hershey, PA: Information Science Reference; doi:10.4018/978-1-61520-967-5.ch094

Kraemer, K., & King, J. L. (2006). Information technology and administrative reform: Will e-government be different? *International Journal of Electronic Government Research, 2*(1), 1–20. doi:10.4018/jegr.2006010101

Kraemer, K., & King, J. L. (2008). Information technology and administrative reform: Will e-government be different? In D. Norris (Ed.), *E-government research: Policy and management* (pp. 1–20). Hershey, PA: IGI Publishing; doi:10.4018/978-1-59904-913-7.ch001

Lampathaki, F., Tsiakaliaris, C., Stasis, A., & Charalabidis, Y. (2011). National interoperability frameworks: The way forward. In Y. Charalabidis (Ed.), *Interoperability in digital public services and administration: Bridging e-government and e-business* (pp. 1–24). Hershey, PA: Information Science Reference; doi:10.4018/978-1-61520-887-6.ch001

Lan, Z., & Scott, C. R. (1996). The relative importance of computer-mediated information versus conventional non-computer-mediated information in public managerial decision making. *Information Resources Management Journal*, *9*(1), 27–0. doi:10.4018/irmj.1996010103

Law, W. (2004). *Public sector data management in a developing economy*. Hershey, PA: IGI Global; doi:10.4018/978-1-59140-259-6.ch034

Law, W. K. (2005). Information resources development challenges in a cross-cultural environment. In M. Khosrow-Pour (Ed.), *Encyclopedia of information science and technology* (pp. 1476–1481). Hershey, PA: Information Science Reference; doi:10.4018/978-1-59140-553-5.ch259

Law, W. K. (2009). Cross-cultural challenges for information resources management. In M. Khosrow-Pour (Ed.), *Encyclopedia of information science and technology* (2nd ed., pp. 840–846). Hershey, PA: Information Science Reference; doi:10.4018/978-1-60566-026-4.ch136

Law, W. K. (2011). Cross-cultural challenges for information resources management. In *Global business: Concepts, methodologies, tools and applications* (pp. 1924–1932). Hershey, PA: Business Science Reference; doi:10.4018/978-1-60960-587-2.ch704

Malkia, M., & Savolainen, R. (2004). eTransformation in government, politics and society: Conceptual framework and introduction. In M. Malkia, A. Anttiroiko, & R. Savolainen (Eds.), eTransformation in governance: New directions in government and politics (pp. 1-21). Hershey, PA: Idea Group Publishing. doi:10.4018/978-1-59140-130-8.ch001

Management Association. I. (2010). Information resources management: Concepts, methodologies, tools and applications (4 Volumes). Hershey, PA: IGI Global. doi:10.4018/978-1-61520-965-1

Management Association. I. (2010). Electronic services: Concepts, methodologies, tools and applications (3 Volumes). Hershey, PA: IGI Global. doi:10.4018/978-1-61520-967-5

Mandujano, S. (2011). Network manageability security. In D. Kar & M. Syed (Eds.), *Network security, administration and management: Advancing technology and practice* (pp. 158–181). Hershey, PA: Information Science Reference; doi:10.4018/978-1-60960-777-7.ch009

Marich, M. J., Schooley, B. L., & Horan, T. A. (2012). A normative enterprise architecture for guiding end-to-end emergency response decision support. In M. Jennex (Ed.), *Managing crises and disasters with emerging technologies: Advancements* (pp. 71–87). Hershey, PA: Information Science Reference; doi:10.4018/978-1-4666-0167-3.ch006

Markov, R., & Okujava, S. (2008). Costs, benefits, and risks of e-government portals. In G. Putnik & M. Cruz-Cunha (Eds.), *Encyclopedia of networked and virtual organizations* (pp. 354–363). Hershey, PA: Information Science Reference; doi:10.4018/978-1-59904-885-7.ch047

Martin, N., & Rice, J. (2013). Evaluating and designing electronic government for the future: Observations and insights from Australia. In V. Weerakkody (Ed.), *E-government services design, adoption, and evaluation* (pp. 238–258). Hershey, PA: Information Science Reference; doi:10.4018/978-1-4666-2458-0.ch014

i. Martinez, A. C. (2008). Accessing administration's information via internet in Spain. In F. Tan (Ed.), *Global information technologies: Concepts, methodologies, tools, and applications* (pp. 2558–2573). Hershey, PA: Information Science Reference; doi:10.4018/978-1-59904-939-7.ch186

Mbarika, V. W., Meso, P. N., & Musa, P. F. (2006). A disconnect in stakeholders' perceptions from emerging realities of teledensity growth in Africa's least developed countries. In M. Hunter & F. Tan (Eds.), *Advanced topics in global information management* (Vol. 5, pp. 263–282). Hershey, PA: Idea Group Publishing; doi:10.4018/978-1-59140-923-6.ch012

Mbarika, V. W., Meso, P. N., & Musa, P. F. (2008). A disconnect in stakeholders' perceptions from emerging realities of teledensity growth in Africa's least developed countries. In F. Tan (Ed.), *Global information technologies: Concepts, methodologies, tools, and applications* (pp. 2948–2962). Hershey, PA: Information Science Reference; doi:10.4018/978-1-59904-939-7.ch209

Means, T., Olson, E., & Spooner, J. (2013). Discovering ways that don't work on the road to success: Strengths and weaknesses revealed by an active learning studio classroom project. In A. Benson, J. Moore, & S. Williams van Rooij (Eds.), *Cases on educational technology planning, design, and implementation: A project management perspective* (pp. 94–113). Hershey, PA: Information Science Reference; doi:10.4018/978-1-4666-4237-9.ch006

Melitski, J., Holzer, M., Kim, S., Kim, C., & Rho, S. (2008). Digital government worldwide: An e-government assessment of municipal web sites. In G. Garson & M. Khosrow-Pour (Eds.), *Handbook of research on public information technology* (pp. 790–804). Hershey, PA: Information Science Reference; doi:10.4018/978-1-59904-857-4.ch069

Memmola, M., Palumbo, G., & Rossini, M. (2009). Web & RFID technology: New frontiers in costing and process management for rehabilitation medicine. In L. Al-Hakim & M. Memmola (Eds.), *Business web strategy: Design, alignment, and application* (pp. 145–169). Hershey, PA: Information Science Reference; doi:10.4018/978-1-60566-024-0.ch008

Meng, Z., Fahong, Z., & Lei, L. (2008). Information technology and environment. In Y. Kurihara, S. Takaya, H. Harui, & H. Kamae (Eds.), *Information technology and economic development* (pp. 201–212). Hershey, PA: Information Science Reference; doi:10.4018/978-1-59904-579-5.ch014

Mentzingen de Moraes, A. J., Ferneda, E., Costa, I., & Spinola, M. D. (2011). Practical approach for implementation of governance process in IT: Information technology areas. In N. Shi & G. Silvius (Eds.), *Enterprise IT governance, business value and performance measurement* (pp. 19–40). Hershey, PA: Information Science Reference; doi:10.4018/978-1-60566-346-3.ch002

Merwin, G. A. Jr, McDonald, J. S., & Odera, L. C. (2008). Economic development: Government's cutting edge in IT. In M. Raisinghani (Ed.), *Handbook of research on global information technology management in the digital economy* (pp. 1–37). Hershey, PA: Information Science Reference; doi:10.4018/978-1-59904-875-8.ch001

Meso, P., & Duncan, N. (2002). Can national information infrastructures enhance social development in the least developed countries? An empirical investigation. In M. Dadashzadeh (Ed.), *Information technology management in developing countries* (pp. 23–51). Hershey, PA: IRM Press; doi:10.4018/978-1-931777-03-2.ch002

Meso, P. N., & Duncan, N. B. (2002). Can national information infrastructures enhance social development in the least developed countries? In F. Tan (Ed.), *Advanced topics in global information management* (Vol. 1, pp. 207–226). Hershey, PA: Idea Group Publishing; doi:10.4018/978-1-930708-43-3.ch014

Middleton, M. (2008). Evaluation of e-government web sites. In G. Garson & M. Khosrow-Pour (Eds.), *Handbook of research on public information technology* (pp. 699–710). Hershey, PA: Information Science Reference; doi:10.4018/978-1-59904-857-4.ch063

Mingers, J. (2010). Pluralism, realism, and truth: The keys to knowledge in information systems research. In D. Paradice (Ed.), *Emerging systems approaches in information technologies: Concepts, theories, and applications* (pp. 86–98). Hershey, PA: Information Science Reference; doi:10.4018/978-1-60566-976-2.ch006

Mital, K. M. (2012). ICT, unique identity and inclusive growth: An Indian perspective. In A. Manoharan & M. Holzer (Eds.), *E-governance and civic engagement: Factors and determinants of e-democracy* (pp. 584–612). Hershey, PA: Information Science Reference; doi:10.4018/978-1-61350-083-5.ch029

Mizell, A. P. (2008). Helping close the digital divide for financially disadvantaged seniors. In F. Tan (Ed.), *Global information technologies: Concepts, methodologies, tools, and applications* (pp. 2396–2402). Hershey, PA: Information Science Reference; doi:10.4018/978-1-59904-939-7.ch173

Molinari, F., Wills, C., Koumpis, A., & Moumtzi, V. (2011). A citizen-centric platform to support networking in the area of e-democracy. In H. Rahman (Ed.), *Cases on adoption, diffusion and evaluation of global e-governance systems: Impact at the grass roots* (pp. 282–302). Hershey, PA: Information Science Reference; doi:10.4018/978-1-61692-814-8.ch014

Molinari, F., Wills, C., Koumpis, A., & Moumtzi, V. (2013). A citizen-centric platform to support networking in the area of e-democracy. In H. Rahman (Ed.), *Cases on progressions and challenges in ICT utilization for citizen-centric governance* (pp. 265–297). Hershey, PA: Information Science Reference; doi:10.4018/978-1-4666-2071-1.ch013

Monteverde, F. (2010). The process of e-government public policy inclusion in the governmental agenda: A framework for assessment and case study. In J. Cordoba-Pachon & A. Ochoa-Arias (Eds.), *Systems thinking and e-participation: ICT in the governance of society* (pp. 233–245). Hershey, PA: Information Science Reference; doi:10.4018/978-1-60566-860-4.ch015

Moodley, S. (2008). Deconstructing the South African government's ICT for development discourse. In A. Anttiroiko (Ed.), *Electronic government: Concepts, methodologies, tools, and applications* (pp. 622–631). Hershey, PA: Information Science Reference; doi:10.4018/978-1-59904-947-2.ch053

Moodley, S. (2008). Deconstructing the South African government's ICT for development discourse. In C. Van Slyke (Ed.), *Information communication technologies: Concepts, methodologies, tools, and applications* (pp. 816–825). Hershey, PA: Information Science Reference; doi:10.4018/978-1-59904-949-6.ch052

Mora, M., Cervantes-Perez, F., Gelman-Muravchik, O., Forgionne, G. A., & Mejia-Olvera, M. (2003). DMSS implementation research: A conceptual analysis of the contributions and limitations of the factor-based and stage-based streams. In G. Forgionne, J. Gupta, & M. Mora (Eds.), *Decision-making support systems: Achievements and challenges for the new decade* (pp. 331–356). Hershey, PA: Idea Group Publishing; doi:10.4018/978-1-59140-045-5.ch020

Mörtberg, C., & Elovaara, P. (2010). Attaching people and technology: Between e and government. In S. Booth, S. Goodman, & G. Kirkup (Eds.), *Gender issues in learning and working with information technology: Social constructs and cultural contexts* (pp. 83–98). Hershey, PA: Information Science Reference; doi:10.4018/978-1-61520-813-5.ch005

Murphy, J., Harper, E., Devine, E. C., Burke, L. J., & Hook, M. L. (2011). Case study: Lessons learned when embedding evidence-based knowledge in a nurse care planning and documentation system. In A. Cashin & R. Cook (Eds.), *Evidence-based practice in nursing informatics: Concepts and applications* (pp. 174–190). Hershey, PA: Medical Information Science Reference; doi:10.4018/978-1-60960-034-1.ch014

Mutula, S. M. (2013). E-government's role in poverty alleviation: Case study of South Africa. In H. Rahman (Ed.), *Cases on progressions and challenges in ICT utilization for citizen-centric governance* (pp. 44–68). Hershey, PA: Information Science Reference; doi:10.4018/978-1-4666-2071-1.ch003

Nath, R., & Angeles, R. (2005). Relationships between supply characteristics and buyer-supplier coupling in e-procurement: An empirical analysis.[IJEBR]. *International Journal of E-Business Research, 1*(2), 40–55. doi:10.4018/jebr.2005040103

Nissen, M. E. (2006). Application cases in government. In M. Nissen (Ed.), *Harnessing knowledge dynamics: Principled organizational knowing & learning* (pp. 152–181). Hershey, PA: IRM Press; doi:10.4018/978-1-59140-773-7.ch008

Norris, D. F. (2003). Leading-edge information technologies and American local governments. In G. Garson (Ed.), *Public information technology: Policy and management issues* (pp. 139–169). Hershey, PA: Idea Group Publishing; doi:10.4018/978-1-59140-060-8.ch007

Norris, D. F. (2008). Information technology among U.S. local governments. In G. Garson & M. Khosrow-Pour (Eds.), *Handbook of research on public information technology* (pp. 132–144). Hershey, PA: Information Science Reference; doi:10.4018/978-1-59904-857-4.ch013

Northrop, A. (1999). The challenge of teaching information technology in public administration graduate programs. In G. Garson (Ed.), *Information technology and computer applications in public administration: Issues and trends* (pp. 1–22). Hershey, PA: Information Science Reference; doi:10.4018/978-1-87828-952-0.ch001

Northrop, A. (2003). Information technology and public administration: The view from the profession. In G. Garson (Ed.), *Public information technology: Policy and management issues* (pp. 1–19). Hershey, PA: Idea Group Publishing; doi:10.4018/978-1-59140-060-8.ch001

Northrop, A. (2007). Lip service? How PA journals and textbooks view information technology. In G. Garson (Ed.), *Modern public information technology systems: Issues and challenges* (pp. 1–16). Hershey, PA: Idea Group Publishing; doi:10.4018/978-1-59904-051-6.ch001

Null, E. (2013). Legal and political barriers to municipal networks in the United States. In A. Abdelaal (Ed.), *Social and economic effects of community wireless networks and infrastructures* (pp. 27–56). Hershey, PA: Information Science Reference; doi:10.4018/978-1-4666-2997-4.ch003

Okunoye, A., Frolick, M., & Crable, E. (2006). ERP implementation in higher education: An account of pre-implementation and implementation phases. *Journal of Cases on Information Technology*, *8*(2), 110–132. doi:10.4018/jcit.2006040106

Olasina, G. (2012). A review of egovernment services in Nigeria. In A. Tella & A. Issa (Eds.), *Library and information science in developing countries: Contemporary issues* (pp. 205–221). Hershey, PA: Information Science Reference; doi:10.4018/978-1-61350-335-5.ch015

Orgeron, C. P. (2008). A model for reengineering IT job classes in state government. In G. Garson & M. Khosrow-Pour (Eds.), *Handbook of research on public information technology* (pp. 735–746). Hershey, PA: Information Science Reference; doi:10.4018/978-1-59904-857-4.ch066

Owsinski, J. W., & Pielak, A. M. (2011). Local authority websites in rural areas: Measuring quality and functionality, and assessing the role. In Z. Andreopoulou, B. Manos, N. Polman, & D. Viaggi (Eds.), *Agricultural and environmental informatics, governance and management: Emerging research applications* (pp. 39–60). Hershey, PA: Information Science Reference; doi:10.4018/978-1-60960-621-3.ch003

Owsiński, J. W., Pielak, A. M., Sęp, K., & Stańczak, J. (2014). Local web-based networks in rural municipalities: Extension, density, and meaning. In Z. Andreopoulou, V. Samathrakis, S. Louca, & M. Vlachopoulou (Eds.), *E-innovation for sustainable development of rural resources during global economic crisis* (pp. 126–151). Hershey, PA: Business Science Reference; doi:10.4018/978-1-4666-4550-9.ch011

Pagani, M., & Pasinetti, C. (2008). Technical and functional quality in the development of t-government services. In A. Anttiroiko (Ed.), *Electronic government: Concepts, methodologies, tools, and applications* (pp. 2943–2965). Hershey, PA: Information Science Reference; doi:10.4018/978-1-59904-947-2.ch220

Pani, A. K., & Agrahari, A. (2005). On e-markets in emerging economy: An Indian experience. In M. Khosrow-Pour (Ed.), *Advanced topics in electronic commerce* (Vol. 1, pp. 287–299). Hershey, PA: Idea Group Publishing; doi:10.4018/978-1-59140-819-2.ch015

Papadopoulos, T., Angelopoulos, S., & Kitsios, F. (2011). A strategic approach to e-health interoperability using e-government frameworks. In A. Lazakidou, K. Siassiakos, & K. Ioannou (Eds.), *Wireless technologies for ambient assisted living and healthcare: Systems and applications* (pp. 213–229). Hershey, PA: Medical Information Science Reference; doi:10.4018/978-1-61520-805-0.ch012

Papadopoulos, T., Angelopoulos, S., & Kitsios, F. (2013). A strategic approach to e-health interoperability using e-government frameworks. In *User-driven healthcare: Concepts, methodologies, tools, and applications* (pp. 791–807). Hershey, PA: Medical Information Science Reference; doi:10.4018/978-1-4666-2770-3.ch039

Papaleo, G., Chiarella, D., Aiello, M., & Caviglione, L. (2012). Analysis, development and deployment of statistical anomaly detection techniques for real e-mail traffic. In T. Chou (Ed.), *Information assurance and security technologies for risk assessment and threat management: Advances* (pp. 47–71). Hershey, PA: Information Science Reference; doi:10.4018/978-1-61350-507-6.ch003

Papp, R. (2003). Information technology & FDA compliance in the pharmaceutical industry. In M. Khosrow-Pour (Ed.), *Annals of cases on information technology* (Vol. 5, pp. 262–273). Hershey, PA: Information Science Reference; doi:10.4018/978-1-59140-061-5.ch017

Parsons, T. W. (2007). Developing a knowledge management portal. In A. Tatnall (Ed.), *Encyclopedia of portal technologies and applications* (pp. 223–227). Hershey, PA: Information Science Reference; doi:10.4018/978-1-59140-989-2.ch039

Passaris, C. E. (2007). Immigration and digital government. In A. Anttiroiko & M. Malkia (Eds.), *Encyclopedia of digital government* (pp. 988–994). Hershey, PA: Information Science Reference; doi:10.4018/978-1-59140-789-8.ch148

Pavlichev, A. (2004). The e-government challenge for public administration. In A. Pavlichev & G. Garson (Eds.), *Digital government: Principles and best practices* (pp. 276–290). Hershey, PA: Idea Group Publishing; doi:10.4018/978-1-59140-122-3.ch018

Penrod, J. I., & Harbor, A. F. (2000). Designing and implementing a learning organization-oriented information technology planning and management process. In L. Petrides (Ed.), *Case studies on information technology in higher education: Implications for policy and practice* (pp. 7–19). Hershey, PA: Idea Group Publishing; doi:10.4018/978-1-878289-74-2.ch001

Planas-Silva, M. D., & Joseph, R. C. (2011). Perspectives on the adoption of electronic resources for use in clinical trials. In M. Guah (Ed.), *Healthcare delivery reform and new technologies: Organizational initiatives* (pp. 19–28). Hershey, PA: Information Science Reference; doi:10.4018/978-1-60960-183-6.ch002

Pomazalová, N., & Rejman, S. (2013). The rationale behind implementation of new electronic tools for electronic public procurement. In N. Pomazalová (Ed.), *Public sector transformation processes and internet public procurement: Decision support systems* (pp. 85–117). Hershey, PA: Engineering Science Reference; doi:10.4018/978-1-4666-2665-2.ch006

Postorino, M. N. (2012). City competitiveness and airport: Information science perspective. In M. Bulu (Ed.), *City competitiveness and improving urban subsystems: Technologies and applications* (pp. 61–83). Hershey, PA: Information Science Reference; doi:10.4018/978-1-61350-174-0.ch004

Poupa, C. (2002). Electronic government in Switzerland: Priorities for 2001-2005 - Electronic voting and federal portal. In Å. Grönlund (Ed.), *Electronic government: Design, applications and management* (pp. 356–369). Hershey, PA: Idea Group Publishing; doi:10.4018/978-1-930708-19-8.ch017

Powell, S. R. (2010). Interdisciplinarity in telecommunications and networking. In *Networking and telecommunications: Concepts, methodologies, tools and applications* (pp. 33–40). Hershey, PA: Information Science Reference; doi:10.4018/978-1-60566-986-1.ch004

Priya, P. S., & Mathiyalagan, N. (2011). A study of the implementation status of two e-governance projects in land revenue administration in India. In M. Shareef, V. Kumar, U. Kumar, & Y. Dwivedi (Eds.), *Stakeholder adoption of e-government services: Driving and resisting factors* (pp. 214–230). Hershey, PA: Information Science Reference; doi:10.4018/978-1-60960-601-5.ch011

Prysby, C., & Prysby, N. (2000). Electronic mail, employee privacy and the workplace. In L. Janczewski (Ed.), *Internet and intranet security management: Risks and solutions* (pp. 251–270). Hershey, PA: Idea Group Publishing; doi:10.4018/978-1-878289-71-1.ch009

Prysby, C. L., & Prysby, N. D. (2003). Electronic mail in the public workplace: Issues of privacy and public disclosure. In G. Garson (Ed.), *Public information technology: Policy and management issues* (pp. 271–298). Hershey, PA: Idea Group Publishing; doi:10.4018/978-1-59140-060-8.ch012

Prysby, C. L., & Prysby, N. D. (2007). You have mail, but who is reading it? Issues of e-mail in the public workplace. In G. Garson (Ed.), *Modern public information technology systems: Issues and challenges* (pp. 312–336). Hershey, PA: Idea Group Publishing; doi:10.4018/978-1-59904-051-6.ch016

Radl, A., & Chen, Y. (2005). Computer security in electronic government: A state-local education information system. *International Journal of Electronic Government Research, 1*(1), 79–99. doi:10.4018/jegr.2005010105

Rahman, H. (2008). Information dynamics in developing countries. In C. Van Slyke (Ed.), *Information communication technologies: Concepts, methodologies, tools, and applications* (pp. 104–114). Hershey, PA: Information Science Reference; doi:10.4018/978-1-59904-949-6.ch008

Ramanathan, J. (2009). Adaptive IT architecture as a catalyst for network capability in government. In P. Saha (Ed.), *Advances in government enterprise architecture* (pp. 149–172). Hershey, PA: Information Science Reference; doi:10.4018/978-1-60566-068-4.ch007

Ramos, I., & Berry, D. M. (2006). Social construction of information technology supporting work. In M. Khosrow-Pour (Ed.), *Cases on information technology: Lessons learned* (Vol. 7, pp. 36–52). Hershey, PA: Idea Group Publishing; doi:10.4018/978-1-59140-673-0.ch003

Ray, D., Gulla, U., Gupta, M. P., & Dash, S. S. (2009). Interoperability and constituents of interoperable systems in public sector. In V. Weerakkody, M. Janssen, & Y. Dwivedi (Eds.), *Handbook of research on ICT-enabled transformational government: A global perspective* (pp. 175–195). Hershey, PA: Information Science Reference; doi:10.4018/978-1-60566-390-6.ch010

Reddick, C. G. (2007). E-government and creating a citizen-centric government: A study of federal government CIOs. In G. Garson (Ed.), *Modern public information technology systems: Issues and challenges* (pp. 143–165). Hershey, PA: Idea Group Publishing; doi:10.4018/978-1-59904-051-6.ch008

Reddick, C. G. (2010). Citizen-centric e-government. In C. Reddick (Ed.), *Homeland security preparedness and information systems: Strategies for managing public policy* (pp. 45–75). Hershey, PA: Information Science Reference; doi:10.4018/978-1-60566-834-5.ch002

Reddick, C. G. (2010). E-government and creating a citizen-centric government: A study of federal government CIOs. In C. Reddick (Ed.), *Homeland security preparedness and information systems: Strategies for managing public policy* (pp. 230–250). Hershey, PA: Information Science Reference; doi:10.4018/978-1-60566-834-5.ch012

Reddick, C. G. (2010). Perceived effectiveness of e-government and its usage in city governments: Survey evidence from information technology directors. In C. Reddick (Ed.), *Homeland security preparedness and information systems: Strategies for managing public policy* (pp. 213–229). Hershey, PA: Information Science Reference; doi:10.4018/978-1-60566-834-5.ch011

Reddick, C. G. (2012). Customer relationship management adoption in local governments in the United States. In S. Chhabra & M. Kumar (Eds.), *Strategic enterprise resource planning models for e-government: Applications and methodologies* (pp. 111–124). Hershey, PA: Information Science Reference; doi:10.4018/978-1-60960-863-7.ch008

Reeder, F. S., & Pandy, S. M. (2008). Identifying effective funding models for e-government. In A. Anttiroiko (Ed.), *Electronic government: Concepts, methodologies, tools, and applications* (pp. 1108–1138). Hershey, PA: Information Science Reference; doi:10.4018/978-1-59904-947-2.ch083

Riesco, D., Acosta, E., & Montejano, G. (2003). An extension to a UML activity graph from workflow. In L. Favre (Ed.), *UML and the unified process* (pp. 294–314). Hershey, PA: IRM Press; doi:10.4018/978-1-93177-744-5.ch015

Ritzhaupt, A. D., & Gill, T. G. (2008). A hybrid and novel approach to teaching computer programming in MIS curriculum. In S. Negash, M. Whitman, A. Woszczynski, K. Hoganson, & H. Mattord (Eds.), *Handbook of distance learning for real-time and asynchronous information technology education* (pp. 259–281). Hershey, PA: Information Science Reference; doi:10.4018/978-1-59904-964-9.ch014

Roche, E. M. (1993). International computing and the international regime. *Journal of Global Information Management, 1*(2), 33–44. doi:10.4018/jgim.1993040103

Rocheleau, B. (2007). Politics, accountability, and information management. In G. Garson (Ed.), *Modern public information technology systems: Issues and challenges* (pp. 35–71). Hershey, PA: Idea Group Publishing; doi:10.4018/978-1-59904-051-6.ch003

Rodrigues Filho, J. (2010). E-government in Brazil: Reinforcing dominant institutions or reducing citizenship? In C. Reddick (Ed.), *Politics, democracy and e-government: Participation and service delivery* (pp. 347–362). Hershey, PA: Information Science Reference; doi:10.4018/978-1-61520-933-0.ch021

Rodriguez, S. R., & Thorp, D. A. (2013). eLearning for industry: A case study of the project management process. In A. Benson, J. Moore, & S. Williams van Rooij (Eds.), Cases on educational technology planning, design, and implementation: A project management perspective (pp. 319-342). Hershey, PA: Information Science Reference. doi:10.4018/978-1-4666-4237-9.ch017

Roman, A. V. (2013). Delineating three dimensions of e-government success: Security, functionality, and transformation. In J. Gil-Garcia (Ed.), *E-government success factors and measures: Theories, concepts, and methodologies* (pp. 171–192). Hershey, PA: Information Science Reference; doi:10.4018/978-1-4666-4058-0.ch010

Ross, S. C., Tyran, C. K., & Auer, D. J. (2008). Up in smoke: Rebuilding after an IT disaster. In H. Nemati (Ed.), *Information security and ethics: Concepts, methodologies, tools, and applications* (pp. 3659–3675). Hershey, PA: Information Science Reference; doi:10.4018/978-1-59904-937-3.ch248

Ross, S. C., Tyran, C. K., Auer, D. J., Junell, J. M., & Williams, T. G. (2005). Up in smoke: Rebuilding after an IT disaster. *Journal of Cases on Information Technology*, *7*(2), 31–49. doi:10.4018/jcit.2005040103

Roy, J. (2008). Security, sovereignty, and continental interoperability: Canada's elusive balance. In T. Loendorf & G. Garson (Eds.), *Patriotic information systems* (pp. 153–176). Hershey, PA: IGI Publishing; doi:10.4018/978-1-59904-594-8.ch007

Rubeck, R. F., & Miller, G. A. (2009). vGOV: Remote video access to government services. In A. Scupola (Ed.), Cases on managing e-services (pp. 253-268). Hershey, PA: Information Science Reference. doi:10.4018/978-1-60566-064-6.ch017

Saekow, A., & Boonmee, C. (2011). The challenges of implementing e-government interoperability in Thailand: Case of official electronic correspondence letters exchange across government departments. In Y. Charalabidis (Ed.), *Interoperability in digital public services and administration: Bridging e-government and e-business* (pp. 40–61). Hershey, PA: Information Science Reference; doi:10.4018/978-1-61520-887-6.ch003

Saekow, A., & Boonmee, C. (2012). The challenges of implementing e-government interoperability in Thailand: Case of official electronic correspondence letters exchange across government departments. In *Digital democracy: Concepts, methodologies, tools, and applications* (pp. 1883–1905). Hershey, PA: Information Science Reference; doi:10.4018/978-1-4666-1740-7.ch094

Sagsan, M., & Medeni, T. (2012). Understanding "knowledge management (KM) paradigms" from social media perspective: An empirical study on discussion group for KM at professional networking site. In M. Cruz-Cunha, P. Gonçalves, N. Lopes, E. Miranda, & G. Putnik (Eds.), *Handbook of research on business social networking: Organizational, managerial, and technological dimensions* (pp. 738–755). Hershey, PA: Business Science Reference; doi:10.4018/978-1-61350-168-9.ch039

Sahi, G., & Madan, S. (2013). Information security threats in ERP enabled e-governance: Challenges and solutions. In *Enterprise resource planning: Concepts, methodologies, tools, and applications* (pp. 825–837). Hershey, PA: Business Science Reference; doi:10.4018/978-1-4666-4153-2.ch048

Sanford, C., & Bhattacherjee, A. (2008). IT implementation in a developing country municipality: A sociocognitive analysis. *International Journal of Technology and Human Interaction*, *4*(3), 68–93. doi:10.4018/jthi.2008070104

Schelin, S. H. (2003). E-government: An overview. In G. Garson (Ed.), *Public information technology: Policy and management issues* (pp. 120–138). Hershey, PA: Idea Group Publishing; doi:10.4018/978-1-59140-060-8.ch006

Schelin, S. H. (2004). Training for digital government. In A. Pavlichev & G. Garson (Eds.), *Digital government: Principles and best practices* (pp. 263–275). Hershey, PA: Idea Group Publishing; doi:10.4018/978-1-59140-122-3.ch017

Schelin, S. H. (2007). E-government: An overview. In G. Garson (Ed.), *Modern public information technology systems: Issues and challenges* (pp. 110–126). Hershey, PA: Idea Group Publishing; doi:10.4018/978-1-59904-051-6.ch006

Schelin, S. H., & Garson, G. (2004). Theoretical justification of critical success factors. In G. Garson & S. Schelin (Eds.), *IT solutions series: Humanizing information technology: Advice from experts* (pp. 4–15). Hershey, PA: CyberTech Publishing; doi:10.4018/978-1-59140-245-9.ch002

Scime, A. (2002). Information systems and computer science model curricula: A comparative look. In M. Dadashzadeh, A. Saber, & S. Saber (Eds.), *Information technology education in the new millennium* (pp. 146–158). Hershey, PA: IRM Press; doi:10.4018/978-1-931777-05-6.ch018

Scime, A. (2009). Computing curriculum analysis and development. In M. Khosrow-Pour (Ed.), *Encyclopedia of information science and technology* (2nd ed., pp. 667–671). Hershey, PA: Information Science Reference; doi:10.4018/978-1-60566-026-4.ch108

Scime, A., & Wania, C. (2008). Computing curricula: A comparison of models. In C. Van Slyke (Ed.), *Information communication technologies: Concepts, methodologies, tools, and applications* (pp. 1270–1283). Hershey, PA: Information Science Reference; doi:10.4018/978-1-59904-949-6.ch088

Seidman, S. B. (2009). An international perspective on professional software engineering credentials. In H. Ellis, S. Demurjian, & J. Naveda (Eds.), *Software engineering: Effective teaching and learning approaches and practices* (pp. 351–361). Hershey, PA: Information Science Reference; doi:10.4018/978-1-60566-102-5.ch018

Seifert, J. W. (2007). E-government act of 2002 in the United States. In A. Anttiroiko & M. Malkia (Eds.), *Encyclopedia of digital government* (pp. 476–481). Hershey, PA: Information Science Reference; doi:10.4018/978-1-59140-789-8.ch072

Seifert, J. W., & Relyea, H. C. (2008). E-government act of 2002 in the United States. In A. Anttiroiko (Ed.), *Electronic government: Concepts, methodologies, tools, and applications* (pp. 154–161). Hershey, PA: Information Science Reference; doi:10.4018/978-1-59904-947-2.ch013

Seufert, S. (2002). E-learning business models: Framework and best practice examples. In M. Raisinghani (Ed.), *Cases on worldwide e-commerce: Theory in action* (pp. 70–94). Hershey, PA: Idea Group Publishing; doi:10.4018/978-1-930708-27-3.ch004

Shareef, M. A., & Archer, N. (2012). E-government service development. In M. Shareef, N. Archer, & S. Dutta (Eds.), *E-government service maturity and development: Cultural, organizational and technological perspectives* (pp. 1–14). Hershey, PA: Information Science Reference; doi:10.4018/978-1-60960-848-4.ch001

Shareef, M. A., & Archer, N. (2012). E-government initiatives: Review studies on different countries. In M. Shareef, N. Archer, & S. Dutta (Eds.), *E-government service maturity and development: Cultural, organizational and technological perspectives* (pp. 40–76). Hershey, PA: Information Science Reference; doi:10.4018/978-1-60960-848-4.ch003

Shareef, M. A., Kumar, U., & Kumar, V. (2011). E-government development: Performance evaluation parameters. In M. Shareef, V. Kumar, U. Kumar, & Y. Dwivedi (Eds.), *Stakeholder adoption of e-government services: Driving and resisting factors* (pp. 197–213). Hershey, PA: Information Science Reference; doi:10.4018/978-1-60960-601-5.ch010

Shareef, M. A., Kumar, U., Kumar, V., & Niktash, M. (2012). Electronic-government vision: Case studies for objectives, strategies, and initiatives. In M. Shareef, N. Archer, & S. Dutta (Eds.), *E-government service maturity and development: Cultural, organizational and technological perspectives* (pp. 15–39). Hershey, PA: Information Science Reference; doi:10.4018/978-1-60960-848-4.ch002

Shukla, P., Kumar, A., & Anu Kumar, P. B. (2013). Impact of national culture on business continuity management system implementation. *International Journal of Risk and Contingency Management*, *2*(3), 23–36. doi:10.4018/ijrcm.2013070102

Shulman, S. W. (2007). The federal docket management system and the prospect for digital democracy in U S rulemaking. In G. Garson (Ed.), *Modern public information technology systems: Issues and challenges* (pp. 166–184). Hershey, PA: Idea Group Publishing; doi:10.4018/978-1-59904-051-6.ch009

Simonovic, S. (2007). Problems of offline government in e-Serbia. In A. Anttiroiko & M. Malkia (Eds.), *Encyclopedia of digital government* (pp. 1342–1351). Hershey, PA: Information Science Reference; doi:10.4018/978-1-59140-789-8.ch205

Simonovic, S. (2008). Problems of offline government in e-Serbia. In A. Anttiroiko (Ed.), *Electronic government: Concepts, methodologies, tools, and applications* (pp. 2929–2942). Hershey, PA: Information Science Reference; doi:10.4018/978-1-59904-947-2.ch219

Singh, A. M. (2005). Information systems and technology in South Africa. In M. Khosrow-Pour (Ed.), *Encyclopedia of information science and technology* (pp. 1497–1502). Hershey, PA: Information Science Reference; doi:10.4018/978-1-59140-553-5.ch263

Singh, S., & Naidoo, G. (2005). Towards an e-government solution: A South African perspective. In W. Huang, K. Siau, & K. Wei (Eds.), *Electronic government strategies and implementation* (pp. 325–353). Hershey, PA: Idea Group Publishing; doi:10.4018/978-1-59140-348-7.ch014

Snoke, R., & Underwood, A. (2002). Generic attributes of IS graduates: An analysis of Australian views. In F. Tan (Ed.), *Advanced topics in global information management* (Vol. 1, pp. 370–384). Hershey, PA: Idea Group Publishing; doi:10.4018/978-1-930708-43-3.ch023

Sommer, L. (2006). Revealing unseen organizations in higher education: A study framework and application example. In A. Metcalfe (Ed.), *Knowledge management and higher education: A critical analysis* (pp. 115–146). Hershey, PA: Information Science Publishing; doi:10.4018/978-1-59140-509-2.ch007

Song, H., Kidd, T., & Owens, E. (2011). Examining technological disparities and instructional practices in English language arts classroom: Implications for school leadership and teacher training. In L. Tomei (Ed.), *Online courses and ICT in education: Emerging practices and applications* (pp. 258–274). Hershey, PA: Information Science Reference; doi:10.4018/978-1-60960-150-8.ch020

Speaker, P. J., & Kleist, V. F. (2003). Using information technology to meet electronic commerce and MIS education demands. In A. Aggarwal (Ed.), *Web-based education: Learning from experience* (pp. 280–291). Hershey, PA: Information Science Publishing; doi:10.4018/978-1-59140-102-5.ch017

Spitler, V. K. (2007). Learning to use IT in the workplace: Mechanisms and masters. In M. Mahmood (Ed.), *Contemporary issues in end user computing* (pp. 292–323). Hershey, PA: Idea Group Publishing; doi:10.4018/978-1-59140-926-7.ch013

Stellefson, M. (2011). Considerations for marketing distance education courses in health education: Five important questions to examine before development. In U. Demiray & S. Sever (Eds.), *Marketing online education programs: Frameworks for promotion and communication* (pp. 222–234). Hershey, PA: Information Science Reference; doi:10.4018/978-1-60960-074-7.ch014

Straub, D. W., & Loch, K. D. (2006). Creating and developing a program of global research. *Journal of Global Information Management, 14*(2), 1–28. doi:10.4018/jgim.2006040101

Straub, D. W., Loch, K. D., & Hill, C. E. (2002). Transfer of information technology to the Arab world: A test of cultural influence modeling. In M. Dadashzadeh (Ed.), *Information technology management in developing countries* (pp. 92–134). Hershey, PA: IRM Press; doi:10.4018/978-1-931777-03-2.ch005

Straub, D. W., Loch, K. D., & Hill, C. E. (2003). Transfer of information technology to the Arab world: A test of cultural influence modeling. In F. Tan (Ed.), *Advanced topics in global information management* (Vol. 2, pp. 141–172). Hershey, PA: Idea Group Publishing; doi:10.4018/978-1-59140-064-6.ch009

Suki, N. M., Ramayah, T., Ming, M. K., & Suki, N. M. (2013). Factors enhancing employed job seekers intentions to use social networking sites as a job search tool. In A. Mesquita (Ed.), *User perception and influencing factors of technology in everyday life* (pp. 265–281). Hershey, PA: Information Science Reference; doi:10.4018/978-1-4666-1954-8.ch018

Suomi, R. (2006). Introducing electronic patient records to hospitals: Innovation adoption paths. In T. Spil & R. Schuring (Eds.), *E-health systems diffusion and use: The innovation, the user and the use IT model* (pp. 128–146). Hershey, PA: Idea Group Publishing; doi:10.4018/978-1-59140-423-1.ch008

Swim, J., & Barker, L. (2012). Pathways into a gendered occupation: Brazilian women in IT. *International Journal of Social and Organizational Dynamics in IT, 2*(4), 34–51. doi:10.4018/ijsodit.2012100103

Tarafdar, M., & Vaidya, S. D. (2006). Adoption and implementation of IT in developing nations: Experiences from two public sector enterprises in India. In M. Khosrow-Pour (Ed.), *Cases on information technology planning, design and implementation* (pp. 208–233). Hershey, PA: Idea Group Publishing; doi:10.4018/978-1-59904-408-8.ch013

Tarafdar, M., & Vaidya, S. D. (2008). Adoption and implementation of IT in developing nations: Experiences from two public sector enterprises in India. In G. Garson & M. Khosrow-Pour (Eds.), *Handbook of research on public information technology* (pp. 905–924). Hershey, PA: Information Science Reference; doi:10.4018/978-1-59904-857-4.ch076

Thesing, Z. (2007). Zarina thesing, pumpkin patch. In M. Hunter (Ed.), *Contemporary chief information officers: Management experiences* (pp. 83–94). Hershey, PA: IGI Publishing; doi:10.4018/978-1-59904-078-3.ch007

Thomas, J. C. (2004). Public involvement in public administration in the information age: Speculations on the effects of technology. In M. Malkia, A. Anttiroiko, & R. Savolainen (Eds.), *eTransformation in governance: New directions in government and politics* (pp. 67–84). Hershey, PA: Idea Group Publishing; doi:10.4018/978-1-59140-130-8.ch004

Treiblmaier, H., & Chong, S. (2013). Trust and perceived risk of personal information as antecedents of online information disclosure: Results from three countries. In F. Tan (Ed.), *Global diffusion and adoption of technologies for knowledge and information sharing* (pp. 341–361). Hershey, PA: Information Science Reference; doi:10.4018/978-1-4666-2142-8.ch015

van Grembergen, W., & de Haes, S. (2008). IT governance in practice: Six case studies. In W. van Grembergen & S. De Haes (Eds.), *Implementing information technology governance: Models, practices and cases* (pp. 125–237). Hershey, PA: IGI Publishing; doi:10.4018/978-1-59904-924-3.ch004

van Os, G., Homburg, V., & Bekkers, V. (2013). Contingencies and convergence in European social security: ICT coordination in the back office of the welfare state. In M. Cruz-Cunha, I. Miranda, & P. Gonçalves (Eds.), *Handbook of research on ICTs and management systems for improving efficiency in healthcare and social care* (pp. 268–287). Hershey, PA: Medical Information Science Reference; doi:10.4018/978-1-4666-3990-4.ch013

Velloso, A. B., Gassenferth, W., & Machado, M. A. (2012). Evaluating IBMEC-RJ's intranet usability using fuzzy logic. In M. Cruz-Cunha, P. Gonçalves, N. Lopes, E. Miranda, & G. Putnik (Eds.), *Handbook of research on business social networking: Organizational, managerial, and technological dimensions* (pp. 185–205). Hershey, PA: Business Science Reference; doi:10.4018/978-1-61350-168-9.ch010

Villablanca, A. C., Baxi, H., & Anderson, K. (2009). Novel data interface for evaluating cardiovascular outcomes in women. In A. Dwivedi (Ed.), *Handbook of research on information technology management and clinical data administration in healthcare* (pp. 34–53). Hershey, PA: Medical Information Science Reference; doi:10.4018/978-1-60566-356-2.ch003

Villablanca, A. C., Baxi, H., & Anderson, K. (2011). Novel data interface for evaluating cardiovascular outcomes in women. In *Clinical technologies: Concepts, methodologies, tools and applications* (pp. 2094–2113). Hershey, PA: Medical Information Science Reference; doi:10.4018/978-1-60960-561-2.ch806

Virkar, S. (2011). Information and communication technologies in administrative reform for development: Exploring the case of property tax systems in Karnataka, India. In J. Steyn, J. Van Belle, & E. Mansilla (Eds.), *ICTs for global development and sustainability: Practice and applications* (pp. 127–149). Hershey, PA: Information Science Reference; doi:10.4018/978-1-61520-997-2.ch006

Virkar, S. (2013). Designing and implementing e-government projects: Actors, influences, and fields of play. In S. Saeed & C. Reddick (Eds.), *Human-centered system design for electronic governance* (pp. 88–110). Hershey, PA: Information Science Reference; doi:10.4018/978-1-4666-3640-8.ch007

Wallace, A. (2009). E-justice: An Australian perspective. In A. Martínez & P. Abat (Eds.), *E-justice: Using information communication technologies in the court system* (pp. 204–228). Hershey, PA: Information Science Reference; doi:10.4018/978-1-59904-998-4.ch014

Wang, G. (2012). E-democratic administration and bureaucratic responsiveness: A primary study of bureaucrats' perceptions of the civil service e-mail box in Taiwan. In K. Kloby & M. D'Agostino (Eds.), *Citizen 2.0: Public and governmental interaction through web 2.0 technologies* (pp. 146–173). Hershey, PA: Information Science Reference; doi:10.4018/978-1-4666-0318-9.ch009

Wangpipatwong, S., Chutimaskul, W., & Papasratorn, B. (2011). Quality enhancing the continued use of e-government web sites: Evidence from e-citizens of Thailand. In V. Weerakkody (Ed.), *Applied technology integration in governmental organizations: New e-government research* (pp. 20–36). Hershey, PA: Information Science Reference; doi:10.4018/978-1-60960-162-1.ch002

Wedemeijer, L. (2006). Long-term evolution of a conceptual schema at a life insurance company. In M. Khosrow-Pour (Ed.), *Cases on database technologies and applications* (pp. 202–226). Hershey, PA: Idea Group Publishing; doi:10.4018/978-1-59904-399-9.ch012

Whybrow, E. (2008). Digital access, ICT fluency, and the economically disadvantages: Approaches to minimize the digital divide. In F. Tan (Ed.), *Global information technologies: Concepts, methodologies, tools, and applications* (pp. 1409–1422). Hershey, PA: Information Science Reference; doi:10.4018/978-1-59904-939-7.ch102

Whybrow, E. (2008). Digital access, ICT fluency, and the economically disadvantages: Approaches to minimize the digital divide. In C. Van Slyke (Ed.), *Information communication technologies: Concepts, methodologies, tools, and applications* (pp. 764–777). Hershey, PA: Information Science Reference; doi:10.4018/978-1-59904-949-6.ch049

Wickramasinghe, N., & Geisler, E. (2010). Key considerations for the adoption and implementation of knowledge management in healthcare operations. In M. Saito, N. Wickramasinghe, M. Fuji, & E. Geisler (Eds.), *Redesigning innovative healthcare operation and the role of knowledge management* (pp. 125–142). Hershey, PA: Medical Information Science Reference; doi:10.4018/978-1-60566-284-8.ch009

Wickramasinghe, N., & Geisler, E. (2012). Key considerations for the adoption and implementation of knowledge management in healthcare operations. In *Organizational learning and knowledge: Concepts, methodologies, tools and applications* (pp. 1316–1328). Hershey, PA: Business Science Reference; doi:10.4018/978-1-60960-783-8.ch405

Wickramasinghe, N., & Goldberg, S. (2007). A framework for delivering m-health excellence. In L. Al-Hakim (Ed.), *Web mobile-based applications for healthcare management* (pp. 36–61). Hershey, PA: IRM Press; doi:10.4018/978-1-59140-658-7.ch002

Wickramasinghe, N., & Goldberg, S. (2008). Critical success factors for delivering m-health excellence. In N. Wickramasinghe & E. Geisler (Eds.), *Encyclopedia of healthcare information systems* (pp. 339–351). Hershey, PA: Medical Information Science Reference; doi:10.4018/978-1-59904-889-5.ch045

Wyld, D. (2009). Radio frequency identification (RFID) technology. In J. Symonds, J. Ayoade, & D. Parry (Eds.), *Auto-identification and ubiquitous computing applications* (pp. 279–293). Hershey, PA: Information Science Reference; doi:10.4018/978-1-60566-298-5.ch017

Yaghmaei, F. (2010). Understanding computerised information systems usage in community health. In J. Rodrigues (Ed.), *Health information systems: Concepts, methodologies, tools, and applications* (pp. 1388–1399). Hershey, PA: Medical Information Science Reference; doi:10.4018/978-1-60566-988-5.ch088

Yee, G., El-Khatib, K., Korba, L., Patrick, A. S., Song, R., & Xu, Y. (2005). Privacy and trust in e-government. In W. Huang, K. Siau, & K. Wei (Eds.), *Electronic government strategies and implementation* (pp. 145–190). Hershey, PA: Idea Group Publishing; doi:10.4018/978-1-59140-348-7.ch007

Yeh, S., & Chu, P. (2010). Evaluation of e-government services: A citizen-centric approach to citizen e-complaint services. In C. Reddick (Ed.), *Citizens and e-government: Evaluating policy and management* (pp. 400–417). Hershey, PA: Information Science Reference; doi:10.4018/978-1-61520-931-6.ch022

Young-Jin, S., & Seang-tae, K. (2008). E-government concepts, measures, and best practices. In A. Anttiroiko (Ed.), *Electronic government: Concepts, methodologies, tools, and applications* (pp. 32–57). Hershey, PA: Information Science Reference; doi:10.4018/978-1-59904-947-2.ch004

Yun, H. J., & Opheim, C. (2012). New technology communication in American state governments: The impact on citizen participation. In K. Bwalya & S. Zulu (Eds.), *Handbook of research on e-government in emerging economies: Adoption, e-participation, and legal frameworks* (pp. 573–590). Hershey, PA: Information Science Reference; doi:10.4018/978-1-4666-0324-0.ch029

Zhang, N., Guo, X., Chen, G., & Chau, P. Y. (2011). User evaluation of e-government systems: A Chinese cultural perspective. In F. Tan (Ed.), *International enterprises and global information technologies: Advancing management practices* (pp. 63–84). Hershey, PA: Information Science Reference; doi:10.4018/978-1-60960-605-3.ch004

Zuo, Y., & Hu, W. (2011). Trust-based information risk management in a supply chain network. In J. Wang (Ed.), *Supply chain optimization, management and integration: Emerging applications* (pp. 181–196). Hershey, PA: Business Science Reference; doi:10.4018/978-1-60960-135-5.ch013

Compilation of References

Aaker, D. (1996). *Building Strong Brands*. New York: The free press.

Aaker, J. (1997). Dimensions of brand personality. *JMR, Journal of Marketing Research*, *34*(3), 347. doi:10.2307/3151897

Abiad, P. (2014). *El abuelo del periodismo de marca cumple 120 años*. Retrieved from http://www.pabloabiad.com/2014/10/el-abuelo-del-periodismo-de-marca.html

Adler, N. J., & Harzing, A. W. (2009). When knowledge wins: Transcending the sense and nonsense of academic rankings. *Academy of Management Learning & Education*, *8*(1), 72–95. doi:10.5465/AMLE.2009.37012181

AERCO & Creative Territory. (n.d.). *The role of the Community Manager*. Retrieved April 7, 2016 from http://www.maestrosdelweb.com/images/2010/04/community-manager.pdf

Aghion, P., & Howitt, P. (1992). A Model of growth through Creative Destruction. *Econometrica*, *60*(2), 323–351. doi:10.2307/2951599

Aghion, P., & Howitt, P. (1997). *Endogenous Growth Theory*. Cambridge, MA: MIT Press.

Aguilar, M. G., Campos, S. R., & Batlle, P. F. (2012). El uso académico de las redes sociales en universitarios. *Comunicar*, *19*(38), 131–138.

Aguillo, I. F., Granadino, B., Ortega, J. L., & Prieto, J. A. (2006). Scientific research activity and communication measured with cybermetrics indicators. *Journal of the American Society for Information Science and Technology*, *57*(10), 1296–1302. doi:10.1002/asi.20433

Ahlquist, J. (2013). 10 Social Media Best Practices in Higher Education. *Josie Ahlquist*. Retrieved February, 15th, 2016, from http://www.josieahlquist.com/2013/10/07/10smbp/

Ahonen, T. (2011). *Main Trends in the Telecommunications Market*. Retrieved April 15, 2015 from http://www.citia.co.uk/content/files/50_44-887.pdf

Alcántara, A. (2014). *Súper profesional: Tómate tu vida profesional como algo personal*. Alienta.

Alonso, S., Cabrerizo, F. J., Herrera-Viedma, E., & Herrera, F. (2009). h-Index: A review focused in its variants, computation and standardization for different scientific fields. *Journal of Informetrics*, *3*(4), 273–289. doi:10.1016/j.joi.2009.04.001

Álvarez de Blanco, R. (2008). *Tu, marca personal: Como conseguir gestionar tu vida con talento y convertirla en una experiencia extraordinadira*. Madrid: Prentice-Hall.

Anderson, C. (2008). The end of theory: the data deluge makes the scientific method obsolete. *Wired, 16*(7). Retrieved June 15, 2015, from http://archive.wired.com/science/discoveries/magazine/16-07/pb_theory

Anderson, C. H. (2008), *The Long Tail: Why the Future of Business is Selling Less of More. Hyperon Books.*

Archambault, A., & Grudin, J. (2012, May). A longitudinal study of facebook, LinkedIn, & twitter use. In *Proceedings of the SIGCHI Conference on Human Factors in Computing Systems* (pp. 2741-2750). New York, NY: ACM. doi:10.1145/2207676.2208671

Ardilla, & Orozco. (n.d.). *El libro de los valores.* El tiempo.

Arnold, D. N., & Fowler, K. K. (2011). Nefarious numbers. *Notices of the AMS*, *58*(3), 434–437.

Arnould, E. J., & Craig, J. (2005). Consumer Culture Theory (CCT): Twenty Years of Research. *The Journal of Consumer Research*, *31*(4), 868–882. doi:10.1086/426626

Arqués, N. (2007). *Y tú qué marca eres?: 12 claves para gestionar tu reputación personal.* Barcelona: Alienta.

Athar, A. (2014). *Sentiment analysis of scientific citations.* University of Cambridge, Computer Laboratory, Technical Report (UCAM-CL-TR-856).

Avalos, C. (2013). *La marca identidad y estrategia.* Argentina: La Crujía.

Bador, P., & Lafouge, T. (2010). Analyse comparative du facteur d'impact et du h-index pour les journaux de pharmacologie. *Therapie*, *65*(2), 129–137. doi:10.2515/therapie/2009061 PMID:20478245

Bailey, I. I. I., & Schechter, A. (1994). The corporation as brand: An identity dilemma. *Chief Executive*, *98*, 42–45.

Bajkiewicz, T. E., Kraus, J. J., & Yeon Hong, S. (2011). The impact of newsroom changes and the rise of social media on the practice of media relations. *Public Relations Review*, *37*(3), 329–333. doi:10.1016/j.pubrev.2011.05.001

Balsebre, A. (1994). *El lenguaje radiofónico.* Madrid: Cátedra.

Barabasi, A. L. (2014). *Linked: the new science of networks.* New York: Perseus Books.

Barabasi, A. L., & Albert, R. (1999). Emergence of Scaling in Random Networks. *Science*, *286*(5439), 509–512. doi:10.1126/science.286.5439.509 PMID:10521342

Barciela, F. (2013a). *El 'boom' del periodismo de marca.* Retrieved from http://economia.elpais.com/economia/2013/02/22/actualidad/1361540029_041048.html

Barciela, F. (2013b). *El último desafío, el 'brand journalism'.* Retrieved from http://www.apmadrid.es/images/stories/125-136.pdf

Bar-Ilan, J. (2004). A microscopic link analysis of academic institutions within a country-the case of Israel. *Scientometrics*, *59*(3), 391–403. doi:10.1023/B:SCIE.0000018540.33706.c1

Beltrán López, G. (2012). Geolocalización y redes sociales: Un munso social, local y móvil (1a edición ed.). Valencia, Spain: Bubok.

Beltrán López, G. (2014). Geomarketing: geolocalización, redes sociales y turismo (1a edición ed.). Valencia, Spain: Bubok.

Bermúdez, A. M. R. (2012). *Profesores 2.0 en la universidad del siglo XXI.* Criterios para la integración educativa de la web social en la universidad.

Berry, D. M., & Pawlik, J. (2008). What is code: A conversation with Deleuze, Guattari and code. In *Libre culture. Meditations on free culture.* Winnipeg: Pygmalion Books. Retrieved June 15, 2015, from http://sro.sussex.ac.uk/46287/1/BERRYMOSS-LibreCulture2008.pdf

Bertot, J., Jaeger, P., & Grimes, J. (2010). Using ICTs to create a culture of transparency: E-government and social media as openness and anti-corruption tools for societies. *Government Information Quarterly*, *27*(3), 264–271. doi:10.1016/j.giq.2010.03.001

Björneborn, L., & Ingwersen, P. (2004). Toward a basic framework for webometrics. *Journal of the American Society for Information Science and Technology*, *55*(14), 1216–1227. doi:10.1002/asi.20077

Bollen, J., Van de Sompel, H., Hagberg, A., & Chute, R. (2009). A principal component analysis of 39 scientific impact measures. *PLoS ONE*, *4*(6), e6022. doi:10.1371/journal.pone.0006022 PMID:19562078

Bonsón, E., & Bednárová, M. (2013). Corporate LinkedIn practices of Eurozone companies. *Online Information Review*, *37*(6), 969–984. doi:10.1108/OIR-09-2012-0159

Bonzi, S., & Snyder, H. W. (1991). Motivations for citation: A comparison of self citation and citation to others. *Scientometrics*, *21*(2), 245–254. doi:10.1007/BF02017571

Borgman, C. L., & Furner, J. (2002). *Scholarly communication and bibliometrics*. Academic Press.

Borgman, C. L. (2000). Digital libraries and the continuum of scholarly communication. *The Journal of Documentation*, *56*(4), 412–430. doi:10.1108/EUM0000000007121

Bornmann, L. (2014). Do altmetrics point to the broader impact of research? An overview of benefits and disadvantages of altmetrics. *Journal of Informetrics*, *8*(4), 895–903. doi:10.1016/j.joi.2014.09.005

Bornmann, L., & Daniel, H. D. (2005). Does the h-index for ranking of scientists really work? *Scientometrics*, *65*(3), 391–392. doi:10.1007/s11192-005-0281-4

Bornmann, L., & Daniel, H. D. (2007). What do we know about the h index? *Journal of the American Society for Information Science and Technology*, *58*(9), 1381–1385. doi:10.1002/asi.20609

Boyd, D., & Ellison, N. (2007). Social network sites: Definition, history, and scholarship. *Journal of Computer-Mediated Communication*, *13*(1), 210–230. doi:10.1111/j.1083-6101.2007.00393.x

Braun, T., Glänzel, W., & Schubert, A. (2006). A Hirsch-type index for journals. *Scientometrics*, *69*(1), 169–173. doi:10.1007/s11192-006-0147-4

Brecht, B. (1927). Teoría de la radio. In *De las ondas rojas a las radios libres*. Barcelona: Gustavo Gili.

Brooks, T. A. (1986). Evidence of complex citer motivations. *Journal of the American Society for Information Science*, *37*(1), 34–36. doi:10.1002/asi.4630370106

Brouer, R. L., Stefanone, M. A., Badawy, R. L., Egnoto, M. J., & Seitz, S. R. (2015). Losing Control of Company Information in the Recruitment Process: The Impact of LinkedIn on Organizational Attraction. In *Proceedings of the 48th Hawaii International Conference on System Sciences* (pp. 1879-1888). Hawaii, HI: IEEE. doi:10.1109/HICSS.2015.226

Brumback, R. A. (2009). Impact factor wars: Episode V-the empire strikes back. *Journal of Child Neurology*, *24*(3), 260–262. doi:10.1177/0883073808331366 PMID:19258283

Buhalis, D. (2015). SoCoMo marketing for travel and tourism: Empowering co-creation of value. *Journal of Destination Marketing & Management*.

Caballo, V. (1998). *Manual de evaluacion y entrenamiento de las habilidades sociales*. Madrid: Siglo XXI de España.

Camilleri, M. A. (2015a). Using Big Data for Customer-Centric Marketing. In *Handbook of Research on Open Data Innovations in Business and Government*. IGI Global. Retrieved April 15, 2015, from http://www.researchgate.net/publication/275349565_Using_Big_Data_for_Customer-Centric_Marketing

Canavilhas, J., & Ivars, B. (2012). Uso y credibilidad de fuentes periodísticas 2.0 en Portugal y España. *El profesional de la información*, *21*(1), 63-69.

Carbellido, C. (2013, November 3). *The profile of the good community manager. Not any serves*. Retrieved from http://www.uncommunitymanager.es/perfil-community-manager/

Carr, N. (2011). *Superficiales ¿Qué está haciendo internet con nuestras mentes?* Madrid: Ed. Taurus.

Casellas, V. (2014). *Marca personal para profesores y marketing educativo*. Retrieved from http://www.artecasellas.es/marca-personal-para-profesores-y-marketing-educativo-por-vicky-casellas/

Castells, M. (2001). *La era de la información: economía, sociedad y cultura*. Madrid: Alianza editorial.

Castells, M. (2001). *La Galaxia Internet*. Barcelona: Random House Mondadori.

Castells, M. (2007). Communication, Power and Counter-power in the Network Society. *International Journal of Communication*, *1*, 238–266.

Castells, M. (2009). *Communication and power*. Oxford, UK: Oxford University Press.

Cebrián Herreros, M. (2008). *La radio en Internet: De la ciberradio a las redes sociales y la radio móvil*. Buenos Aires: La Crujía.

Chamorro-Premuzic, T. (2013, July). The Future of You. *Harvard Business Review*.

Chasco, C. (2003). El geomárketing y la distribución comercial. *Investigación y Marketing*, *79*, 6–13.

Chicheri, J. (2012). *31 historias de branding personal contadas en primera persona*. Bubok.

CINAIC 2015

Cloutier, J. (1973). *La communication audio-scripto-visuelle à heure des self-media, ou l'ère d'Emerec*. Montreal: Presses de l'Université de Montreal.

Codina, L. (2010, October 11). *Science 2.0: Social networks and online applications for scholars*. Retrieved April 7, 2016 from http://www.upf.edu/hipertextnet/numero-7/ciencia-2-0.html

Corazón, A. (2014). *El País*. Retrieved from http://cultura.elpais.com/

Cornellá, A. (1999). *Cómo sobrevivir a la infoxicación*. Academic Press.

Costas, R., & Bordons, M. (2007). The h-index: Advantages, limitations and its relation with other bibliometric indicators at the micro level. *Journal of Informetrics*, *1*(3), 193–203. doi:10.1016/j.joi.2007.02.001

Courtault, J. M., & Hayek, N. (2008). On the Robustness of the h-index: A mathematical approach. *Economic Bulletin*, *3*(78), 1–9.

Courtault, J. M., Hayek, N., Rimbaux, E., & Zhu, T. (2010). Research in economics and management in France: A bibliometric study using the h-index. *Journal of Socio-Economics*, *39*(2), 329–337. doi:10.1016/j.socec.2009.12.006

Cowen, T. (2004). *Creative Destruction: How Globalization Is Changing the World's Cultures*. Princeton, NJ: Princeton University Press.

Cummings, H. (2015). *Finding Your Brand's voice.* Retrieved from https://www.distilled.net/t

Dai, Y. (2014). *Análisis comunicacional del uso que las universidades españolas están haciendo de Twitter: evaluación e implicaciones para la práctica.* (Unpublished Master's thesis). Universidad Carlos III de Madrid, Madrid, Spain.

Dai, K., Nespereira, C. G., Vilas, A. F., & Redondo, R. P. D. (2015). Scraping and Clustering Techniques for the Characterization of LinkedIn Profiles. *Computer Science and Information Technology*, *5*(1), 1–15.

Dans, E. (2014). *Eres lo que compartes.* Retrieved 06/25, 2014, from http://www.enriquedans.com/2014/06/eres-lo-que-compartes.html

Datchary, C. (2005). Se disperser avec les tics une nouvelle compétence? In *Le travail avec les technologies de l'information* (pp. 157-173). Paris: Hermès-Lavoisier. Retrieved June 15, 2015, from http://hal.archives-ouvertes.fr/docs/00/33/17/82/PDF/_Datchary.pdf

Davidson, P. M., Newton, P. J., Ferguson, C., Daly, J., Elliott, D., Homer, C., & Jackson, D. et al. (2014). Rating and Ranking the Role of Bibliometrics and Webometrics in Nursing and Midwifery. *The Scientific World Journal*. PMID:24550691

Del Fresno, M. (2011b). Infosociabilidad: monitorización e investigación en la web 2.0 para la toma de decisiones. *El profesional de la información*, *20*(5), 548-554.

Del Fresno, M. (2014). Haciendo visible lo invisible: Visualización de la estructura de las relaciones en red en Twitter por medio del Análisis de Redes Sociales. *El profesional de la información*, *23*(3), 246-252.

Del Fresno, M. (2015). We are the media: the ongoing disruption of social space via social media. In *#Commoncore Project. How social media is changing the politics in education.* Consortium for Policy Research in Education Pennsylvania. Retrieved June 15, 2015, from http://www.hashtagcommoncore.com

Del Fresno, M. (2011a). *Netnografía.* Barcelona: UOC.

Del Fresno, M. (2012 a). *El consumidor social. Reputación online y social media.* Barcelona: UOC.

Del Fresno, M. (2012 b). Comprendiendo los social media y mass media un modelo para el estudio de la comunicación interpersonal colectiva en tiempos de Internet. *Derecom*, *11*, 99–109.

Del Fresno, M., & López-Peláez, A. (2014). Social work and Netnography: The case of Spain and generic drugs. *Qualitative Social Work: Research and Practice*, *13*(1), 85–107. doi:10.1177/1473325013507736

Del Río Pereda, P. (1996). *Psicología de los medios de comunicación: hacia el diseño sociocultural en comunicación audiovisual.* Madrid: Síntesis.

Delgado López-Cózar, E., Robinson-García, N., & Torres-Salinas, D. (2014). The Google Scholar Experiment: How to index false papers and manipulate bibliometric indicators. *Journal of the Association for Information Science and Technology*, *65*(3), 446–454. doi:10.1002/asi.23056

Deutsch, L. (2012). *Can you hear your Brand.* Retrieved April 13, 2015, from http://www.larsdeutsch.net/doc/CanYouHearYourBrand.pdf

Duart, J. M. (2009). Internet, redes sociales y educación.RUSC. Universities and Knowledge Society Journal, 6(1).

Durón, M. E. (2013). *Audiobranding and its importance to your personal Brand.* Retrieved April 13, 2015, from http://www.personalbrandingblog.com/audio-branding-and-its-importance-to-your-personal-brand/

Ebner, M., Lienhardt, C., Rohs, M., & Meyer, I. (2010). Microblogs in Higher Education. A chance to facilitate informal and process-oriented learning? *Computers & Education*, *55*(1), 92–100. doi:10.1016/j.compedu.2009.12.006

Echavarria Ramirez, A. F. (n.d.). *Academic Social Networks ... the boom of the academic Web 2.0 | so-called gatic* [Blog of the support group in the use of ICTS for the Universidad ICESI]. Retrieved April 7, 2016 from http://www.icesi.edu.co/blogs/egatic/2010/11/18/redes-sociales-academicas/

Erikson, E. (1950). *Childhood and society*. New York: Norton.

Eyrich, N., Padman, M., & Sweetser, K. (2008). PR practitioners' use of social media tools and communication technology. *Public Relations Review*, *34*(4), 412–414. doi:10.1016/j.pubrev.2008.09.010

Eysenbach, G. (2011). Can tweets predict citations? Metrics of social impact based on Twitter and correlation with traditional metrics of scientific impact. *Journal of Medical Internet Research*, *13*(4), e123. doi:10.2196/jmir.2012 PMID:22173204

Facebook Messenger reaches 900 million users. (2016, July 4). Retrieved April 13, 2016, from http://www.abc.es/tecnologia/redes/abci-facebook-messenger-alcanza-900-millones-usuarios-201604071756_noticia.html

Favaloro, E. J. (2009). The journal impact factor: Don't expect its demise any time soon. *Clinical Chemistry and Laboratory Medicine*, *47*(11), 1319–1324. doi:10.1515/CCLM.2009.328 PMID:19769542

Feather, J. (2008). *In The information society: A study of continuity and change*. London: Library Association.

Forkosh, A., & Hershkovitz, A. (2011). The use of social networks by higher education institutes in Israel. In *Proceedings of the Chais Conference on instructional technologies research 2011: Learning in the technological era* (pp. 14-20). Raanana: The Open University of Israel.

Foster, R., & Kaplan, S. (2001). *Creative Destruction: Why Companies that are Built to Last Underperform the Market – And how to Successfully Transform Them*. New York: Crown Business.

Foucault, M. (1998). *Technologies of the Self. A seminar with Micher Foucault*. The University of Massachussets Press.

Fournier, S. (1998). Consumers and their brands: Developing Relationship theory in consumer research. *The Journal of Consumer Research*, *24*(4), 343–373. doi:10.1086/209515

Franceschet, M. (2010). Ten good reasons to use the Eigenfactor™ metrics. *Information Processing & Management*, *46*(5), 555–558. doi:10.1016/j.ipm.2010.01.001

Franck, K. (1995). When I Enter Virtual Reality, What Body Will I Leave Behind? *Architectural Design*, *65*(118), 20-22. Retrieved June 15, 2015, from digitalarts.bgsu.edu/faculty/wmadsen/.../in_VR_Where_is_my_body.pdf

Fruchterman, T. M., & Reingold, E. M. (1991). Graph drawing by force-directed placement. *Software, Practice & Experience*, *21*(11), 1129–1164. doi:10.1002/spe.4380211102

Gallego Pérez, J.I. (2010). *Podcasting. Nuevos modelos de distribución para los contenidos sonoros*. Barcelona: UOC Press.

García Matilla, A. (2009). *El concepto de Emirec. Los medios para la comunicación educativa*. Retrieved August 5, 2014, from http://ayura.udea.edu.co/medios/documentos/EL%20CONCEPTO%20DE%20EMIREC

García-Santiago, L., & De Moya-Anegón, F. (2009). Using co-outlinks to mine heterogeneous networks. *Scientometrics*, *79*(3), 681–702. doi:10.1007/s11192-007-1855-0

Garfield, E. (1972). *Citation analysis as a tool in journal evaluation*. American Association for the Advancement of Science. Retrieved April 15, 2015, from http://www.elshami.com/Terms/I/impact%20factor-Garfield.pdf

Garfield, E. (1979). Is citation analysis a legitimate evaluation tool? *Scientometrics*, *1*(4), 359–375. doi:10.1007/BF02019306

Glänzel, W. (2006). On the h-index-A mathematical approach to a new measure of publication activity and citation impact. *Scientometrics*, *67*(2), 315–321. doi:10.1007/s11192-006-0102-4

Glänzel, W., & Moed, H. F. (2002). Journal impact measures in bibliometric research. *Scientometrics*, *53*(2), 171–193. doi:10.1023/A:1014848323806

Glass, L. (1994). *Cómo expresarse correctamente*. Barcelona: Paidós.

Goldhaber, M. (1997). The attention economy on the net. *First Monday*, *2*(4). doi:10.5210/fm.v2i4.519

Gonzalez, G. (2016, August 4). *85% of adults in the world uses Facebook services*. Retrieved April 13, 2016, from http://hipertextual.com/2016/04/uso-de-facebook-en-el-mundo

González-Díaz, C., Iglesias-García, M., & Codina, L. (2015, September - October). Presence of Spanish universities on scientific digital social networks: case of communication studies. *The Information Professional, 24*, 640 - 647.

Govers, R., & Go, F. (2009). *Place branding: Glocal, virtual and physical identities, constructed, imagined and experienced*. Hampshire: Palgrave Macmillan. doi:10.1057/9780230247024

Gowers, T., & Nielsen, M. (2009). Massively collaborative mathematics. *Nature*, *461*(7266), 879–881. doi:10.1038/461879a PMID:19829354

Graham, P. (2005). *Web 2.0 Want to start a startup?* Retrieved June 15, 2015, from http://www.paulgraham.com/web20.html

Grosseck, G., & Holotescu, C. (2010). Microblogging multimedia-based teaching methods best practices with Cirip. Eu. *Procedia: Social and Behavioral Sciences*, *1*, 495–501.

Grossman, G. (1994). Carefully crafted identity can build brand equity. *The Public Relations Journal*, *50*, 18–21.

Guillory, J., & Hancock, J. T. (2012). The effect of LinkedIn on deception in resumes. *Cyberpsychology, Behavior, and Social Networking*, *15*(3), 135–140. doi:10.1089/cyber.2011.0389 PMID:22335544

Guzmán, A. P., Del Moral, M. E., & González, F. (2012). Usos de Twitter en las universidades iberoamericanas. *Revista Latinoamericana de Tecnología Educativa*, *11*(1), 27–39.

Hall, N. (2014). The Kardashian Index: A measure of discrepant social media profile for scientists. *Genome Biology*, *15*(7), 424. doi:10.1186/s13059-014-0424-0 PMID:25315513

Hanna, R., Rohm, A., & Crittenden, V. L. (2011). Were all connected: The power of the social media ecosystem. *Business Horizons*, *54*(3), 265–273. doi:10.1016/j.bushor.2011.01.007

Haraway, D. (1989). *The Cyborg Manifesto*. Retrieved June 15, 2015, from http://faculty.georgetown.edu/irvinem/theory/Haraway-CyborgManifesto-1.pdf

Harzing, A. W., & Van Der Wal, R. (2009). A Google Scholar h-index for journals: An alternative metric to measure journal impact in economics and business. *Journal of the American Society for Information Science and Technology*, *60*(1), 41–46. doi:10.1002/asi.20953

Haustein, S., Peters, I., Bar-Ilan, J., Priem, J., Shema, H., & Terliesner, J. (2014). Coverage and adoption of altmetrics sources in the bibliometric community. *Scientometrics*, *101*(2), 1145–1163. doi:10.1007/s11192-013-1221-3

Heath, R. L., & Ni, L. (2008). *Corporate Social Responsibility*. Retrieved September 2, 2015, from http://www.instituteforpr.org/corporate-social-responsibility

Heim, M. (1998). *The virtual reality of the tea ceremony*. Retrieved June 15, 2015, from http://www.mheim.com/files/vrtea.pdf

Henning, V., & Reichelt, J. (2008, December). Mendeley-a last. fm for research? In *eScience, 2008. eScience'08. IEEE Fourth International Conference on* (pp. 327-328). Retrieved January 22, 2015, from http://blog.mendeley.com/wp-content/uploads/2009/05/henning-reichelt-2008-mendeley-a-lastfm-for-research-ieee-e-science.pdf

He, Q., Pei, J., Kifer, D., Mitra, P., & Giles, L. (2010). Context-aware citation recommendation. In *Proceedings of the 19th international conference on World wide web* (pp. 421-430). ACM. doi:10.1145/1772690.1772734

Herce, D. (2012). *Lo que no se mide no se mejora.* Retrieved 06/25, 2012, from http://dherce.es/2012/12/23/lo-que-no-se-mide-no-se-mejora/

Hirsch, J. E. (2005). An index to quantify an individuals scientific research output. *Proceedings of the National Academy of Sciences of the United States of America, 102*(46), 16569–16572. doi:10.1073/pnas.0507655102 PMID:16275915

Hodge, D. R., & Lacasse, J. R. (2010). Evaluating journal quality: Is the H-index a better measure than impact factors? *Research on Social Work Practice.*

Hoekman, J., Frenken, K., & Tijssen, R. J. (2010). Research collaboration at a distance: Changing spatial patterns of scientific collaboration within Europe. *Research Policy, 39*(5), 662–673. doi:10.1016/j.respol.2010.01.012

Holmberg, K. J. (2015). *Altmetrics for Information Professionals: Past, Present and Future.* Chandos Publishing.

Holotescu, C., & Grosseck, G. (2010). Learning to microblogging to learn. A case study on learning scenarios in a microblogging context. In *Proceedings of the 6th international scientific conference e learning and software for educational Bucharest* (pp. 15-16). Bucharest: RomaniaUniversity Press.

Hong, S. (2012). Online news on Twitter: Newspapers' social media adoption and their online readership. *Information Economics and Policy, 24*(1), 69–74. doi:10.1016/j.infoecopol.2012.01.004

Ibáñez, J.A., & Baraybar, A. (2011). Fuentes 2.0 y periodistas. Transformaciones en la comunicación museística. *El profesional de la información, 20*(6), 634-638.

Ingwersen, P. (1998). The calculation of web impact factors. *The Journal of Documentation, 54*(2), 236–243. doi:10.1108/EUM0000000007167

Invat.tur. (2015). *Big Data: retos y oportunidades para el turismo.* Retrieved from http://www.slideshare.net/invattur/estudio-big-data-retos-y-oportunidades-para-el-turismo

Jenkins, H. (2006). *Convergence Culture: Where Old and New Media Collide.* New York: New York University Press.

Junco, R. (2012). The relationship between frequency of Facebook use, participation in Facebook activities, and student engagement. *Computers & Education, 58*(1), 162–171. doi:10.1016/j.compedu.2011.08.004

Junco, R., Heiberger, G., & Loken, E. (2010). The effect of Twitter on college student engagement and grades. *Journal of Computer Assisted Learning, 27*(2), 119–132. doi:10.1111/j.1365-2729.2010.00387.x

Kapferer, J. (1992). *Startegic Brand Management.* Kogan.

Kaplan, A., & Haenlein, M. (2010). Users of the world, unite! The challenges and opportunities of Social Media. *Business Horizons, 53*(1), 59–68. doi:10.1016/j.bushor.2009.09.003

Katz, R. N. (2010). Scholars, scholarship and scholarly enterprise in the digital age. *EDUCAUSE Review, 45*(2), 44–56.

Keller, K. (1993). Conceptualizing, measuring, and managing customer- based brand equity. *Journal of Marketing, 57*(1), 1–22. doi:10.2307/1252054

Kelly, B., & Delasalle, J. (2012). Can LinkedIn and Academia.edu enhance access to open repositories? In *OR2012: The 7th International Conference on Open Repositories*. University of Bath.

Kelly, C. D., & Jennions, M. D. (2006). The h index and career assessment by numbers. *Trends in Ecology & Evolution*, *21*(4), 167–170. doi:10.1016/j.tree.2006.01.005 PMID:16701079

Kenekayoro, P., Buckley, K., & Thelwall, M. (2014). Hyperlinks as inter-university collaboration indicators. *Journal of Information Science*, *40*(4), 514–522. doi:10.1177/0165551514534141

Kessous, E., Mellet, K., & Zouinar, M. (2010). L'économie de l'attention: Entre protection des ressources cognitives et extraction de la valeur. *Sociologie du Travail*, *52*(3), 359–373. doi:10.1016/j.soctra.2010.06.009

Kietzmann, J., Hermkens, K., McCarthy, I., & Silvestre, B. (2011). Social media? Get serious! Understanding the functional building blocks of social media. *Business Horizons*, *54*(3), 241–251. doi:10.1016/j.bushor.2011.01.005

Kim, D., Kim, J. H., & Nam, Y. (2014). How does industry use social networking sites? An analysis of corporate dialogic uses of Facebook, Twitter, YouTube, and LinkedIn by industry type. *Quality & Quantity*, *48*(5), 2605–2614. doi:10.1007/s11135-013-9910-9

Klamm, D. (2011). 6 Best Practices for Universities Embracing Social Media. *Mashable*. Retrieved February, 15th, 2016, from http://mashable.com/2011/10/10/universities-social-media/#lyqXRDcSp5qa

Knothe, G. (2006). Comparative citation analysis of duplicate or highly related publications. *Journal of the American Society for Information Science and Technology*, *57*(13), 1830–1839. doi:10.1002/asi.20409

Knoth, P., & Herrmannova, D. (2014). Towards Semantometrics: A New Semantic Similarity Based Measure for Assessing a Research Publication's Contribution. *D-Lib Magazine*, *20*(11), 8.

Kousha, K., & Thelwall, M. (2015). Web indicators for research evaluation. Part 3: books and non standard outputs. *El profesional de la información, 24*(6), 724-736.

Kousha, K., & Thelwall, M. (2007). Google Scholar citations and Google Web/URL citations: A multi-discipline exploratory analysis. *Journal of the American Society for Information Science and Technology*, *58*(7), 1055–1065. doi:10.1002/asi.20584

Kozinets, R. (2005). *Desert Pilgrim.* Available online http://www.youtube.com/watch?v=BM3jrNzt3YI&feature=plcp

Kozinets, R. V. (1999). Desert Pilgrim. In Multimedia Ethnography, Heretical Consumer Research (HCR) 1999 Conference, Columbus, OH.

Kozinets, R. V. (2012). Marketing Netnography: Prom/ot(ulgat)ing a New Research Method. *Methodological Innovations Online, 1*(7), 37-45. Retrieved from http://www.pbs.plym.ac.uk/mi/index.html

Kozinets, R. V. (2002). Desert Pilgrim. *Consumption Markets & Culture*, *5*(September), 171–186. doi:10.1080/10263860290015602

Kozinets, R. V. (2015). *Netnography: Redefined.* London: Sage. doi:10.1002/9781118767771.wbiedcs067

Kretschmer, H., & Aguillo, I. (2004). Visibility of collaboration on the Web. *Scientometrics*, *61*(3), 405–426. doi:10.1023/B:SCIE.0000045118.68430.fd

Labbé, C. (2010). Ike Antkare one of the great stars in the scientific firmament. *ISSI Newsletter, 6*(2), 48-52.

Labrecque, L. I., Markos, E. C., & Milne, G. R. (2011). Online Personal Branding: Processes, Challenges, and Implications. *Journal of Interactive Marketing*, *25*(1), 37–50. doi:10.1016/j.intmar.2010.09.002

Lambrechts, D. (2011a, August 30). *Guia community manager*. Retrieved March 1, 2016, from Http://cursosnet.es/wp-content/uploads/2014/05/community-manager.pdf

Lambrechts, D. (2011b, September 13). *Analytical tools for social networks*. Retrieved from http://www.maestrosdelweb.com/guia-community-manager-herramientas-redes-sociales/

Lanham, R. (2007). *The Economics of Attention: Style and Substance in the Age of Information*. Chicago: University Of Chicago Press.

Lanier, J. (2010). *You are not a gadget: A manifesto*. New York: Vintage.

Lasswell, H. D. (1938). *Propaganda techniques in the world war*. New York: Peter Smith. Retrieved June 15, 2015, from http://babel.hathitrust.org/cgi/pt?id=mdp.39015000379902,view=1up,seq=9

Lasswell, H. D. (1948). The structure and function of communication in society. In *The communication of ideas*. New York: Harper. Retrieved June 15, 2015, from www.irfanerdogan.com/dergiweb2008/24/12.pdf

Latorre, J. (2015). *Innovando en formación pública*. Retrieved 06/26, 2015, from https://josanlatorre.wordpress.com/2015/06/24/innovando-en-formacion-publica/

Lazarsfeld, P. (1940) *Radio and the Printed Page: An Introduction to the Study of Radio and Its Role in the Communication of Ideas*. New York: Duell, Sloan and Pearce. Retrieved June 15, 2015, from https://archive.org/stream/radiotheprintedp00lazarich/radiotheprintedp00lazarich_djvu.txt

Lazarsfeld, P., & Merton, K. (1943). *Studies in radio and film propaganda*. Retrieved June 15, 2015, from http://onlinelibrary.wiley.com/doi/10.1111/j.2164-0947.1943.tb00897.x/abstract

Leary, T. (1988). Foreword to *The Scientist*. In *The Scientist: A Metaphysical Autobiography* (pp. 4–8). Berkeley, CA: Ronin Press.

Lee, J., Kraus, K. L., & Couldwell, W. T. (2009). Use of the h index in neurosurgery: Clinical article. *Journal of Neurosurgery*, *111*(2), 387–392. doi:10.3171/2008.10.JNS08978 PMID:19392590

Lessig, L. (2009). *Code: versión 2.0*. Retrieved June 15, 2015, from http://codev2.cc/download+remix/Lessig-Codev2.pdf

Levine, R., & Locke, C. (2009). *El manifiesto cluetrain: El ocaso de la empresa*. Deusto SA ediciones.

Lewis, D. (1999). *Information Overload*. London: Penguin.

Li, C., & Bernoff, J. (2008). *Groundswell: Winning in a World Transformed by Social Technologies*. Boston: Harvard Business Press.

Likert, R. (1932). A technique for the measurement of attitudes. *Archives de Psychologie*.

Lilly, J. C. (1967). *Programming and Metaprogramming in the Human Biocomputer*. New York: Julian Press.

Lilly, J. C. (1988). *The Scientist: A Metaphysical Autobiography*. Berkeley, CA: Ronin Press.

Lindstrom, M. (2005). *Brands Sense. Build Powerful Brands through Touch, Taste,Smell, Sight and Sound*. New York: The Free Press.

Lin, J., & Fenner, M. (2013). Altmetrics in evolution: Defining and redefining the ontology of article-level metrics. *Information Standards Quarterly*, *25*(2), 20. doi:10.3789/isqv25no2.2013.04

LinkedIn. (2014). *LinkedIn outcome rankings under the hood*. [Technical sheet]. Retrieved February 15, 2016, from: https://university.linkedin.com/content/dam/university/global/en_US/site/pdf/linkedin_outcome_graduate_rankings.pdf

Lippmann, W. (1922). *Public opinión*. London: Allen and Unwin. Retrieved June 15, 2015, from http://monoskop.org/images/b/bf/Lippman_Walter_Public_Opinion.pdf

Lister, M., Dovey, J., Giddings, S., Grant, I., & Kelly, K. (2009). *New Media: A Critical Introduction*. London: Routledge.

Lycett, M. (2013). Datafication: Making Sense of (Big) Data in a Complex World. *European Journal of Information Systems*, *22*(4), 381–386. doi:10.1057/ejis.2013.10

Maciá, F. (2011). *Técnicas avanzadas de posicionamiento en buscadores*. Anaya Multimedia.

Mack, D., Behler, A., Roberts, B., & Rimland, E. (2007). Reaching students with Facebook: data and best practices. *Electronic Journal of Academic and Special Librarianship, 8*(2), 4.

Marcos, T. (2008). Marca personal, máximo valor de las personas. In Marca Personal: Cómo convertirse en la opción preferente. ESIC.

Marcos, T. (2009). *Como crear y mantener una marca personal*. Madrid.

Martínez Rizo, F. (2011). Los rankings de universidades: una visión crítica. *Revista de la educación superior*.

Martínez Zarandona, I. (2005). *Tres pilares de la educación para los medios*. Retrieved August 2, 2014, from http://www.uned.es/ntedu/espanol/master/primero/modulos/teorias-del-aprendizaje-y-comunicacion-educativa/artirene.htm

Martín, R., Sánchez de la Nieta, L., & Hernández, M. A. (2014). *Es periodismo el "brand journalism"?: análisis de los portales informativos creados por multinacionales de referencia en los últimos dos años. Asociación Española de Investigación de la Comunicación (AE-IC)*. Universidad del País Vasco /Euskal Herriko Unibertsitatea.

Mas-Bleda, A., & Aguillo, I. F. (2015). *La web social como nuevo medio de comunicación y evaluación científica*. Barcelona: Editorial UOC.

Mas-Bleda, A., Thelwall, M., Kousha, K., & Aguillo, , I. (2014). Successful researchers publicizing research online: An outlink analysis of European highly cited scientists' personal websites. *The Journal of Documentation*, *70*(1), 148–172. doi:10.1108/JD-12-2012-0156

Mas-Bleda, A., Thelwall, M., Kousha, K., & Aguillo, I. F. (2014). Do highly cited researchers successfully use the social web? *Scientometrics*, *101*(1), 337–356. doi:10.1007/s11192-014-1345-0

Mateos, C. V., & Labra, J. P. (2008). Evaluación del uso y manejo de las tecnologías de la información y la comunicación en los docentes universitarios. Un caso mexicano. *Revista Latinoamericana de Tecnología Educativa*, *7*(1), 13–32.

Mattelart, A. (2003). *The Information Society: An Introduction*. London: Sage.

Mayer-Schönberger, V., & Cukier, K. (2014). *Big Data: A Revolution That Will Transform How We Live, Work, and Think*. London: John Murray Publishers.

McCarthy, L., & Lindsay, K. (2014). The 23 Things for Research. *23 Things blog*. Retrieved October 14, 2015, from http://blogs.bodleian.ox.ac.uk/23things/schedule/

McLeod, J., & Chaffee, S. (1973). Interpersonal approaches to communication research. *The American Behavioral Scientist*, *16*(4), 469–500. doi:10.1177/000276427301600402

McLuhan, M. (1964). *Understanding media: The extensions of man*. Cambridge, MA: MIT Press.

McQuail, D. (Ed.). (2002). *McQuail's reader in mass communication theory*. London: Sage.

McQuail, D., & Windahl, S. (1993). *Communication Models for the Study of Mass Communications*. New York: Longman.

Megias, J. (2014). *Manifiesto contra la mediocridad.* Retrieved 06/26, 2015, from http://javiermegias.com/blog/2014/07/manifiesto-contra-la-mediocridad/

Mejia Armijo, R. (2011, July 18). *Tools to work together without having!!* Retrieved from http://www.maestrosdelweb.com/herramientas-para-trabajar-titanpad-googledocs-basecamp/

Mejide, R. (2014). *Urbrands*. Espasa Calpe.

Metcalf, M., & Palmer, M. (2011). *Innovative Leadership Fieldbook*. Tucson, AZ: Integral Publishers.

Mingers, J. (2009). Measuring the research contribution of management academics using the Hirsch-index. *The Journal of the Operational Research Society*, *60*(9), 1143–1153. doi:10.1057/jors.2008.94

Minguillo, D., & Thelwall, M. (2012). Mapping the network structure of science parks: An exploratory study of cross-sectoral interactions reflected on the web. *Aslib Proceedings*, *64*(4), 332–357. doi:10.1108/00012531211244716

Mitchell, W. (1996). *City of Bits: Space, Place, and the Infobahn*. Cambridge, MA: MIT Press.

Moed, H. F. (2005). Statistical relationships between downloads and citations at the level of individual documents within a single journal. *Journal of the American Society for Information Science and Technology*, *56*(10), 1088–1097. doi:10.1002/asi.20200

Moed, H. F. (2006). *Citation analysis in research evaluation* (Vol. 9). Springer Science & Business Media.

Moed, H. F., Burger, W. J. M., Frankfort, J. G., & Van Raan, A. F. (1985). The use of bibliometric data for the measurement of university research performance. *Research Policy*, *14*(3), 131–149. doi:10.1016/0048-7333(85)90012-5

Monjas, M. (1993). *Programa de enseñanza de habilidades de interacción social. Para niños y niñas en edad escolar.* Valladolid: MIMC.

Moore, E., & Garzón, C. (2010). Social cartography: The art of using maps to build community power. Academic Press.

Moussa, S., & Touzani, M. (2010). Ranking marketing journals using the Google Scholar-based hg-index. *Journal of Informetrics*, *4*(1), 107–117. doi:10.1016/j.joi.2009.10.001

Moyers, B. (2008). *Bill moyers rewind: Isaac Asimov (1988).* Retrieved 06/17, 2015, from http://www.pbs.org/moyers/journal/blog/2008/03/bill_moyers_rewind_isaac_asimo_1.html

Mutom, D. S. (2011). *Social media for researchers and online personal branding.* University of Warwick. Retrieved October 13, 2013, from https://www.academia.edu/2612640/SOCIAL_MEDIA_FOR_RESEARCHERS_and_ONLINE_PERSONAL_BRANDING_e-book_

Negroponte, N. (2000). *El mundo digital*. Barcelona: Ediciones B.

Newcomb, T. (1953). An Approach to the Study of Communicative Acts. *Psychological Review*, *60*(6), 393–404. doi:10.1037/h0063098 PMID:13112341

Nicolaisen, J. (2007). Citation analysis. *Annual Review of Information Science & Technology*, *41*(1), 609–641. doi:10.1002/aris.2007.1440410120

Nie, N., & Erbring, L. (2002). Internet and Society: A Preliminary Report. *IT & Society*, 275-283. Retrieved June 15, 2015, from www.stanford.edu/group/siqss/itandsociety/v01i01/v01i01a18.pdf

Obispo, M. (2015). *Una historia detrás de cada paciente. Cómo aplicar el periodismo de marca a la comunicación de salud.* Retrieved from http://periodismodemarca.llorenteycuenca.com/una-historia-detras-de-cada-paciente-como-aplicar-el-periodismo-de-marca-a-la-comunicacion-de-salud/

Obispo, M., Sánchez, H., & Méndez, S. (2015). Medios sociales en el periodismo español: Demanda, expectativas y oferta universitaria. *Cuadernos.info*, (36), 155-171. doi: 10.7764/cdi.36.636

Olden, J. D. (2007). How do ecological journals stack-up? Ranking of scientific quality according to the h index. *Ecoscience*, *14*(3), 370–376. doi:10.2980/1195-6860(2007)14[370:HDEJSR]2.0.CO;2

Orduna-Malea, E., Serrano-Cobos, J., & Lloret-Romero, N. (2009). Las universidades públicas españolas en Google Scholar: presencia y evolución de su publicación académica web. *El profesional de la información, 18*(5), 493-500.

Orduna-Malea, E. (2011). Personalización e interactividad en los rankings de universidades publicados en la Web. *Anuario ThinkEPI*, *5*, 216–222.

Orduna-Malea, E. (2013a). Aggregation of the web performance of internal university units as a method of quantitative analysis of a university system: The case of Spain. *Journal of the American Society for Information Science and Technology*, *64*(10), 2100–2114. doi:10.1002/asi.22912

Orduna-Malea, E. (2013b). Espacio universitario español en la Web (2010): Estudio descriptivo de instituciones y productos académicos a través del análisis de subdominios y subdirectorios. *Revista Espanola la de Documentacion Cientifica*, *36*(3), e017. doi:10.3989/redc.2013.3.958

Orduna-Malea, E., Ayllón, J. M., Martín-Martín, A., & López-Cózar, E. D. (2015). Methods for estimating the size of Google Scholar. *Scientometrics*, *104*(3), 931–949. doi:10.1007/s11192-015-1614-6

Orduna-Malea, E., & Aytac, S. (2015). Revealing the online network between university and industry: The case of Turkey. *Scientometrics*, *105*(3), 1849–1866. doi:10.1007/s11192-015-1596-4

Orduna-Malea, E., Delgado López-Cózar, E., Serrano-Cobos, J., & Romero, N. L. (2015). Disclosing the network structure of private companies on the web: The case of Spanish IBEX 35 share index. *Online Information Review*, *39*(3), 360–382. doi:10.1108/OIR-11-2014-0282

Orduña-Malea, E., & López-Cózar, E. D. (2014). Google Scholar Metrics evolution: An analysis according to languages. *Scientometrics*, *98*(3), 2353–2367. doi:10.1007/s11192-013-1164-8

Orduna-Malea, E., & Ontalba-Ruipérez, J. A. (2013). Selective linking from social platforms to university websites: A case study of the Spanish academic system. *Scientometrics*, *95*(2), 593–614. doi:10.1007/s11192-012-0851-1

O'Reilly, T. (2005). *What Is Web 2.0*. Retrieved June 15, 2015, from http://oreilly.com/web2/archive/what-is-web-20.html

Ortega, J. L., Orduna-Malea, E., & Aguillo, I. F. (2014). Are web mentions accurate substitutes for inlinks for Spanish universities? *Online Information Review*, *38*(1), 59–77. doi:10.1108/OIR-10-2012-0189

Pashler, H. (1998). *The Psychology of Attention*. Cambridge, MA: MIT Press.

Peña Jiménez, P. (2012). Interactividad y redes sociales en la radio española, nuevas formas de participación. *Telos*, (92), 105–117.

Peña Jiménez, P., & Ventero Velasco, M. (2011). Nuevas formas de participación en radio. In *Radio 3.0: Una nueva radio para una nueva era*. Madrid: Fragua.

Pérez Ortega, A. (2008). Marca Personal: Cómo convertirse en la opción preferente. Ed. ESIC.

Pérez Ortega, A. (2012). Marca personal, cómo convertirse en la opción preferente (2nd ed.) ESIC.

Pérez Rufí, J. P. (Ed.). (2013). *Nuevos tiempos para la industria radiofónica en España*. Málaga: EUMED - Universidad de Málaga.

Perez, A. (2014). *Marca Pesonal para Dummies*. Barcelona: Wiley Publishing.

Pérez, F. J., Portilla, I., & Sánchez, C. (2011). Social Networks, Media and Audiences: A Literature Review. *Comunicación y Sociedad*, *24*(1), 63–74.

Peter, J. P., & Olson, J. C. (1983). Is Science Marketing? *Journal of Marketing*, *47*(Autumn), 111–125. doi:10.2307/1251404

Peters, T. (1997). *The Brand Called You*. Retrieved December 15, 2015, from http://www.fastcompany.com/28905/brand-called-you

Peters, T. (1997, August/September). *The Brand Called You. Fast Company*.

Pino, I. (2014). *Hacia un hexálogo del Periodismo de Marca. In Periodismo de marca y reputación corporativa*. Madrid: LLORENTE & CUENCA.

Piscitelli, A., Adaime, I., & Binder, I. (2010). *El Proyecto Facebook y la Posuniversidad. Sistemas operativos sociales y entornos abiertos de aprendizaje*. Colección Fundación Telefónica.

Plasencia, A. (2015). *Recordar el future*. Retrieved from http://adolfoplasencia.es/blog/recordar-el-futuro/

Plasencia, A. (2015). *Wikipedia: Un diálogo con Jimmy Wales*. Retrieved from http://adolfoplasencia.es/blog/wikipedia-un-dialogo-con-jimmy-wales/

Poza-Luján, J. L., Calduch-Losa, Á., Albors, A., Cabrera, M., Teruel, D., Rebollo, M., & Somavilla, R. D. (2015). Propuesta de parámetros y caracterización de los grupos de las redes sociales orientados a la docencia universitaria: experiencia y resultados. *Revista de Educación a Distancia*, (44).

Prentice, R. (2005). The anatomy of surgical simulations: The mutual articulation of bodies in and through the machine. *Social Studies of Science*, *35*(6), 837–866. doi:10.1177/0306312705053351

Priem, J., Piwowar, H. A., & Hemminger, B. M. (2012). *Altmetrics in the wild: Using social media to explore scholarly impact*. Cornell University Working Paper. Retrieved May 2, 2016 from http://arxiv.org/abs/1203.4745

Priem, J., & Hemminger, B. H. (2010). Scientometrics 2.0: New metrics of scholarly impact on the social Web. *First Monday*, *15*(7). doi:10.5210/fm.v15i7.2874

Qualman, E. (2009). *Socialnomics: How Social Media Transforms the Way We Live and Do Business*. New York: John Wiley.

Rampersad, H. K. (2008). A new blueprint for powerful and authentic personal branding. *Performance Improvement*, *47*(July), 34–37. doi:10.1002/pfi.20007

Rasmussen, D. (2012). *Forget Job Security: Build your Marketability!* CreateSpace Independent Publishing Platform.

Rebiun, R. B. U. E. (2010). *Science 2.0: Implementation of the social web to research* (Report). Retrieved from http://eprints.rclis.org/3867/

Reed, R. (2011). *SOLOMO manifesto: Just about everything marketers need to know about de convergence of social, local, and mobile (SoLoMo)*. Academic Press.

Reidenberg, J. (1998). Lex Informatica: The Formulation of Information Policy Rules Through Technology. *Texas Law Review*, *76*(3), 553–593.

Rich, A., & Schroeder, H. (1976). *Research issues in assertiveness training*. Academic Press.

Rimbaux, E. (2010). *Research in Economics and Management in France*. Retrieved April 10, 2015, from http://core.ac.uk/download/pdf/6508361.pdf

Roblyer, M., McDaniel, M., Webb, M., Herman, J., & Witty, J. V. (2010). Findings on Facebook in higher education: A comparison of college faculty and student uses and perceptions of social networking sites. *The Internet and Higher Education*, *13*(3), 134–140. doi:10.1016/j.iheduc.2010.03.002

Roca, G. (2014). *Estrategias digitales para hacer crecer el negocio*. Congreso de hoteleros españoles, Sta Cruz de Tenerife.

Roel, M. (2010). Desafíos de la televisión ante la consolidación del ecosistema digital. *Ámbitos, 19*, 25-42.

Ruiz King, F. J. (n.d.). *Web 2.0. A New Learning Environment in the Network*. Retrieved April 7, 2016 from http://dim.pangea.org/revistaDIM13/Articulos/pacoruiz.pdf

Ruiz, R. M., López, F. C., & Millán, J. (2013). Experimentando con las redes sociales en la enseñanza universitaria en ciencias. *Revista Eureka sobre enseñanza y divulgación de las ciencias, 10*(3), 394-405.

Russo, R. (1998). *Straight Man: A Novel*. New York: Random House.

Sanabria Mesa, A. L., & Hernández Jorge, C. M. (2011). Percepción de los estudiantes y profesores sobre el uso de las TIC en los procesos de cambio e innovación de la educación superior. *Aloma: Revista de Psicologia, Ciències de l'Educació i de l'Esport*, (29).

Sapiña, L. (2013). *Aprender a divulgar la ciencia*. Metode.

Sarkar, A., Agarwal, S., Ghosh, A., & Nath, A. (2015). Impacts of Social Networks: A Comprehensive Study on Positive and Negative Effects on Different Age Groups in a Society. *International Journal (Toronto, Ont.)*, *3*(5).

Schlesinger Díaz, M., & Cervera, A. (2008). Estudio comparativo entre personalidad de marca ideal vs. percebida. *Innovar*, 31.

Schwabel, D. (2009). *Me 2.0: A Powerful way to Achieve Brand Success*. New York: Kaplan Publishers.

SciVerse Scopus. (2015). *About Scopus*. Retrieved April 15, 2015, from http://www.elsevier.com/solutions/scopus

Seglen, P. O. (1989). From bad to worse: Evaluation by Journal Impact. *Trends in Biochemical Sciences*, *14*(8), 326–327. doi:10.1016/0968-0004(89)90163-1 PMID:2799903

Seglen, P. O. (1992). The skewness of science. *Journal of the American Society for Information Science*, *43*(9), 628–638. doi:10.1002/(SICI)1097-4571(199210)43:9<628::AID-ASI5>3.0.CO;2-0

Seglen, P. O. (1997a). Why the impact factor of journals should not be used for evaluating research. *BMJ (Clinical Research Ed.)*, *314*(7079), 497. doi:10.1136/bmj.314.7079.497 PMID:9056804

Seglen, P. O. (1997b). Citations and journal impact factors: Questionable indicators of research quality. *Allergy*, *52*(11), 1050–1056. doi:10.1111/j.1398-9995.1997.tb00175.x PMID:9404555

Sellas Güell, T. (2011). *El podcasting. La (r)evolución sonora*. Barcelona: UOC.

Senft, T. (2008). *Camgirls: Celebrity and Community in the Age of Social Networks*. New York: Peter Lang.

Sherry, J. F., Jr. (2012). Fuck you very much, Reviewer C. In Clarence Clobbers Tenderly. St. Bathans, New Zealand: University of St. Bathans Press.

Sherry, J. F., Jr. (2015). Sitting in the back of a crowded lecture hall. In Chickasaw Craft Threnody. South Bend, IN: Aire Lyre Media.

Shirky, C. (2009). *Here Comes Everybody: The Power of Organizing Without Organizations*. New York: Penguin.

Shuai, X., Pepe, A., & Bollen, J. (2012). *How the scientific community reacts to newly submitted preprints: Article downloads, twitter mentions, and citations*. Academic Press.

Silver, N. (2013). *The signal and the noise: why so many predictions fail — but some don't*. New York: Penguin Books.

Simonite, T. (2012). What Facebook knows. *Technology Review*, (July-August), 42–49.

Skeels, M. M., & Grudin, J. (2009). When social networks cross boundaries: a case study of workplace use of facebook and LinkedIn. In *Proceedings of the ACM 2009 International Conference On Supporting Group Work* (pp. 95-104). New York: ACM. doi:10.1145/1531674.1531689

Smith, A. G. (1999). A tale of two web spaces: Comparing sites using web impact factors. *The Journal of Documentation, 55*(5), 577–592.

Sobejano, J. (2012). *Microinnovación a la vuelta de la esquina*. capítulo del libro "Nunca te fies de un economista que no dude", Deusto ediciones.

Stacey, J. (1988). Can there be a feminist ethnography? *Women's Studies International Forum, 11*(1), 21–27. doi:10.1016/0277-5395(88)90004-0

Stanford. (n.d.). *Science 2.0: The Design Science of collaboration*. Retrieved from https://www.youtube.com/watch?v=nxBmKkLJCEc#t=50

Stepanyan, K., Borau, K., & Ullrich, C. (2010). A social network analysis perspective on student interaction within the Twitter microblogging environment. In *Proceedings of the Advanced Learning Technologies (ICALT), 2010 IEEE 10th International Conference* (pp. 70-72), Sousse. doi:10.1109/ICALT.2010.27

Stieger, S., & Burger, C. (2010). Let's go informative: Continuous student ratings with web 2.0 application Twitter. *Cyberpsychology, Behavior, and Social Networking, 13*(2), 163–167. doi:10.1089/cyber.2009.0128 PMID:20528272

Stuart, D., & Thelwall, M. (2006). Investigating triple helix relationships using URL citations: A case study of the UK West Midlands automobile industry. *Research Evaluation, 15*(2), 97–106. doi:10.3152/147154406781775968

Styhre, A. (2009). The cinematic mode of organizing: Media and the problem of attention in organization theory. *Information and Organization, 19*(1), 47–58. doi:10.1016/j.infoandorg.2008.06.001

Suler, J. (2004). The online disinhibition effect. *Cyberpsychology & Behavior, 7*(3), 321–326. doi:10.1089/1094931041291295 PMID:15257832

Swash, G. (1998). UK business information on the Internet. *New Library World, 99*(1144), 238–242. doi:10.1108/03074809810236793

Sweeney, J., & Brandon, C. (2006). Brand personality: Exploring the potential to move from factor analytical to circumplex models. *Psychology and Marketing, 23*(8), 639–663. doi:10.1002/mar.20122

Swot Analysis. (2011). Retrieved from http://www.washington.edu/

Tague-Sutcliffe, J. (1992). An introduction to informetrics. *Information Processing & Management, 28*(1), 1–3. doi:10.1016/0306-4573(92)90087-G

Takaki, M., Bravo, R., & Martínez, E. (2015). La gestión de la identidad corporativa en la Universidad: Análisis y consecuencias desde la perspectiva del profesorado. *Revista Europea de Dirección y Economía de la Empresa, 24*(1), 25–34. doi:10.1016/j.redee.2014.05.001

Tapscott, D., & Williams, A. (2006). *Wikinomics: How Mass Collaboration Changes Everything*. New York: Portfolio.

Tartt, D. (1992). *The Secret History*. New York: Alfred A. Knopf.

Teruel, L., Poza-Lujan, J.-L., Cabrera, M., Calduch-Losa, A. Rebollo, M., Diez-Somavilla, R., & Lloret-Romero, N. (2015). *Is it possible to use social media in university teaching?* CINAIC 2015. III Congreso Internacional sobe Aprendizaje, Innovación y Competitividad

The World University Rankings. (n.d.). Retrieved from www.timeshighereducation.co.uk

Thelwall, M. (2008). Bibliometrics to webometrics. *Journal of Information Science*. Retrieved April 24, 2016 from http://jis.sagepub.com/content/early/2008/06/13/0165551507087238.short

Thelwall, M., & Kousha, K. (2015a). Web indicators for research evaluation. Part 1: Citations and links to academic articles from the Web. *El profesional de la información, 24*(5), 587-606.

Thelwall, M., & Kousha, K. (2015b). Web indicators for research evaluation. Part 2: Social media metrics. *El profesional de la información, 24*(5), 607-620.

Thelwall, M., Haustein, S., Larivière, V., & Sugimoto, C. R. (2013). *Do altmetrics work? Twitter and ten other social web services*. Academic Press.

Thelwall, M. (2001). Results from a web impact factor crawler. *The Journal of Documentation, 57*(2), 177–191. doi:10.1108/EUM0000000007081

Thelwall, M. (2002). Conceptualizing documentation on the Web: An evaluation of different heuristic-based models for counting links between university Web sites. *Journal of the American Society for Information Science and Technology, 53*(12), 995–1005. doi:10.1002/asi.10135

Thelwall, M. (2004). *Link analysis: An information science approach*. Elsevier, UK: Academic Press.

Thelwall, M. (2004). *Link analysis: an information science approach*. San Diego: Academic Press.

Thelwall, M. (2009). *Introduction to webometrics: Quantitative web research for the social sciences*. Morgan & Claypool.

Thelwall, M., & Harries, G. (2004). Can personal web pages that link to universities yield information about the wider dissemination of research? *Journal of Information Science, 30*(3), 240–253. doi:10.1177/0165551504044669

Thelwall, M., Klitkou, A., Verbeek, A., Stuart, D., & Vincent, C. (2010). Policy-relevant Webometrics for individual scientific fields. *Journal of the American Society for Information Science and Technology, 61*(7), 1464–1475. doi:10.1002/asi.21345

Thelwall, M., & Kousha, K. (2015). ResearchGate: Disseminating, communicating, and measuring Scholarship? *Journal of the Association for Information Science and Technology, 66*(5), 876–889. doi:10.1002/asi.23236

Thelwall, M., & Price, L. (2003). Disciplinary differences in academic web presence–a statistical study of the UK. *Libri, 53*(4), 242–253. doi:10.1515/LIBR.2003.242

Thelwall, M., & Sud, P. (2011). A comparison of methods for collecting web citation data for academic organisations. *Journal of the American Society for Information Science and Technology, 62*(8), 1488–1497. doi:10.1002/asi.21571

Thelwall, M., & Tang, R. (2003). Disciplinary and linguistic considerations for academic Web linking: An exploratory hyperlink mediated study with Mainland China and Taiwan. *Scientometrics, 58*(1), 155–181. doi:10.1023/A:1025487826207

Thelwall, M., Vaughan, L., & Björneborn, L. (2005). Webometrics. *Annual Review of Information Science & Technology, 39*(1), 81–135. doi:10.1002/aris.1440390110

Thomson Reuters ISI Web of Science. (2010). *Web of Science*. Retrieved April 17, 2015, from http://thomsonreuters.com/en/products-services/scholarly-scientific-research/scholarly-search-and-discovery/web-of-science.html

Toffler, A. (1970). *Future shock*. New York: Bantam Books. Retrieved June 15, 2015, from http://goo.gl/KDGWwn

Toffler, A., & Toffler, H. (1994). Surfing the third wave. *Microtimes*, *118*. Retrieved June 15, 2015, from http://www.yoyow.com/marye/tofflers94.html

Torres Salinas, D., & Docampo, D. (2013). The new Highly cited researchers list from Thomson Reuters and the Shanghai: Spain´s position and university map. *El profesional de la Información*, 265.

Tufts University. (n.d.). *Social Media Best Practices*. Retrieved February, 15th, 2016, from http://webcomm.tufts.edu/social-media-best-practices13/

Turkle, S. (1984). *The Second Self. Computers and the human spirit*. Cambridge, MA: MIT Press. Retrieved June 15, 2015, from http://mitpress.mit.edu/books/chapters/0262701111intro1.pdf

Turkle, S. (1994). *Constructions and reconstructions of the Self in Virtual Reality*. Cambridge, MA: MIT Press. Retrieved June 15, 2015, from http://web.mit.edu/sturkle/www/constructions.html

Turkle, S. (1995). *Life on the Screen: Identity in the Age of the Internet*. New York: Simon and Schuster.

University of Leicester. (n.d.). *Twitter best practice*. Retrieved February, 15th, 2016, from http://www2.le.ac.uk/offices/cap/marcomms/communications/social/handbook/twitter/twitter-best-practise

Unkelos-Shpigel, N., Sherman, S., & Hadar, I. (2015). Finding the missing link to industry: LinkedIn professional groups as facilitators of empirical research. In *Proceedings of the Third International Workshop on Conducting Empirical Studies in Industry* (pp. 43-46). Piscataway, NJ: IEEE Press. doi:10.1109/CESI.2015.14

Urueña, A., Annie, F., Blanco, D., & Valdecasa, E. (2011). *Social network in internet*. Retrieved from http://www.ontsi.red.es/ontsi/sites/default/files/redes_sociales-documento_0.pdf

Usluel, Y. K., & Mazman, S. G. (2009). Adoption of web 2.0 tools in distance education. *Procedia: Social and Behavioral Sciences*, *1*(1), 818–823. doi:10.1016/j.sbspro.2009.01.146

Van Dijck, J. (2013). *The Culture of Connectivity: A Critical History of Social Media*. New York: Oxford University Press. doi:10.1093/acprof:oso/9780199970773.001.0001

Van Dijck, J. (2013). You have one identity: Performing the self on Facebook and LinkedIn. *Media Culture & Society*, *35*(2), 199–215. doi:10.1177/0163443712468605

Vanclay, J. K. (2008). Ranking forestry journals using the h-index. *Journal of Informetrics*, *2*(4), 326–334. doi:10.1016/j.joi.2008.07.002

Vander Hofstadt, C. (2003). *El libro de las habilidades de comunicación*. Academic Press.

Vaughan, L. (2004). Exploring website features for business information. *Scientometrics*, *61*(3), 467–477. doi:10.1023/B:SCIE.0000045122.93018.2a

Vaughan, L., & Romero-Frías, E. (2014). Web search volume as a predictor of academic fame: An exploration of Google trends. *Journal of the Association for Information Science and Technology*, *65*(4), 707–720. doi:10.1002/asi.23016

Vaughan, L., & Shaw, D. (2005). Web citation data for impact assessment: A comparison of four science disciplines. *Journal of the American Society for Information Science and Technology*, *56*(10), 1075–1087. doi:10.1002/asi.20199

Vaughan, L., & Shaw, D. (2007). A new look at evidence of scholarly citation in citation indexes and from web sources. *Scientometrics*, *74*(2), 317–330. doi:10.1007/s11192-008-0220-2

Vaughan, L., & Wu, G. (2004). Links to commercial web sites as a source of business information. *Scientometrics*, *60*(3), 487–496. doi:10.1023/B:SCIE.0000034389.14825.bc

Veletsianos, G. (2011). Higher education scholars' participation and practices on Twitter. *Journal of Computer Assisted Learning*. Retrieved February, 15th, 2016, from http://onlinelibrary.wiley.com/doi/10.1111/j.1365-2729.2011.00449.x/full

Verhoeven, P., Tench, R., Zerfass, A., Moreno, A., & Verčič, D. (2012). How European PR practitioners handle digital and social media. *Public Relations Review*, *38*(1), 162–164. doi:10.1016/j.pubrev.2011.08.015

Vessuri, H. (2010). *El rol de la investigación en la Educación Superior*. Academic Press..

Vidal, C. E., Martínez, J. G., Fortuño, M. L., & Cervera, M. G. (2011). Actitudes y expectativas del uso educativo de las redes sociales en los alumnos universitarios. RUSC. *Universities and Knowledge Society Journal*, *8*(1), 171–185.

Vignelli, M. (2010). The Vignelli canon. New York: Academic Press.

Virilio, P. (1991). The aesthetics of disappearance. New York: Semiotext(e).

Virilio, P. (1995). *Speed and Information: Cyberspace Alarm!* Retrieved June 15, 2015, from http://www.ctheory.net/printer.asp?id=72

Virilio, P. (1996). Speed Pollution. *Wired*. Retrieved June 15, 2015, from http://archive.wired.com/wired/archive/4.05/virilio.html

Weaver, R., Johnson, E., Sweetser, K., & Howes, P. (2009). An examination of the role of online social media in journalists' source mix. *Public Relations Review*, *35*(3), 314–316. doi:10.1016/j.pubrev.2009.05.008

Wertheim, M. (1999). *The pearly gates of cyberspace. In Cyber_reader. Critical writtings for the digital era*. New York: Phaidon Press.

Wilkinson, D., Harries, G., Thelwall, M., & Price, L. (2003). Motivations for academic Web site interlinking: Evidence for the Web as a novel source of information on informal scholarly communication. *Journal of Information Science*, *29*(1), 49–56. doi:10.1177/016555150302900105

Wilson, C. S. (1999). Informetrics. *Annual Review of Information Science & Technology*, *34*, 107–247.

Woollaston, V. (2014). *How often do you look at your phone? The average user now picks up their device more than 1,500 times a week*. Retrieved April 15, 2015, from http://www.dailymail.co.uk/sciencetech/article-2783677/How-YOU-look-phone-The-average-user-picks-device-1-500-times-day.html

Yasuhiro, M. (1996). *El Just in time hoy en Toyota*. Deusto SA ediciones.

Yuste, B., & Cabrera, M. (2014). *Emprender en Periodismo. Nuevas oportunidades para el profesional de la información*. Barcelona: UOC.

About the Contributors

Margarita Cabrera Mendez has a PhD in Audiovisual Communication. She currently works as a full-time professor at the Polytechnic University of Valencia in Spain. Since 2003, she has taught both undergraduate and graduate students. Currently she is developing a research project that involves hispanic social media and communication through various social networks. She primarily coordinated and edited the book: 'Escribir en Internet: guía para los nuevos medios y las redes sociales' for The Fundéu BBVA (formerly The Foundation of Urgent Spanish). She also coordinated the Master degree "Digital Comunication and Contents Management. CALSI" and co-directed the social media workshop: Comunica 2.0. Her new book is about entrepreneurship in journalism: "Emprender en periodismo". She was also a founding member of MASmedios & TheLineBTWN. Both companies specialize in developing effective digital platforms in communication and brand experiences.

Nuria Lloret-Romero is a Consultant in Digital Comunication, Web Analytics and Personal Branding and Online Identity. Professor in Digital Comunication and collaborative online Projects in the Polytechnic University of Valencia Director Master degree "Digital Comunication and Contents Management. CALSI" Founder of company Masmedios Ltd and The Line Between Ltd. Researcher focused the use of technology to create and enable online collaborative projects . Director Research group CALSI inside the IDF "The Institute of Design" Editor and autor Book "Systems Science and Collaborative Information Systems: Theories, Practices and New Research" Director Master degree in Music and PhD Music Coordinator. Arts lover in special of the Photography, Video-art and Music.

* * *

Gersón Beltrán López is a geographer, specialist in Geolocation, Geomarketing, Social Media and Tourism. PhD in Geography by the University of Valencia, Postgraduate in Regional Geographical Analysis by the University of Valencia and postgraduate in Geographic Information Systems by the University of Girona. + 20 years of experience in Education and Consulting. Professor at the Department of Geography at the University of Valencia, Manager in Geoturismo SLU, Cofounder of Marketingeo and Social Media Manager at the Valencian Cartographic Institute.

Raúl Terol Bolinches studied Audiovisual Communication at Polytechnic University of Valencia. In 2003 he founded the local radio station Llosa FM, of which he is a director from that time until today. Llosa FM was the first municipal radio station to broadcast online in Valencia. He published in 2007 "El papel de la radio municipal en la Sociedad de la Información: El caso de Llosa FM". Later Raúl Terol studied a Masters of Social Media in the UPV. Where is currently completing his PhD on the Radio 3.0. Raul Terol is also a member of ECREA.

Ángeles Calduch-Losa achieved the B.S. degree in mathematical sciences (1995) and the B.S. degree in statistics (2003), both of them from Valencia University. She got her Ph. D. degree in mathematics from the Jaume I University of Castellón de la Plana in 2004. She has been working since 2001 at Technical Valencia University, where she is currently a lecturer and a researcher, in the fields of data mining and web metrics. Her educational research interests include social networks and students' competency management.

Mark Anthony Camilleri is a resident academic lecturer in the Department of Corporate Communication. He lectures in an international masters programme run by the University of Malta in collaboration with King's College, University of London. Dr Camilleri has published his research in peer-reviewed journals and chapters. He is a frequent speaker and reviewer at the American Marketing Association's (AMA) Marketing & Public Policy conference and in the Academy of Management's (AoM) Annual Meeting.

Miguel del Fresno, Ph.D. (Sociology), Associate professor at UNED (Universidad Nacional de Educación a Distancia) is a senior marketing and communication consultant and researcher in social network analysis and netnography. He is a professor in the Masters of Communication program at Universidad Navarra (Pamplona), Universidad País Vasco (Bilbao), Universidad Politécnica (Valencia), Universidad Complutense (Madrid), and Universidad Sevilla, Universidad Jaume I (Castellón) in Spain, as well as a visiting professor at University of the Republic of Uruguay (Montevideo) and University of California, Berkley. He is a visiting researcher at the University of California, Berkley and University of California, San Diego. Del Fresno earned an MBA and an Executive Masters in e-Business from IE Business School (Madrid), as well as a Masters in Information Society and Knowledge (Universitat Oberta de Catalunya) and a Ph.D. in Sociology (Universidad Nacional de Educacion a Distancia).

Ariana Daniela Del Pino Espinoza is a university professor and graphic designer. She has 25 years and has taught at several universities in Ecuador. She wrote the Book Sectorial Marks for Ecuador. Currently teaches at Escuela Superior Politécnica del Litoral, (ESPOL), the best university in Ecuador and Practices Coordinator, she is doing his doctoral studies at the Polytechnic University of Valencia. She has a Master in Design and Branding.

Rebeca Díez-Somavilla, PhD in Communication Studies, Professor at the Polytechnic University of Valencia (Campus de Gandia). Coordinator of the Masters in social media CALSI. Comunica2 co-director of the International Congress on social networks.

Cristina I. Font is a student at the Polytechnic University of Valencia.

Susana Herrera Damas is a Full Journalism Professor at the University Carlos III of Madrid.

Robert Kozinets is a globally recognized expert on social media, marketing, branding and innovation. Besides inventing netnography, he has authored and co-authored over 100 pieces of research on the intersection of technology, media, brands, and consumers. This includes four books—three of them Sage Method books. Currently, he is Associate Editor of the Journal of Consumer Research and the Journal of Retailing, an Academic Trustee of the Marketing Science Institute, and is on the Board of Directors of the Association for Consumer Research. On the industry side, he has extensive speaking, training, and consulting experience with a range of global companies and organizations, including HSBC, TD Banking and Financial Group, American Express, Merck, Sony, Nissan, eBay, Campbell Soup, and L'Oréal. He is the Hufschmid Professor of Strategic Public Relations and Business Communication at the University of Southern California's Annenberg School for Communication and Journalism.

José-Antonio Ontalba-Ruipérez is a professor of methodology and techniques of scientific research at Polytechnic University of Valencia. Specializing in informetrics and webometrics. Member of Trademetrics research group.

Enrique Orduna-Malea works as a postdoctoral researcher at the Polytechnic University of Valencia (UPV). He is specialized in the design and application of methods and metrics to the evaluation of the web performance of entities belonging to specific environments, such as academic (authors, journals, universities) and business (companies, organizations). He pertains to the EC3 Research Group (University of Granada) and Trademetrics Research Group (UPV).

Jose-Luis Poza-Lujan received the B.S. degree in education from Valencia University in 1990 and the B.S. degree in computing from Technical Valencia University, Valencia (UPV), in 1999. He received his Ph.D. degree in informatics from the Technical University of Valencia in 2012, where he has worked since 2000. He is, currently, a lecturer and a researcher at the UPV, in the fields of smart devices and intelligent distributed control networks. His educational research interests include social networks, soft skills for engineering, and students' competency management.

Ruth Matovelle Villamar is a Phd candidate for the Doctoral program in Design from the Universitat Politécnica de Valéncia, a Master´s degree in Branding from Escuela Superior Politécnica del Litoral, Bachellor of Information technologies/Arts from Sheridan College in Ontario, Canada. Currently Vice-president of REDIS (National Network of Academic Design), Tenure Professor of ESPOL at EDCOM (Design and Communication Faculty). Former Dean and former investigation department head for EDCOM in Guayaquil, Ecuador.

Index

J

L

M

N

O

P

R

S

T

U

V

W

www.ingramcontent.com/pod-product-compliance
Ingram Content Group UK Ltd.
Pitfield, Milton Keynes, MK11 3LW, UK
UKHW050950290926
15470UKWH00005B/1750

9 781522 509172